T0076137

LARGE SCALE NETWORKS

MODELING AND SIMULATION

LARGE SCALE NETWORKS

MODELING AND SIMULATION

Radu Dobrescu • Florin Ionescu

CRC Press is an imprint of the
Taylor & Francis Group, an **informa** business

CRC Press
Taylor & Francis Group
6000 Broken Sound Parkway NW, Suite 300
Boca Raton, FL 33487-2742

Printed on acid-free paper
Version Date: 20160719

International Standard Book Number-13: 978-1-4987-5017-2 (Hardback)

Visit the Taylor & Francis Web site at
http://www.taylorandfrancis.com

and the CRC Press Web site at
http://www.crcpress.com

Contents

Preface

This book offers a rigorous analysis of the achievements in the field of traffic control in large networks, oriented on two main aspects: the self-similarity in traffic behavior and the scale-free characteristic of a complex network. Additionally, the authors provide a new insight into understanding the inner nature of things and the cause and effect based on the identification of relationships and behavior within a model, using the study of the influence of the topological characteristics of a network on traffic behavior. The effects of this influence are then discussed in order to find new solutions for traffic monitoring and diagnosis and for the prediction of traffic anomalies.

The international seminar "Interdisciplinary Approaches of Fractal Analysis" (IAFA) in 2003 inspired research in a new direction, which led to the writing of this book. As cochairmen of the seminar, organized by the Politehnica University of Bucharest with the kind support of Alexander von Humboldt-Foundation, Radu Dobrescu and Florin Ionescu opened a novel direction of research on the fractal analysis of information in large-scale networks, which continued and was enriched with each new edition of IAFA (held every 2 years, the last—the seventh—in 2015). Numerous scientific papers were jointly published and several PhD theses were written on the sample topic, where the research for PhD was cosupervised by the two authors, both university professors, for which research grants came through international funding. The appearance of this book is a synthesis of the results obtained so far through the expertise gained on the topic.

This book refers to complex systems that have a network-like structure. Complexity should express the level of interconnectedness and interdependencies of a system. In a complex system, it is often the case that the utility of a structure or process is expressed at the next higher level of organization relative to the process itself. This book offers a new insight into the identification of relationships and behavior of Internet traffic, based on the study of the influence of the topological characteristics of a network upon the traffic behavior. The key property to illustrate this interdependency is the self-similarity, which can be expressed both in the fractal-like topological structure of scale-free networks and in the characteristics of traffic flow in a packet-switched environment. In practice, this approach leads to developing good predictors of network performance.

The reader can follow the development, validation, and use of self-similar and multifractal models, queuing, control and performance evaluation, assessing the incremental utility of various models, hierarchical models based on aggregation, analytic approximation models for various performance metrics, tradeoff, and sensitivity analysis using a multiobjective optimization framework. Therefore, this book, aimed at undergraduates and graduate students as well as experienced researchers in the field of computer networks, is very timely and it covers topics ranging from the classical approaches right up to present-day research applications. For the experimentalist, the book may serve as an introduction to mathematical modeling topics, while the theoretician will particularly profit from the description of key problems in the context of self-similar dynamic processes. The book provides the perfect background for researchers wishing to pursue the goal of multiscale modeling in complex networks, perhaps one of the most challenging and important tasks for researchers interested in the progress of information technology and communication.

The authors are very well aware of the fact that the desire to explain the new and complicated ideas in this area, where even the terminology is hardly settled, may not be rewarding. However, the problems of optimal traffic control and the investigation of new traffic features caused by the huge users and services volume in the networks offer important insights for a wide audience (students, engineers, researchers, and academics) focused on the main idea that in future it will be impossible to analyze any complicated systems without using the fractal self-similar approach. Also, the authors hope that the bold assumption that the topological characteristics of a network influence the traffic behavior will be confirmed convincingly.

Radu Dobrescu
Florin Ionescu

Authors

Radu Dobrescu earned a diploma in engineering in automatic control from the Faculty of Control and Computers at the Polytehnical Institute of Bucharest, Romania, in 1968. He then earned his PhD in automatic control of electrical processes from the Polytehnical Institute of Bucharest, Romania, in 1976. He currently serves as a professor in the Department of Automation and Industrial Informatics of the Faculty of Automatic Control and Computers, Politehnica University of Bucharest. From 1992, he has advised more than 50 students in their doctoral research in the field of systems engineering. His scientific works span three main domains: data acquisition, processing, and transmission; networked control systems; and modeling of complex systems. He has published more than 30 books and university courses, about 80 scientific papers in journals and more than 120 papers in volumes of international conferences. He has served as a project manager for around 30 research projects. He was the general chairman of seven biennial International Symposiums on Interdisciplinary Approaches in Fractal Applications (IAFA 2003–2015). He has been an IEEE member since 1991 and an IEEE senior member since 2005. He was the past chair of the IEEE Romania Section (2010–2014). The monograph *Complexity and Information* published by the Romanian Academy Publishing House in 2010 was awarded the Grigore Moisil Academy Prize.

Florin Ionescu was born in Romania and earned a Diploma in Engineering in 1968 and a PhD (thesis on "Machine-Tools and Hydraulic Drives") in 1981, both at the University Politehnica Bucharest, where he worked between 1969 and 1987 as an assistant professor and as senior assistant professor, respectively. In 1987, he left Romania and joined the Alexander von Humboldt Foundation (AvH) as a research fellow and worked on "Modelling and Simulation of Hydraulic Drive Systems" at the IHP, Rheinisch Westfalische Hochschule für Technik (RWTH) Aachen, Germany (Professor Backé). Between 1989 and 1991, he was a researcher and a visiting professor at the Institute of Fluid Systems, Hydraulic Engines and Plants, TU Darmstadt, Germany (Professor Stoffel), where he also realized a postdoctoral scientific achievement equivalent to a "habilitation." From 1991 to 2010, he was a full professor at the University of Applied Sciences in Konstanz, Germany Hochschule für Technik, Wirtschaft und Gestalltung-Konstanz (HTWG-KN), and a visiting professor in other universities. In 1992, he was appointed by Steinbeis Foundation (STW), Stuttgart, to lead—besides his teaching and research activity—the Steinbeis Transfer Centre in the specialization "Dynamics of Machines, Hydraulics and Pneumatics" and in 2005, he worked at the Steinbeis Transfer Centre's Project Consulting for Eastern Europe-Konstanz in order to promote research, technology development, and innovation and consultancy for the industry. In 1994, Professor Ionescu was appointed by (HTWG) as the director of Mechatronics Research Department.

He retired in 2010 and was appointed by Steinbeis University Berlin as a full professor in 2011, along with a position as senate member and director of Steinbeis Transfer Institute Dynamic Systems. His activities were teaching, research, and, in the framework of STW, development and technology transfer for the Industry. He delivered lectures in mechanics, applied mathematics, hydraulics, and pneumatics (he is the author of Hydraulics Pneumatics Analysis and Synthesis (HYPAS), an object-oriented multitasking interactive software for hybrid drive systems and control), vibrations of machines, robotics and comparison of mathematical models and solid bodies, and modeling and simulation of systems. The topics of research were centered around the major fields of object-oriented modeling and simulation of large-scale (virtual manufacturing) systems, and was focused on machine tools, hybrid drives and control, robotics, fault detection and prediction, robotics, which involved micro–nano robotics and neuro-fuzzy in acupuncture and man–machine systems. Professor Ionescu led over 150 research, development, and innovation projects. He has organized and directed research stages in Konstanz of PhD and postdoctoral scientists from many countries, with the support of federal and land ministries of research and education, of the German industry, as well as of national research organizations such as AvH, Conference of German Academies of Sciences (KDAW), German Research Association (DFG), German Academic Exchange Service (DAAD), Industrial Associations "Otto von Guericke" (AiF), the European Union, and NATO (the North Atlantic Treaty Organization). Professor Ionescu is (co)author of several patents and prototypes of micro–nano robots, cutting tools, electrohydraulic devices and automatic systems, and automatic control of sewerage installations. He is (co)author of over 400 papers, prints in proceedings of international conferences, organized by IFAC, IEEE, IMEKO, WCNA, IASTED, ISMA, KES, ASME, AMSE, JAMS, ICMAs, conferences held in more than 20 countries from Europe, Asia, Australia and North America. Professor Ionescu is (co)author, and/or editor of 45 books or multiauthor-chapters, published by Springer, Elsevier, CRC Press, Taylor & Francis, de Gruyter, Pergamon Press, Artech House, Indersciences, Eds. of RO and BG Academies. Professor Ionescu was vice-president of the American Romanian Academy (ARA) and editor of *ARA Journal*. He also was a member of the national (scientific) councils of RO, of professional (inter)national associations, technical committees, editorial boards, reviewer for conferences, journals and publishing houses, auditor, evaluator of scientific and education activities and R & D projects or programs. Professor Ionescu was awarded several diplomas, honorary doctorates, and honorary professorships.

chapter one

State of the art and trends in information networks modeling

1.1 Self-similarity and fractals in traffic modeling

1.1.1 Building of a framework

One of the key issues in measurement-based network control is to predict traffic in the next control time interval based on the online measurements of traffic characteristics. The goal is to forecast future traffic variations as precisely as possible, based on the measured traffic history. Traffic prediction requires accurate traffic models that can capture the statistical characteristics of the actual traffic. Inaccurate models may overestimate or underestimate the network traffic.

In a landmark paper, Leland et al. (1993) reported the discovery of self-similarity (SS) in the local-area network (LAN) traffic, more precisely the Ethernet traffic. Specifically, they studied two separate processes: the number of bytes arriving per time interval (a.k.a. byte count process), and the number of IP packets (a.k.a. packet count process).

Since packets can vary widely in size (40–1500 bytes in the case of Ethernet), it is in principle conceivable that the two processes would be quite different. However, the paper shows that they are both self-similar and also display fractal behavior according to the changing network conditions. Since then, self-similarity, fractals, and long-range dependence (LRD) have emerged as powerful tools for modeling traffic behavior and have brought a significant change in the understanding of network traffic.

It has been found in numerous studies (Crovella and Bestavros 1997, Willinger et al. 1997, Grossglauser and Bolot 1999, Ostring and Sirisena 2001 are among the most representative) that data traffic in high-speed networks exhibits self-similarity that cannot be captured by classical models, necessitating the development of self-similar models.

The problem with self-similar models is that they are computationally complex. Their fitting procedure is very time-consuming while their parameters cannot be estimated based on the on-line measurements. The main objective of this book is to discuss the problem of traffic prediction in the presence of self-similarity and particularly to describe a number of short-memory and long-memory stochastic traffic models and to associate them to non-model-based predictors, based on the following criteria:

1. *Accuracy*: The most important criterion for choosing a predictor is the quality of its predictions, since the goal of the predictor is to closely model the future.
2. *Simplicity*: In order to achieve a real-time predictor, a certain level of simplicity is necessary. Simplicity has an intrinsic value because of the ease of use and implementation.
3. *On-line*: Most traffic modeling has been done for off-line data. In reality, for network control, we want to use on-line measurements to forecast the future. We do not know anything about the underlying traffic structure instead we should estimate the predictor parameters based on the on-line measurements.
4. *Adaptability*: A good predictor should adapt to the changing traffic. As time progresses, more samples become available. Therefore, more information is known about the traffic characteristics. An adaptive traffic predictor should use new information for improvement and updation of its parameters.

The statistical characteristics of the network traffic were studied especially for understanding the factors that influence the performance and the scalability of large systems. Hence, there exist a number of studies related to the evidence of the self-similarity in modern communication networks (Willinger et al. 1996, Cao et al. 2000, Floyd and Paxson 2001, Sikdar and Vastola 2001), such as ON/OFF models for heavy tails distributions, users behavior, back-off algorithms, routers buffering, TCP algorithms for congestion avoidance, etc.

1.1.2 Mathematical background of self-similar processes

1.1.2.1 Stable distributions and semistable processes

Let us begin with the central limit theorem (due to DeMoivre and Laplace):

X_j is a sequence of independent and identically distributed (i.i.d.) random variables (RVs), having finite first moment (μ) and second moment (σ^2). Denote $S_n = X_1 + \cdots + X_n$, and let $N(0,1)$ be the normal distribution with mean 0 and variance 1. Then, the RV $Z_n = (S_n - n\mu)/\sigma\sqrt{n}$ approaches $N(0,1)$ in distribution as $n \to \infty$.

Note that the shifting and scaling parameters depend on n. A compact and general form of the theorem can be written as

$$Z_n = a_n S_n + b_n. \tag{1.1}$$

This is not very interesting because no matter what the initial distribution, the limit is the same: $N(0,1)$. However, if we drop the assumption of finiteness of σ, Z_n can have as its limit any member of an entire family of distributions, the so-called stable distributions. The explanation of this occurrence is: if the common distribution X of the i.i.d. RVs X_j is a stable one, there exists at least one choice of sequences $\{a_n\}$, $\{b_n\}$ such that $Z_n = a_n S_n + b_n = a_n \Sigma_n X + b_n \rightarrow X$. In other words, X is stable under the "sum, shift, scale" algorithm defined above. The theory of stable distributions attained maturity in the 1950s.

The next step in generalization was to drop the independence assumption, and here the stochastic processes enter the picture. A stochastic process (SP) can be thought of as a set of RVs $\{X_i\}$, where the index i is referred to as the "time" parameter. In this context, the index will be either real (we say that the process has continuous time, and use the widespread notation X_t), real bidimensional ($X_{t,s}$), or an integer (discrete time, denoted X_n). The novelty consists in the fact that the individual RV need not be independent—in fact, the dependency itself is what makes a SP interesting. In order to completely characterize a SP, in general, we need to specify all the finite-dimensional distributions of its RVs: unidimensional (individual distributions of X_1, X_2, etc.), bidimensional (joint distributions of X_1 and X_2, X_1 and X_3, X_2 and X_3, etc.), …, n-dimensional, … Two SPs $\{X_t\}$ and $\{Y_t\}$ are said to be equal if all their finite-dimensional distributions are equal.

The connection with the stable laws is achieved through the so-called *increment process*. Let $\{X_t\}$ be a SP in continuous time. For any pair $s \leq t$, the difference $X_t - X_s = X_{s,t}$ is a RV indexed by s and t. The set $\{X_{s,t}\}$ is therefore a bidimensional SP in continuous time, called the increment process associated with $\{X_t\}$. Obviously, a given unidimensional SP uniquely defines a (bidimensional) increment process. The converse requires a little care, because more than one SP can have the same increment process. The simplest example is $\{X_t\}$ and $\{X_t + C\}$, where C is a deterministic constant. To remove this impediment, one usually "fixes the origin" by assuming $X_0 = 0$, and this permits the exact recovery of $\{X_t\}$ when $\{X_{s,t}\}$ is known: $X_t = X_{t,0}$. The SP analogous to a stable distribution is a SP with independent increments, all of which have the same stable distribution.

The analogy goes even further: every time interval $[0,t]$ can be divided into an arbitrary number n of equal subintervals, and X_t becomes a sum of n increments, just like S_n from the central limit theorem (the difference is that the increments are not independent any more). This type of argument led Lamperti in a much-quoted paper (Lamperti 1962) to define semistable processes as those satisfying:

$$\{X_{rt}\} = \{a_r X_t + b_r\} \tag{1.2}$$

where the notation emphasizes that a and b depend on r. Lamperti himself explains the term semistable: "The name is intended to suggest the analogy with the theory of stable laws, and is rendered more appropriate by the fact [...] that a semistable process, if it is assumed to have independent increments, must actually be a stable one." He then goes on to prove the fundamental property of this class of SPs: for any semistable process, the scaling parameter must be of the form:

$$a_r = r^{\alpha}, \quad \alpha \geq 0 \tag{1.3}$$

This power law has had a long, illustrious career.

1.1.2.2 First empirical processes with power-law statistics

In 1951, the British physicist-turned-hydrologist H. E. Hurst described the first instance of a natural process exhibiting power-law behavior (Hurst 1951). The analysis was based on several records of levels of water in the river Nile, the longest of which extends over no less than 663 years (622–1284 AD).

By the very nature of the measurement process, such empirical data series are discrete and finite. As an analytical tool, Hurst used the rescaled adjusted range (R/S) statistic. A brief description of R/S will immediately reveal its relevance for hydrology: Let a data series be represented by $\{x_0, x_1, x_2, \ldots, x_n\}$. The lowercase letters are meant to emphasize that this is not (yet!) a stochastic process (SP), but rather an ordered set of deterministic, fixed values. For each index $n \leq N$, the following steps are performed:

- Compute the sample mean μ_n and sample variance $S(n)$ of the truncated series $\{x_0, x_1, x_2, \ldots, x_n\}$. We expect these values to converge as $n \to \infty$. If we imagine the data series to be a sample path of an underlying SP, we say that the SP is *stationary*.
- For each $k \leq n$, compute the rescaled value of the data series up to k: $A_k = x_0 + \cdots + x_k - k\mu_n$. This step "extracts" the average from the data, leaving only the variations around the average. If x_k has the physical interpretation of the net annual amount of water that flows into a reservoir (positive or negative), the sum $x_0 + \cdots + x_k$ is the total quantity of water in the reservoir after k years, and A_k itself is the random variation of this quantity around its mean. In real life, the outflow of water from the reservoir is controllable, so the mean will ideally be kept equal to zero.
- Compute $\text{MAX}_n = \max[0, A_1, \ldots, A_n]$ and $\text{MIN}_n = \min[0, A_1, \ldots, A_n]$. These are the extreme quantities of water that have to be stored in a period of n years.
- Define $R(n) = \text{MAX}_n - \text{MIN}_n$. This is the adjusted range, which is the capacity needed in the reservoir in order for it to either underflow or overflow due to the random variations in water input. Unlike $S(n)$, $R(n)$ does not in general converge. The intuitive explanation is that,

as time (n) goes by, more and more "unlikely" excursions of A_k will occur, thus ever-increasing the capacity $R(n)$ needed to cover them.
- Define $R/S_n = R(n)/S(n)$.

From what was said above, it is clear that R/S_n is not bounded as a function of n. Referring again to the underlying SP, R/S_n is itself a RV, so it makes sense to assume that $E[R/S_n]$. If $E[R/S_n]$ is not bounded, the next question is: What is a function that closely resembles its dependency on n? Hurst's discovery was that $E[R/S_n]$ behaves like a power law in n, that is, there are constants C and H such that the ratio $E[R/S_n]/(Cn^H) \to 1$ as $n \to \infty$. We write this as follows:

$$E\left[\frac{R}{S_n}\right] \sim n^H \tag{1.4}$$

H is now called the Hurst parameter of the respective SP. When trying to fit the process with any of the "classical" models (e.g., Poisson processes, ARMA time-series, etc.), it was relatively easy to prove that the H resulting from such models is exactly 0.5—at odds with the experimental value. When represented in logarithmic coordinates, Equation 1.4 is a curve with an asymptote of slope H. This can be used to estimate H.

The second example was given by the economist Adelman (1965), who used Fourier transforms to analyze series of U.S. economic indicators collected since 1889. He found that all power spectra have a singularity (pole) at the origin, that is, the power at small frequencies is very high. (In time domain, this corresponds to long cycles, the longest of which have periods close to the sample size N.) Again there was discrepancy, since "classical" models would only yield power spectra which are flat at the origin.

1.1.2.3 *From semistability to self-similarity*

In 1965, Benoit Mandelbrot introduced self-similar SPs (the original French name was *processus homothétiques à soi*, while another widely used term is fractal processes). This is a model which finally captures the power-law behavior described earlier. (His most cited paper is the one published 3 years later in the *SIAM Review* [Mandelbrot and Van Ness 1968].) Referring directly to the increment process $X_{s,t} = X_t - X_s$, he defines stochastic self-similarity as

$$X_{t_0,t_0+rt} = r^H X_{t_0,t_0+t}, \quad \forall t_0, t, \quad \forall r > 0. \tag{1.5}$$

Compared to Equation 1.2, it is clear that Mandelbrot's self-similar processes are semistable in the sense of Lamperti. Taking advantage of the similar initials, we can abbreviate both as SS. The similarity between the two concepts goes even further, since Mandelbrot constructs his SS process (fractional Brownian motion, *fBm* for short) starting with the usual Brownian motion (*Bm*), which is used as an example by Lamperti in his already cited paper.

Assuming that the reader is familiar with the properties of *Bm*, we only mention two properties, which are important for our discussion:

- *Bm* has independent increments.
- *Bm* is self-similar, with Hurst parameter $H = 0.5$.

Denoting *Bm* as $B(t)$ and *fBm* as $B_H(t)$, here is a simplified version of Mandelbrot's definition of the *fBm*: $B_H(0) = 0$, $H \in [0,1]$ and

$$B_H(t) - B_H(0) = \frac{1}{\Gamma(H+1/2)} \left\{ \int_{-\infty}^{0} \left[(t-s)^{H-1/2} - (-s)^{H-1/2} \right] dB(s) + \int_{0}^{t} (t-s)^{H-1/2} dB(s) \right\} \tag{1.6}$$

The reader need not worry now about the technicalities of this definition (gamma function, stochastic integration), but rather needs to concentrate on the big picture. The integral from $-\infty$ causes the current value of the process ($B_H(t)$) to depend on the *entire* history of the process, and this is the reason why, unlike the *Bm*, *fBm* does not have independent increments. For $H > 0.5$, the integrand goes to zero as $t \to -\infty$, so the "influence" of the distant past on the present is less than the "influence" of the more recent past (in this respect, we have something like a weighted average). The case $H < 0.5$ is much less understood, but, fortunately, without practical relevance as far as we know (Figure 1.1).

Returning to the definition of the *fBm*, we define *fBm* in Equation 1.6 as indeed an SS process in the sense of Equation 1.5, and that its *R/S* statistic obeys the power law of Equation 1.4. For a deeper mathematical treatment of the *fBm*, the reader is directed to Embrechts and Maejima (2002).

1.1.2.4 *Self-similarity versus long-range dependence*

The integral in Equation 1.6 points to another important property of SS processes: *long-range dependence* (LRD). Intuitively, LRD can be viewed as Adelman's "long cycles"—correlations which, instead of "dying out" quickly, extend over long periods of time.

Mathematically it is not as easy to agree to what LRD means, because several definitions (not all equivalents) exist. One of them uses the correlation

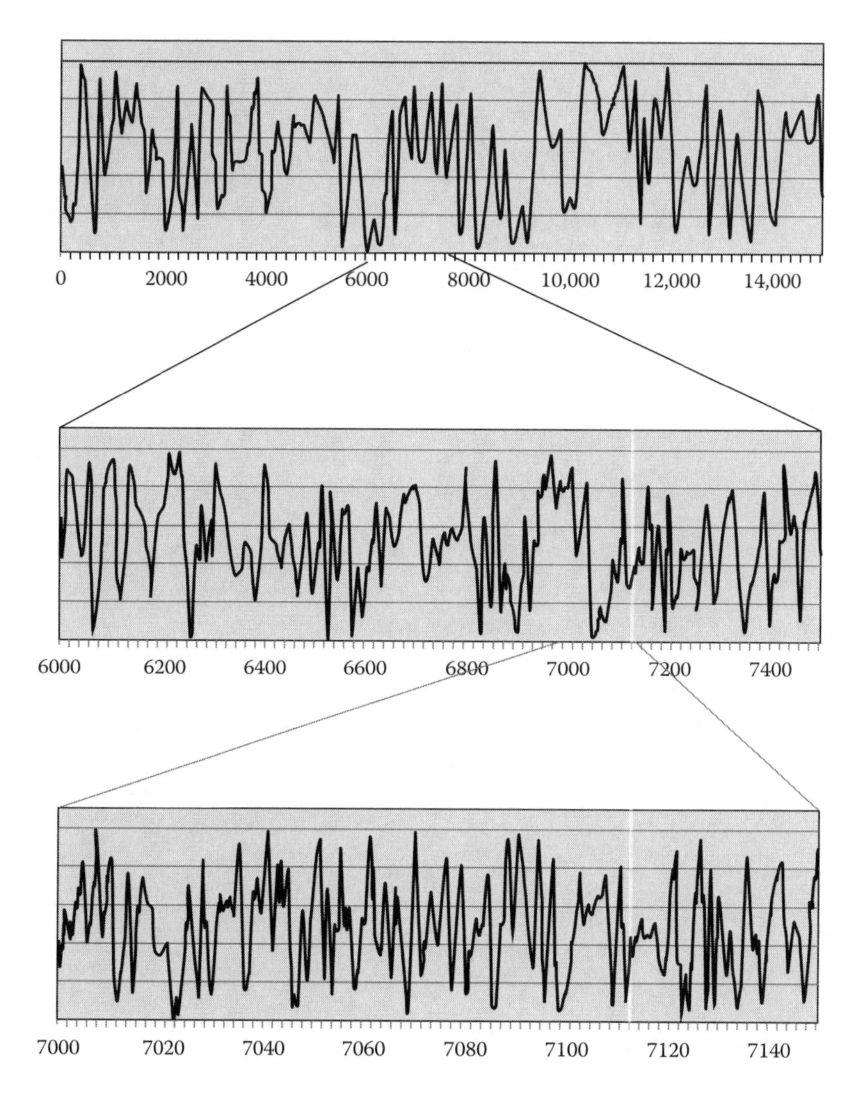

Figure 1.1 Self-similar aspects for unidimensional stochastic processes.

function of the discrete SP: $\rho(k) = \mathrm{Cov}[X_t, X_{t+k}]/\sigma^2$ is the autocorrelation of lag k. The SP is called LRD if there are constants $\alpha \in (0,1)$ and $C > 0$ such that

$$\frac{\rho(k)}{Ck^{-\alpha}} \xrightarrow{k \to \infty} 1, \quad \text{which we write simply as } \rho(k) \sim k^{-\alpha} \tag{1.7}$$

For comparison, note that the AR(1) process $X_i = aX_{i-1} + \epsilon_i$, with $a < 1$ has the polynomial autocorrelation $\rho(k) \sim a^k$. It is known that such a polynomial decreases faster than any power law of the type (1.7). More dramatically, the autocorrelation series $\sum_{k=1}^{N} \rho(k)$ is convergent for the polynomial form, but divergent for the power law. It is proved that a *fBm* with parameter H is LRD in the above sense, and that $H = 1 - \alpha/2 \in (1/2, 1)$, or, equivalently $\alpha = 2 - 2H \in (0,1)$ (Mandelbrot and Van Ness 1968). When represented in logarithmic coordinates, Equation 1.7 is called the correlogram of the process and has an asymptote of slope—α. This property can be used to estimate α (and therefore H), either graphically or through some analytic fitting procedure (e.g., least squares).

There are other definitions (see Beran 1994 for a thorough introduction), but we only mention another one, which uses the variance of the sample mean:

$$\frac{\mathrm{Var}\left[\sum_{1}^{k} X_i / k\right]}{Ck^{2H-2}} \xrightarrow{k \to \infty} 1, \tag{1.8}$$

or, denoting the sample mean by X^k, $Var[X^k] \sim k^{2H-2}$.

The graphical representation of Equation 1.8 in logarithmic coordinates is called the variance–time plot and can also be used to estimate H. At this point we have three known methods for estimating H: R/S plot (1.4), correlogram (1.7), and variance–time plot (1.8).

A word of caution: SS and LRD are not overlapping properties. There are SS processes which are not LRD (the simple *Bm* is an example), and, conversely, there are LRD processes that are not SS. However, the *fBm* with $H > 0.5$ is both SS and LRD (according to Equations 1.4, 1.7, and 1.8).

All examples of power laws given in Equations 1.3 through 1.5, 1.7, and 1.8 can be presented in two ways:

- Exact laws: Valid for all values of time (the equation has the "=" sign).
- Asymptotic laws: Valid only in the limit as t or $k \to \infty$ (the equation has the "~" sign).

FBm is exactly SS (equality in Equation 1.5) and also exactly LRD (equality in Equations 1.4, 1.7, and 1.8).

1.1.2.5 *Self-similarity discovered in network traffic*

In the already mentioned paper (Leland et al. (1993), reported the discovery of self-similarity in a local-area network (LAN) traffic, more precisely Ethernet traffic. Specifically, they studied two separate processes: number of *bytes* arriving per time interval (a.k.a. byte count process), and number of IP packets (a.k.a. packet count process). Since packets can vary widely in size (40–1500 bytes in the case of Ethernet), it is, in principle, conceivable that the two processes would be quite different. However, the paper shows that they are both SS, with H estimated in the range (0.6–0.9) according to the changing network conditions.

As a general critique of this approach, we observed that all methods used in Beran (1994) (and in numerous papers that followed) detect and estimate LRD rather than SS. Indeed, the only "proof" offered for SS *per se* is the visual inspection of the time series at different timescales. In the same paper, *fBm* is proposed as a model for the byte/packet count processes. Our opinion is that this is just an approximation, which tends to be off the mark at low utilizations of the LAN under test: the periodogram, correlogram and R/S plot all depart considerably from linearity for small t. In contrast, a *fBm* process should in theory generate perfectly straight lines in all these tests.

"Self-similarity" (actually LRD) has since been reported in various types of data traffic: LAN, WAN (Paxson and Floyd 1994), Variable-Bit-Rate video, SS7 control, HTTP, etc. (see Park and Willinger 2000 for an extensive bibliography). An overview of the above studies shows link speeds ranging from 10 Mbps (Ethernet) to 622 Mbps (OC-12), with the link type being "access" (typically connecting a university campus or research lab to an Internet service provider), average bandwidths between 1.4 and 42 Mbps, minimum timescale of 1 ms (in only one instance, usually above 10–100 ms), and at most six orders of magnitude for timescales. Lack of access to high speed, high-aggregation links, and lack of devices capable of measuring such links have prevented similar studies from being performed on Internet backbone links until recently. In principle, backbone traffic could be qualitatively different from the types enumerated earlier, due to factors such as much higher level of aggregation, traffic conditioning (policing and shaping) performed at the edge, and much larger round-trip-time (RTT) for TCP sessions.

Among the first proofs of the self-similarity in the traffic on the Internet backbone (OC-12 ATM and OC-48 POS links) are the correlograms shown in Figure 1.2 (Yao et al. 2003). They demonstrated that this traffic is indeed SS asymptotically and also reported as a new autocorrelation structure for short lags. The autocorrelation function for short lags has the same power form as that of the long lags, that is, $\rho(k) \sim k^{-\alpha}$, but the parameter α turns out to assume values that are significantly larger: $\alpha \in [0.55, 0.71]$ for $k \in [50\ \mu s, 10\ ms]$, compared to $\alpha \in [0.1, 0.18]$ for $k \in [100\ ms, 500\ s]$.

The first plot in Figure 1.2 shows the correlogram for the shortest time unit used in the analysis. Although the plot is clearly linear, its dependence is too chaotic to be of much use. For the second plot, the bytes arrived are aggregated at 0.4 ms time intervals, and the two slopes corresponding to the two values of α are easily observed. The third is a variance–time plot, which is just another way of looking at LRD. The straight line corresponds to a Hurst parameter $H = 0.5$ and therefore the asymptote of the function

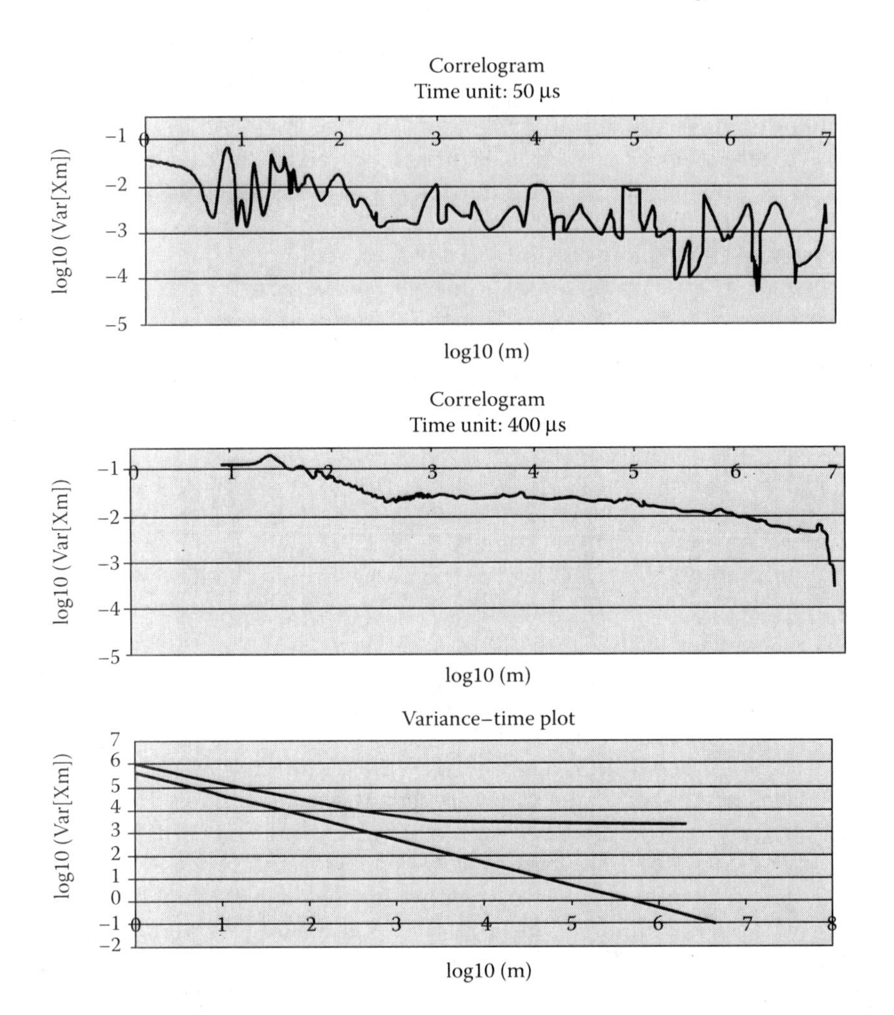

Figure 1.2 Correlograms for different types of Internet traffic.

represented clearly has a larger slope (between 0.84 and 0.96, to be precise). As the arrival process occurs with an average speed of about 700 Mbps, the hypothesis that at high speeds the traffic becomes Poissonian ($H \rightarrow 0.5$) is rejected. The conclusion is that the small timescales of Internet backbone traffic are *not* modelled accurately by *fBm*.

So, although the traffic processes in high-speed Internet links exhibit asymptotic self-similarity, their correlation structure at short timescales renders their modeling to be exact self-similar processes (like the fractional Brownian motion) inaccurate. Based on statistical and queuing analysis of data traces from high-speed Internet links, we conclude that Internet traffic retains its self-similar properties even under high aggregation.

1.2 *Models of complex networks*

It is easy to accept that the World Wide Web is a huge, complex network. It remains more difficult to define the complexity or more exactly, the main characteristics of a complex system. We expect that its definition should be richer than that of algorithmic complexity, and should express the level of interconnectedness and interdependencies of a system. In a complex system, the utility of a structure or a process is often expressed at the next higher level of organization relative to the process itself. Unlike entropy and the related concept of information, complexity is neither extensive nor entirely intensive. What is clear though is that complexity involves a specific description, which, is of course, dependent on the technology and subjective capabilities of the observer. Anyway, we can consider that a complex system is a system with a large number of elements, building blocks, or agents, capable of interacting with each other and with their environment. The interaction between elements may occur only with immediate neighbors or with distant ones; the agents can be all identical or different; they may move in space or occupy fixed positions, and can be in one of the two states or of multiple states. The common characteristic of all complex systems is that they display organization without the application of any external organizing principle. Today, we encounter many more examples of complex systems: physical, mechanical, biological, and social. The stock market, cities, metabolic pathways, ecosystems, the Internet, or the human brain, are all complex. The question naturally arises as to what is in common all these systems. That they all share similar network architectures has emerged as the answer in the last few years. Network theory has become one of the most visible pieces of the body of knowledge that can be applied to the description, analysis, and understanding of complex systems and is now an essential ingredient of their study.

A network is a system of nodes with connecting links. Once this viewpoint is accepted, networks appear everywhere. Consider some examples from two main fields: (a) biological networks: autonomous nervous systems of complex organisms, a network of neurons connected by synapses, gene regulation networks, a network of genes connected by cross-regulation interactions or metabolic networks, or a network of metabolites connected by chemical reactions and (b) social networks, like e-mails services, Internet, and the World Wide Web. The structure of such social networks was formalized initially by using random graphs, in which the existence of a link between any pair of nodes has probability p. Erdos, in collaboration with Renyi, undertook the theoretical analysis of the properties of random graphs and obtained a number of important results, including the identification of the percolation threshold, that is, the average number of links per node necessary in order for a random graph to be fully connected, or the typical number of intermediate links in the shortest path between any two nodes in the graph.

Paul Erdos and Alfred Renyi initiated the study of random graphs in the 1960s (Erdos and Renyi 1960). Random graph theory is, in fact, not the study of graphs, but the study of an ensemble of graphs (or, as mathematicians prefers to call it, a probability space of graphs). The ensemble is a group consisting of many different graphs, where each graph has a probability attached to it. They proposed a graph with N nodes in which each of the $N(N+1)/2$ possible edges is present with a probability p. A property Q is said to exist with a probability P if the total probability of all the graphs in the ensemble having that property is P. Two well-studied graph ensembles are $G_{N,M}$, which is the ensemble of all graphs having N vertices and M edges, and $G_{N,p}$ consists of graphs with N vertices, where each possible edge is realized with a probability p. These families are known to be similar if

$$M = \binom{N}{2} p,$$

so long as p is not too close to 0 or 1 and are referred to as ER graphs.

An important attribute of a graph is the average degree, that is, the average number of edges connected to each node. We shall denote the degree of the i-th node by k_i and the average degree by $\langle k \rangle$. N-vertex graphs with $\langle k \rangle = O(N^0)$ are called sparse graphs. In what follows, we exclusively focus on sparse graphs.

An interesting characteristic of the ensemble $G_{N,p}$ is that many of its properties have a related threshold function, $p_t(N)$, such that if $p < p_t$ the property exists with probability **0**, in the "thermodynamic limit" of $N \to \infty$, and with probability **1** if $p > p_t$. This phenomenon is similar to the physical notion of a phase transition. For a property Q, we have

$$\lim_{N \to \infty} P_{N,p}(Q) = \begin{cases} 0, & \text{if } \dfrac{p(N)}{p_t(N)} \to 0 \\ 1, & \text{if } \dfrac{p(N)}{p_t(N)} \to \infty \end{cases} \tag{1.9}$$

An example of such a property is the existence of a giant component, that is, a set of connected nodes, in the sense that a path exists between any two of them, whose size is proportional to N. Erdos and Renyi showed that for ER graphs such a component exists if $\langle k \rangle > 1$. If $\langle k \rangle < 1$ only small components exist, and the size of the largest component is proportional to $\ln N$. Exactly at the threshold, $\langle k \rangle = 1$, a component of size proportional to $N^{2/3}$ emerges. This phenomenon was described by Erdos as the "double jump." Another property is the average path length between any two sites, which in almost every graph of the ensemble is of order ln N.

Recently, several studies of real-world networks have indicated that the ER model fails to reproduce many of their observed properties. One of the simplest properties of a network that can be measured directly is the degree distribution, or the fraction $P(k)$ of nodes having k connections (degree k). A well-known result for ER networks is that the degree distribution is Poissonian, $P(k) = e^{-z}z^k/k!$, where $z = \langle k \rangle$ is the average degree. Direct measurements of the degree distribution for networks of the Internet (Faloutsos et al. 1999), WWW (Broder et al. 2000), metabolic networks (Jeong et al. 2000), network traffic control (Dobrescu et al. 2004b) and many more, show that the Poisson law does not apply. Most often these nets exhibit a scale-free degree distribution: $P(k) = k^{-\lambda}$, $k = m, \ldots, K$ where $c \approx (\lambda - 1)m^{(\lambda-1)}$ is a normalization factor, and m and K are the lower and upper cutoffs for the connectivity of a node, respectively.

One of the first network models that presents scale-free degree distribution is represented by the small-world network model, which has the so-called small-world phenomenon as main characteristic. This phenomenon is defined by the coexistence of two apparently incompatible conditions: (i) the number of intermediaries between any pair of nodes in the network is quite small, typically referred to as the six degrees of separation phenomenon and (ii) the large local redundancy of the network, that is, the large overlap of the circles of neighbors of two network neighbors. The latter property is typical of ordered lattices, while the former is typical of random graphs. Watts and Strogatz (1998) proposed a minimal model for the emergence of the small-world phenomenon in simple networks. In their model, small-world networks emerge as the result of randomly rewiring a fraction p of the links in a d-dimensional lattice (Figure 1.3).

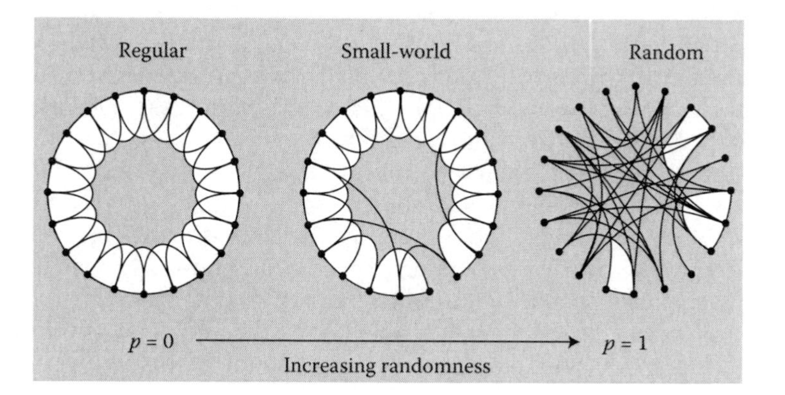

Figure 1.3 Small-world networks model generation. (First published in Watts, D.J. and S.H. Strogatz. *Nature*, 393, 440–442, 1998.)

The parameter p enables one to continuously interpolate between the two limiting cases of a regular lattice ($p = 0$) and a random graph ($p = 1$).

Watts and Strogatz probed the structure of their small-world network model using two quantities: (i) the mean shortest distance L between all pairs of nodes in the network and (ii) the mean clustering coefficient C of the nodes in the network. For a d-dimensional lattice, $L \sim N^{1/d}$ and $C = O(1)$, where N is the number of nodes in the network; for a random graph, $L \sim \ln N$ and $C \sim 1/N$.

An important characteristic of a graph that was not taken into consideration in the small-world model of Watts and Strogatz is the degree of distribution, that is, the distribution of number of connections of the nodes in the network. The Erdos–Renyi class of random graphs has a Poisson degree distribution, while lattice-like networks have even more strongly peaked distributions, which means they are a perfectly ordered lattice having a delta-Dirac degree of distribution. Similarly, the small-world networks generated by the Watts and Strogatz model also have peaked, single-scale, degree distributions, that is, one can clearly identify a typical degree of the nodes comprising the network. Against this theoretical background, Barabasi and coworkers found that a number of real-world networks have a scale-free degree of distribution with tails that decay following the power law (Barabási and Albert 1999). These networks were called scale-free networks (SFNs).

1.3 Scale-free networks

1.3.1 Basic properties of SFNs

Scale-free networks are complex networks in which some nodes are very well connected while most nodes have a very small number of connections. An important characteristic of scale-free networks is that they are size-independent, that is, they preserve the same characteristics regardless of the network size N. Scale-free networks have a degree of distribution that follows a power relationship, $P(k) = k^{-\lambda}$. Many real networks have a scale-free degree distribution, including the Internet.

The Internet is the primary example of a self-organizing complex system, having grown mostly in the absence of centralized control or direction. In this network, information is transferred in the form of packets from the sender to the receiver via routers, which are computers specialized in the transfer packets to another router "closer" to the receiver. A router decides the route of the packet using only local information obtained from its interaction with neighboring routers, not by following instructions from a centralized server. A router stores packets in its finite queue and processes them sequentially. However, if the queue overflows due to excess demand, the router will discard incoming packets, leading to a congestion-like situation. A number of studies have probed the topology of the Internet and its implications for traffic dynamics. In Section 1.1, it has been already mentioned that Internet traffic fluctuations are statistically self-similar and that the traffic displays two separate phases: congested and noncongested. It was also shown that time series of a number of connections are nonstationary and are characterized by different mean values depending on the observation period and that the Internet displays a number of properties that distinguishes it from random graphs: wiring redundancy and clustering, nontrivial eigenvalue spectra of the connectivity matrix and a scale-free degree distribution.

The degree of distribution does not characterize the graph or ensemble in full. There are other quantities, such as the degree–degree correlation (between connected sites), the spatial correlations, etc. Several models have been presented for the evolution of scale-free networks, each of which may lead to a different ensemble. The first suggestion was the preferential attachment model proposed by Barabasi and Albert, which came to be known as the Barabasi–Albert (BA) model. Several variants have been suggested to this model. One of them, known as the Molloy–Reed construction (Molloy and Reed 1998), which ignores the evolution and assumes only the degree of distribution and no correlations between the nodes, will be considered in the following. Thus, reaching a site by following a link is independent of the origin.

The most representative property of a SFN is the power-law distribution $P(k)$ for the number of nodes k, which are connected to a randomly chosen node. Considering the degree of node as a number of edges that start from a given node, one can observe that the node degree of distribution is well represented by $P(k) \sim k^{-\lambda} e^{-(k/K)}$, where the coefficient λ may vary approximately from 1 to 3 for most real networks, and $K \gg 10$ shows an exponential cutoff.

Another important result is that, in spite of the huge dimension of a network like the World Wide Web, having $\langle k \rangle \equiv \sum kP(k) \sim O(1)$, the average length ℓ between two nodes is very small. For this reason SFNs can be considered as small-world networks, in particular.

The third main property of SFNs is clustering, which is characterized by the grouping coefficient C_i. The grouping coefficient of the node i of degree k_i is $C_i = (2E_i/k_i(k_i - 1))$, where E_i is the number of edges between the neighbors of the node i, and $k_i(k_i - 1)/2$ is the maximum number of edges. The grouping coefficient of the whole graph is the average of the grouping coefficients of all nodes.

1.3.2 *Distances and bounds in SFN*

In most random network models the structure is locally tree-like (since most loops occur only for $n(l) \sim N$), and, since the number of sites grows as $n(l) \sim (k-1)^l$, they are also infinite-dimensional. As a consequence, the diameter of such graphs (i.e., the minimal path between the most distant nodes) scales like $D \sim \ln N$. This small diameter is to be contrasted with that of finite-dimensional lattices, where $D \sim N^{1/d_l}$. Watts and Strogatz (1998) have suggested a model that retains the local high clustering of lattices while reducing the diameter to $D \sim \ln N$. This so-called small-world network is achieved by replacing a fraction z of the links in a regular lattice with random links, to random distant neighbors.

We now aim to show that scale-free networks with the exponential degree $2 < \lambda < 3$ has a diameter $D \sim \ln \ln N$, which is smaller than that of ER and small-world networks. If the network is fragmented, we will only be interested in the diameter of the largest cluster (assuming there is one). We consider the diameter of a Molloy–Reed scale-free network defined as the average distance between any two sites on the graph. Actually, it is easier still to focus on the radius of a graph, $L \equiv \langle l \rangle$ as the average distance of all sites from the site of highest degree in the network (if there is more than one, we pick one arbitrarily). The diameter of the graph, D, is restricted to $L \leq D \leq 2L$ and thus scales like L.

Cohen et al. show that the radius of any scale-free graph with $\lambda > 2$ has a rigorous lower bound that scales as $\ln \ln N$ (Cohen et al. 2000). It is easy to understand that the smallest diameter of a graph, of a given degree of distribution, is achieved by the following construction: Start with the highest degree site and then connect to each successive layer the existing sites of the highest degree, until the layer is full. By construction, loops will occur only in the last layer.

To bind the radius L of the graph, we will assume that the low degree sites are connected randomly to the giant cluster. We pick a site of degree $1 \ll k^* \ll (\ln \ln N)^{1/(\lambda-1)}n$. If $l_1 \approx \ln \ln N/\ln(\lambda - 2)$ then $K_{l_1} < k^*$. Therefore, all sites of degree $k \geq k^*$ with probability 1 lie within the l_1 layers from the site we chose. On the other hand, if we start uncovering the graph from any site—provided it belongs to the giant component—then within a distance l_2 from this site, there are at least l_2 bonds. Since $l = l_1 + l_2$, all sites are at a distance of order $\ln \ln N$ from the highest degree site, and $L = \ln \ln N$ is a rigorous lower bound for the diameter of scale-free networks with $\lambda > 2$.

In a similar way, one can demonstrate that the scaling of $D \sim \ln \ln N$ is actually realized in the general case of random scale-free graphs with $2 < \lambda < 3$. For $\lambda > 3$ and $N \gg 1$, k is independent of N, and the radius of the network is $L \sim \ln N$ (Newman et al. 2001). The lower bound is obtained from the highest degree site for $\lambda = 3$, with $K = m\sqrt{N}$. Then, assuming $\ln \ln N \gg 1$, the upper bound becomes $L \sim \ln N/\ln \ln N$. This result has been obtained rigorously for the maximum distance in the BA model where $\lambda = 3$, for $m \geq 2$. For $m = 1$, the graphs in the BA model turn into trees, and the behavior of $D \sim \ln N$ is obtained. It should be noted that for $m = 1$ the giant component in the random model contains only a fraction of the sites (while for $m \geq 2$ it contains all sites at least in the leading order). This might explain why exact trees and BA trees are different from Molloy–Reed random graphs.

Figure 1.4 shows a scale-free network that grows increment by increment from 2 to 11 nodes. When a new node (gray color) makes a new connection, it attaches preferentially to the most connected existent node (black color). These two mechanisms—incremental step-by-step growth and preferential attachment—lead finally to a system dominated by hubs, that is, nodes having a huge number of connections.

After the researchers established that $P(k)$ has a scale-free nature, they tried to generalize the concept of random graph by a deterministic construct of the degree distributions. That leads to the theory of random graphs with arbitrary degree distributions, in which one can compute the average of all graphs with N possible nodes, each graph being weighted with $\prod_{i=1}^{n} P(k_i)$, where $\{k_1, \ldots, k_N\}$ are the nodes degrees. One of the first results in this field was proposed by Molloy and Reed, which demonstrated that if $\sum_k K(k-2)P(k) > 0$, then there exists a phase transition in the distribution space $\{P(k)\}$ (Molloy and Reed 1998). After that, Dorogovtsev et al. (2000a,b) used a "master-equation" to obtain an exact solution for a class of growing network models, Krapivsky et al. (2000) examined the effect of a nonlinear preferential attachment on network dynamics and topology. Models that incorporate aging and cost and capacity constraints were studied by Amaral et al. (2000) to explain deviations from the power-law behavior in several real-life networks. Bianconi and Barabasi (2001) introduced a model addressing the competitive aspect of many real networks such as the World Wide Web. Additionally, in real systems microscopic events affect the network evolution, including the addition or rewiring of new edges or the removal of vertices or edges. Albert and Barabasi (2000) discussed a model that incorporates new edges between existing vertices and the rewiring of old edges. It is now well established that preferential attachment can explain the power-law characteristic of networks, but some other alternative mechanisms affecting the evolution of growing networks can also lead to the observed scale-free topologies. Kleinberg et al. (1999) and

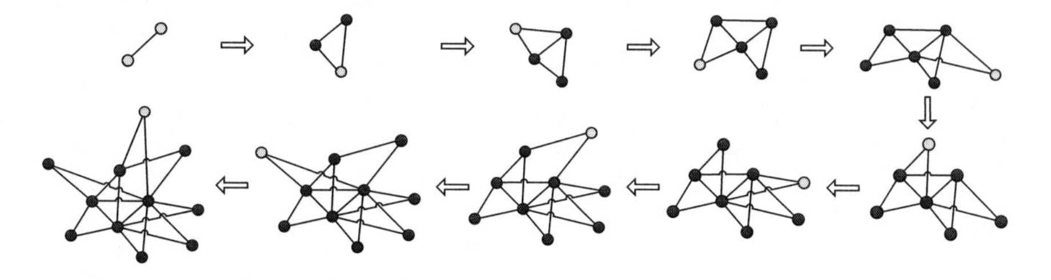

Figure 1.4 Birth of a scale-free network.

Kumar et al. (2000) suggested certain copying mechanisms in an attempt to explain the power-law degree distribution of the World Wide Web.

1.4 *Current trends in traffic flow and complex networks modeling*

The interest of the researchers for studying the traffic in large-scale informatics networks has growth in permanence starting from 1990. The capture and analysis of the traffic behavior in these networks were facilitated by the increased capacity in modeling offered by digital processing techniques (Crovella and Bestavros 1997, Floyd and Paxson 2001, Jiang et al. 2001, Willinger et al. 2001, Dobrescu et al. 2004b, Field et al. 2004).

After a large number of experiments, two main results toward characterizing and quantifying the network traffic processes have been achieved:

- First, self-similarity is an adaptability of traffic in networks. Many factors are involved in creating this characteristic. A new view of this self-similar traffic structure is provided. This view is an improvement over the theory used in most current literature, which assumes that the traffic self-similarity is solely based on the heavy-tailed file-size distribution.
- Second, the scaling region for traffic self-similarity is divided into two timescale regimes: short-range dependence (SRD) and long-range dependence (LRD). Experimental results show that a delay in the network transmission separates the two scaling regions. This gives us a physical source of the periodicity in the observed traffic. Also, bandwidth, TCP window size, and packet size have impacts on SRD. The statistical heavy tailedness (Pareto shape parameter) affects the structure of LRD. In addition, a formula to quantify traffic burstiness can be derived from the self-similarity property.

Starting 1999, the scale-free models were considered most suitable to characterize huge networks like Internet and the World Wide Web. Like the phenomenon of self-similarity was accepted as typical for traffic flow, network topologies considered initially to be random has shown a totally different structure when considering the topologic scaling and repeatability of the scale-free graphs, which are close to the concept of fractality, also introduced and discussed at the end of the twentieth century. Furthermore, studies of fractal traffic with multifractal analysis have given more interesting and applicable results:

- At large timescales, increasing bandwidth does not improve throughput (or network performance). The two factors affecting traffic throughput are network delay and TCP window size. On the other hand, more simultaneous connections smooth traffic, which could result in an improvement of network efficiency.

- At small timescales, traffic burstiness varies. In order to improve network efficiency, we need to control bandwidth, TCP window size, and network delay to reduce traffic burstiness. There are the tradeoffs from each other, but the effect is nonlinear.
- In general, the statistics of the network traffic processes differ from Poisson processes. To apply this prior knowledge from traffic analysis and to improve network efficiency, the notion of the efficient bandwidth can be derived to represent the fractal concentration set. Above that bandwidth, traffic appears bursty and cannot be reduced by multiplexing. But, below the bandwidth, traffic congestion occurs. An important finding is that the relationship between the bandwidth and the transfer delay is nonlinear.

The past few decades have seen an exponential growth in the amount of data being carried across packet-switched networks, and particularly the Internet. In recent analyses of traffic measurements, evidence of non-Markovian effects, such as burstiness across multiple timescales, long-range dependence, and self-similarity have been observed in a wide variety of traffic sources. Given the evidence of long-range dependence and self-similarity in such traffic sources, it is clear that any general model for data traffic must account for these properties. This has led to the development of a number of new models, which allow the characterization of phenomenon like betweenness centrality or traffic burstiness.

It is generally accepted that in many networks nodes having a larger degree also have a larger betweenness centrality. Indeed, the larger the degree of a node, the larger the chance that many of the shortest paths will pass through this node; the chance of many shortest paths passing a law degree node is presumably small.

The betweenness centrality of many small-degree nodes can be comparable to that of the largest hubs of the network. For nonfractal networks, on the other hand, degree and betweenness centrality of nodes are strongly correlated. To demonstrate the difference in the relation between degree and betweenness centrality in real networks we can compare original networks with their random (uncorrelated) counterparts. The random counterpart network can be constructed by rewiring the edges of the original network, yet preserving the degrees of the nodes and enforcing its connectivity. As a result, we obtain a random network with the same degree of distribution, which is always nonfractal regardless of the original network. One can observe that a random network obtained by rewiring edges of the World Wide Web network is much stronger compared to that of the original network (Kitsak et al. 2007).

The ranges of betweenness centrality values for a given degree decrease significantly as we randomly rewire edges of a fractal SF network. Thus, the betweenness centrality of nodes of fractal networks is significantly less correlated with degree than in nonfractal networks. This can be understood as a result of the repulsion between hubs found in fractals: large-degree nodes prefer to connect to nodes of small degree and not to each other. Therefore, the shortest path between the two nodes must necessarily pass through the small-degree nodes, which are found at all scales of a network. Thus, in fractal networks, small-degree nodes have a broad range of values of betweenness centrality while in nonfractal networks nodes of small degree generally have small betweenness centrality.

The problem of traffic burstiness raised a lot of controversies in the literature. Traffic that is bursty on many or all timescales can be characterized statistically using the concept of self-similarity. Self-similarity is often associated with objects in fractal geometry, that is, objects that appear to look alike regardless of the scale at which they are viewed. In the case of stochastic processes, like time series, self-similarity refers to the process distribution; when viewed at varying timescales, the process distribution remains the same. A self-similar time series has noticeable bursts—long periods with extremely high values on all timescales. Characteristics of network traffic, such as interarrival times or length of frames, can be considered as stochastic time series. Therefore, measuring traffic burstiness is the same as characterizing the self-similarity of the corresponding time series.

Various papers discuss the impact of burstiness on network congestion. Their conclusions can be formulated as follows:

- Congested periods can be quite long with losses that are heavily concentrated.
- Linear increases in buffer size do not result in large decreases in packet drop rates.
- A slight increase in the number of active connections can result in a large increase in the packet loss rate.

Many previous works also analyzed the burstiness and the correlation structure of Internet traffic in various timescales in terms of the protocol mechanisms of the TCP, such as timeouts, congestion avoidance, self-clocking, etc. It was shown that large transfers over high-capacity links produced non-Gaussian traffic, while the low-volume transmissions produced Gaussian and long-range-dependent traffic. Long sequences of back-to-back packets can cause significant correlations in short timescales.

A first generation of papers, approximately from 1994 to 2004, argued that the traditionally used Poisson models oversimplified the characteristics of network traffic and were not appropriate for modeling bursty, local-area, and wide-area network traffic. Since 2004, a second generation of papers has challenged the suitability of these results in networks of the new century and has claimed that the traditional Poisson-based and other models are still more appropriate for characterizing today's Internet traffic burstiness. A possible explanation was that as the speed and amount of Internet traffic grow spectacularly, any irregularity of the network traffic, such as self-similarity, might cancel out as a consequence of high-speed optical connections, new communications protocols, and the vast number of multiplexed flows. These papers analyzed the traffic traces of Internet backbone. Terdik and Gyires (2009) applied the theory of smoothly truncated Levy flights and the linear fractal model in examining the variability of Internet traffic from self-similar to Poisson and demonstrated that the series of interarrival times was still close to a self-similar process, but the burstiness of the packet lengths observed from the Internet backbone traces captured in 2008 significantly decreased compared to earlier traces.

This book offers a rigorous analysis of the achievements in the field of traffic control in large networks, oriented on two main aspects: the self-similarity in traffic behavior and the scale-free characteristic of a complex network. Additionally, the authors propose a new insight in understanding the inner nature of things and the cause and effect based on identification of relationships and behaviors within a model, based on the study of the influence of the topological characteristics of a network upon the traffic behavior. The effects of this influence are then discussed in order to find new solutions for traffic monitoring and diagnosis and for prediction of traffic anomalies.

chapter two

Flow traffic models

2.1 Background in traffic modeling

2.1.1 Definition of the informational traffic

One important research area in the context of networking focuses on developing traffic models that can be applied to the Internet and, more generally, to any communication network. The interest in such models is twofold. First, traffic models are needed as an input in network simulations. In turn, these simulations must be performed in order to study and validate algorithms and protocols that can be applied to real traffic, and to analyze how traffic responds to particular network conditions (e.g., congestion). Thus, it is essential that the proposed models reflect as much as possible the relevant characteristics of the traffic it is supposed to represent. Second, a good traffic model may lead to a better understanding of the characteristics of the network traffic itself. This, in turn, can prove to be useful in designing routers and devices which handle network traffic.

A traffic model represents a stochastic process, which can be used also to predict the behavior of a real traffic stream. Ideally, the traffic model should accurately represent all of the relevant statistical properties of the original traffic, but such a model may become overly complex. A major application of traffic models is the prediction of the behavior of the traffic as it passes through a network. In this context, the response of the individual network elements in the traditional Internet can be modeled using one or more single server queues (SSQs). Hence, a useful model for network traffic modeling applications is the one that accurately predicts queuing performance in an SSQ. Matching the first- and second-order statistics provides us with confidence that such a performance matching is not just a lucky coincidence. In order to keep our modeling parsimonious, we aim to typify a given traffic stream using as few parameters as possible.

Usually the traffic is considered as a sequence of arrivals of discrete entities (packets, cells, etc.). Mathematically, it is a point process, which resides in a set of arrival moments $T_1, T_2, \ldots, T_n, \ldots$ measured from the origin 0, that is, $T_0 = 0$. There are other two possible descriptions of the point processes: counting processes and interarrival time processes. A counting process $\{N(t)\}_{t=0}^{\infty}$ is a continuous-time, integer-valued stochastic process, where $N(t) = \max\{n{:}T_n \leq t\}$ expresses the number of arrivals in the time interval $(0,t]$. An interarrival time process is a nonnegative random sequence $\{A_n\}_{n=1}^{\infty}$, where $A_n = T_n - T_{n-1}$ indicates the length of the interval separating arrivals $n - 1$ and n. The two kinds of processes are related through the following equation:

$$\{N(t) = n\} = \{T_n \leq t < T_{n+1}\} = \left\{ \sum_{k=1}^{n} A_k \leq t < \sum_{k=1}^{n+1} A_k \right\} \tag{2.1}$$

The equivalence resides in the cumulative effect $T_n = \sum_{k=1}^{n} A_k$ and in the equality of the events, supposing that the intervals between two arrivals $\{A_n\}$, form a stationary sequence. An alternate characterization of the point processes based on the theory of the stochastic intensity is presented later in Section 2.7.

In case of compound traffic, arrivals may happen in batches, that is, several arrivals can happen at the same instant T_n. This fact can be modeled by using an additional nonnegative random sequence of real values $\{B_n\}_{n=1}^{\infty}$, where B_n is the cardinality of the nth batch (it may be a random number). The traffic model is largely defined by the nature of the stochastic processes $\{N(t)\}$ and $\{A_n\}$ chosen, with the condition that the random variables A_n can have only integer values, that is, the random variables $N(t)$ can grow only when T_n are integers.

One important issue in the selection of the stochastic process is its ability to describe traffic *burstiness*. In particular, a sequence of arrival times will be bursty if the T_n tend to form clusters, that is, if the corresponding $\{A_n\}$ becomes a mix of relatively long and short interarrival times. Mathematically speaking, traffic burstiness is related to short-terms autocorrelations between the interarrival times. However, there is no single widely accepted notion of burstiness (Frost and Melamed 1994); instead, several different measures are used, some of which ignore the effect of second-order properties of the traffic. A first measure is the ratio of peak rate to mean rate, which though has the drawback of being dependent upon the interval used to measure the rate. A second measure is the coefficient of variation $c_A = \sigma[A_n]/E[A_n]$ of the interarrival times. A metric considering second-order properties of the traffic is the index of dispersion for counts (IDC). In particular, given an interval of time τ, $IDC(\tau) = Var[N(\tau)]/E[N(\tau)]$. Finally, as will be better detailed later, the Hurst parameter can be used as a measure of burstiness in case of self-similar traffic.

Another useful notion is the workload process $\{W_n\}_{n=1}^{\infty}$. It is described by the amount of work W_n brought to the system by the nth arrival, assuming that it is independent of the interarrival times and the dimension of the groups. An example is the sequence of the requests for service times of the arrivals in a queue. In such cases, if the traffic is deterministic, only the description of the workload is necessary.

The following notation will be used: the distribution function of A_n is denoted by $F_A(x)$. Similarly, $\lambda_A = 1/E[A_n]$ denotes the traffic rate, $\sigma_A^2 = \text{Var}[A_n]$ and $c_A = \lambda_A\sigma_A$ assuming also that $0 < \sigma_A < \infty$, and that $\{A_n\}$ is a simple one, that is, $P\{A_n = 0\} = 0$. A traffic flow is denoted by X when other particular description (by A, N, or T) is not necessary.

2.1.2 *Internet teletraffic modeling*

2.1.2.1 *Introduction in teletraffic theory*

Teletraffic theory (Akimaru and Kawashima 1999) is the basis for performance evolution and dimensioning of telecommunication networks. It was developed alongside the momentous changes of switching and networking technology in the last decades. The theory has incorporated the recent advances in operation research and queuing theory. Teletraffic theory deals with the application of mathematical modeling of the traffic demand, network capacity and realized performance relationships. The traffic demand is statistical in nature, resulting in the generation of relevant models derived from the theory of stochastic processes. The nature of traffic in today's data networks (e.g., Internet) is completely different from classical telephone traffic and the main difference can be explained by the fact that in traditional telephony the traffic is highly static in nature. The static nature of telephone traffic resulted in "universal laws" governing telephone networks like the Poisson nature of call arrivals. This law states that call arrivals are mutually independent and exponentially distributed with the same parameter. The great success of the Poissonian model is due to the parsimonious modeling, which is a highly desirable property in practice.

A similar "universal law" is that the call holding times follow more or less an *exponential distribution*. This model was also preferred due to its simplicity and analytical tractability in spite of the fact that the actual telephone call duration distribution sometimes deviates significantly from the exponential distribution. However, these deviations did not yield to major errors in the network design due to the nature of the Poisson arrival process. This is because several performance measures do not depend on the distribution but only on the average holding time.

A dramatic change happened concerning the validity of these laws when telephone networks were used not only for voice conversations but also for FAX transmissions and Internet access. The statistical characteristics of these services are significantly different from voice calls. As the popularity of the Internet increased due to the success of Web, more people started to use the classical telephone networks for Internet access. These changes call for reviewing the old laws and present a challenge for all teletraffic researchers.

The picture is completely different in case of data networks. All the expectations of finding similar universal laws for data traffic failed.

It is because data traffic is much more variable than voice traffic. Roughly speaking, it is impossible to find a general model because the individual connections of data communication can change from extremely short to extremely long and the data rate can also span a huge range. There is no static and homogenous nature of data traffic as it was found in case of the voice traffic. This extremely bursty nature of data traffic is mainly caused by the fact that this traffic is generated by machine-to-machine communication in contrast to the human-to-human communication.

This high variability of data traffic in both time (traffic dependencies do not decay exponentially fast as it was the case in voice traffic but long-term dependencies are present, e.g., in the autocorrelation of the traffic) and space (distributions of traffic-related quantities do not have exponential tails as it was the case of voice traffic) call for new models and techniques to be developed. Statistically, the long-term dependencies can be captured by long-range dependence (LRD), that is, autocorrelations that exhibit power-law decay. The extreme spatial variability can be described by heavy-tailed distributions with infinite variance, which is typically expressed by the Pareto distributions. The power-law behavior in both time and space of some statistical descriptors often cause the corresponding traffic process to exhibit fractal characteristics. The fractal properties often manifest themselves in self-similarity. It means that several statistical characteristics of the traffic are the same over a range of timescales. Self-similar traffic models seem to be successful parsimonious models to capture this complex fractal nature of network traffic in the previous decade. However, recent research indicates that the actual data traffic has a more refined burstiness structure, which is better captured by multifractality rather than only self-similarity, which is a special case of mono-fractality.

Besides the very variable characteristics of data traffic there are other factors that make predictions about data traffic characteristics more unreliable. The Internet traffic is doubling each year. The picture is even more complicated if we think of quality of service (QoS) requirements of data services which can be very different from one application to the other. Different QoS requirements generate different traffic characteristics. To describe these different traffic characteristics in case of both stream and elastic traffic flows a number of traffic models and traffic characterization techniques have been developed. Based on a successful traffic modeling, successful traffic dimensioning methods for resource allocation can also be developed.

2.1.2.2 Basic concepts of teletraffic theory

A demand for a connection in a network is defined as a call, which is activated by a customer. The call duration is defined as holding time or service time. The traffic load is the total holding time per unit time. The unit of traffic load is called erlang (erl) after the Danish mathematician Agner

Krarup Erlang (1878–1929), also known as the father of teletraffic theory. The traffic load has the following important properties:

1. The traffic load (offered traffic) a is given by $a = ch$ (erl) where c is the number of calls originating per unit time and h is the mean holding time.
2. The traffic load (offered traffic) is equal to the number of calls originating in the mean holding time.
3. The traffic load (carried traffic) carried by a single trunk is equivalent to the probability (fraction of time) that the trunk is used (busy).
4. The traffic load (carried traffic) carried by a group of trunks is equivalent to the mean (expected) number of busy trunks in the group.

A switching system is defined as a system connecting between inlets and outlets. A system is called a full availability system if any inlet can be connected to any idle outlet. Congestion is a state of the system when a connection cannot be made because of busy outlets or internal paths. The system is called a waiting or delay system if an incoming call can wait for a connection in case of congestion. If no waiting is possible in congestion state the call is blocked and the system is called as loss system or nondelay system. A full availability system can be described by the following:

1. *Input process*: This describes the way of call arrival process.
2. *Service mechanism*: This describes the number of outlets, service time distributions, etc.
3. *Queue discipline*: This specifies ways of call handling during congestion. In delay systems the most typical queuing disciplines are the first-in first-out (FIFO), last-in first-out (LIFO), priority systems, processor sharing, etc.

The Kendall notation A/B/C/D/E-F is used for classification of full availability systems where *A* represents the interarrival time distribution, *B* the service time distribution, *C* the number of parallel servers, *D* the system capacity, *E* the finite customer population, and *F* is the queuing discipline. The following notations are used: *M*, exponential (Markov); E_k, phase k Erlangian; H_n, order n hyperexponential; *D*, deterministic; *G*, general; *GI*, general independent; *MMPP*, Markov-modulated Poisson process; *MAP*, Markov arrival process.

For a Poisson arrival process (exponential interarrival times) in steady state the distribution of existing calls at an arbitrary instant is equal to the distribution of calls just prior to call arrival epochs. This relationship is called Poisson Arrivals See Time Averages (PASTA) because this probability is equal to the average time fraction of calls existing when observed over a sufficiently long period.

If the interarrival time is exponentially distributed, the residual time seen at an arbitrary time instant is also exponential with the same parameter. A model in which the interarrival time and the service time both exponentially distributed is called the Markovian model, otherwise it is called non-Markovian model.

2.1.2.3 *Teletraffic techniques*

Beyond the classical queuing methods there are numerous approximations, bounds, techniques to handle teletraffic systems.

The fluid flow approximation is a useful technique when we have lots of traffic units (packets) in the timescale under investigation. In this case, we can treat it as a continuous flow-like fluid entering a piping system. We can define $A(t)$ and $D(t)$ to be the random variables describing the number of arrivals and departures, respectively, in $(0,t)$. The number of customers in the system at time t is $N(t) = A(t) - D(t)$, assuming that the system is empty initially. By the weak law of large numbers, when $A(t)$ gets large it gets close to its mean and this is the same for $D(t)$. The fluid flow approximation simply replaces $A(t)$ and $D(t)$ by the their means, which are continuous deterministic processes.

The fluid flow approximation uses mean values and the variability in the arrival and departure processes is not taken into account. The diffusion approximation extends this model by modeling this variability (motivated by the central limit theorem) by normal distribution around the mean. Diffusion approximations are also applied to solve difficult queuing systems. For example, in the complex *G/G/1* system the queue length distribution can be obtained by diffusion methods.

An approach based on the information theory called the maximum entropy method is often useful in solving teletraffic systems. The basis for this method is Bernoulli's principle of insufficient reasons which states that all events over a sample space should have the same probability unless there is evidence to the contrary. The entropy of a random variable is minimum (zero) when its value is certain. The entropy is maximum when its value is uniformly distributed because the outcome of an event has maximum uncertainty. The idea is that the entropy be maximized subject to any additional evidence. The method is successfully used for example in the queuing theory.

A number of other methods have also been developed like queuing networks with several solving techniques, fixed point methods, decomposition techniques, etc.

2.1.3 *Internet teletraffic engineering*

Currently network provisioning is based on some rules of the thumb and teletraffic theory has no major impact on the design of the Internet.

The nature of the data traffic is significantly different from the nature of voice traffic. New techniques and models were developed in teletraffic theory of the Internet to respond to these challenges. In the following, we review the most possible two alternatives of Internet teletraffic engineering. The first is called the big bandwidth philosophy and the other is called managed bandwidth philosophy (Molnar and Miklos 1997).

2.1.3.1 Big bandwidth philosophy

There is a certain belief that there is no real need for some advanced teletraffic engineering in the Internet because the overprovisioning of resources can solve the problems. This is the big bandwidth philosophy. People from this school say that in spite of the dramatic increase in the Internet traffic volume each year, the capacity of links and also the switching and routing devices will be so cheap that overprovisioning of resources will be possible. It is worth investigating a little bit more deeply how realistic the "big bandwidth philosophy" is. It is assumed that the transmission and information technology can follow the trend of "Internet traffic doubling each year" trend and can provide cheap solutions. From a technological point of view it seems that this expectation is not unrealistic at least in the near future. Indeed, if you imagine today's Internet and you just increase the capacity of links you could have a network that supports even real-time communications without any QoS architectures. On the other hand, the locality of data in the future will also be dominant, which makes caching an important technical issue in future networks. Even today if you want to transmit all the bits that are stored on hard drives it would take over 20 years for completing the process. This trend probably gives a relative decrease in the total volume of transmitted information.

Another important factor is that the streaming traffic, which really requires some QoS support, is not dominant in the Internet. It was believed that it would become dominant but none of these expectations have been fulfilled so far, which can also be predicted for the future. The demand for this traffic type is not growing as fast as the capacity is increasing. Consider the following example: we have 1% streaming traffic so it needs some QoS support. We have two options. We can introduce some QoS architecture or we can increase the capacity by 5%. Advocates of the "big bandwidth philosophy" school argue that the second option is cheaper. They also say that multimedia applications will use store-and-reply technique instead of real-time streaming. They argue that the capacity of storage is increasing at about the same rate as transmission capacity. Moreover, due to transmission bottlenecks (e.g., wireless link) it makes sense to store information in local.

It is also interesting if we investigate the reason for capacity increase in the previous years. For example, we can see that people are not paying for cable modems or ADSL not because their modem links could not bring them more data, but because when they click on a hyperlink they want that page on their screen immediately. So they need the big capacity, not for downloading lots of bits, but rather for achieving a low latency when they initiate a file download. This is also the reason for the fact that the average utilization of LANs have been decreased by about a factor of 10 over the last decade: people want high bandwidth to provide low latency.

Will overprovisioning be the solution? Nobody knows at this time. It is rather difficult to predict what will happen mainly because this is not only a technical issue but rather depends on political and economic factors. However, as a modest prediction we might say that even if overprovisioning can be a solution for backbone networks it is less likely that it will happen also in access networks. For cases where overprovisioning cannot be applied we have a limited capacity which should be managed somehow. This leads us to the second alternative which is the "managed bandwidth philosophy."

2.1.3.2 *Managed bandwidth philosophy*

In the case of limited network resources some kind of traffic control should be implemented to provide appropriate capacity and router memory for each traffic class or stream to fulfill its QoS requirements. Basically, there are three major groups of QoS requirements: transparency, accessibility, and throughput. *Transparency* expresses the time and semantic integrity of the transferred data. As an example for data transfer semantic integrity is usually required but delay is not so important. *Accessibility* measures the probability of refusal of admission and also the delay of setup in case of blocking. As an example the blocking probability is in this class, which is a well-known and frequently used measure in telephone networks. The *throughput* is the main QoS measure in data networks. As an example a throughput of 100 kbit/s can ensure the transfer of most of the Web pages quasi-instantaneously (<1 s).

Considering the traffic types by nature two main groups can be identified: stream traffic and elastic traffic. The stream traffic is composed of flows characterized by their intrinsic duration and rate. Typical examples of stream traffic are the audio and video real-time applications: telephone, interactive video services, and videoconferencing. The time integrity of stream traffic must be preserved. The negligible loss, delay and jitter are the generally required QoS measures. The *elastic traffic* usually consists of digital objects (documents) transferred from one place to another. The traffic is elastic because the flow rate can vary due to external causes (e.g., free capacity). Typical elastic applications are the Web, e-mail, or file transfers. In case of elastic traffic the semantic integrity must be preserved. Elastic traffic can be characterized by the arrival process of requests and the distribution of object sizes. The throughput and the response time are the typical QoS measures in this class.

2.1.3.3 Open-loop control of stream traffic

The stream traffic is usually controlled by an *open-loop preventive traffic control* based on the notion of traffic contract. Traffic contract is a successful negotiation between the user and the network in which the user requests are described by a set of traffic parameters and required QoS parameters. Based on these requests the network performs an admission control accepting the communication and the traffic contract only if QoS requirements can be satisfied. The effectiveness of this control highly depends on how accurately the performance can be predicted based on the traffic descriptors. In practice, it turned out that it is not simple to define practically useful traffic descriptors. It is because it should be simple (understandable by the user), useful (for resource allocation), and controllable (verifiable by the network). The results of intensive research on finding such traffic descriptors with all these properties showed that it is practically impossible. As an example the standardized token bucket-type descriptors (both in ATM and Internet research bodies) are good controllable descriptors but they are less useful for resource allocation. The users are encouraged to use mechanisms (e.g., traffic shaping) to ensure declared traffic descriptors. Mechanisms can also be implemented at the network ingress to police traffic descriptors (traffic policing). Both shaping and policing are frequently based on the mentioned token bucket-type mechanisms.

The major types of open-loop traffic control (admission control) strategies depend on whether statistical multiplexing gain is aimed to be utilized and to what extent. Table 2.1 shows the main categories.

If no multiplexing gain is targeted to be achieved, we have the simplest case and we can simply allocate the maximal rate (peak rate) of all the connections, which is called the peak rate allocation. The advantage of this approach is that the only traffic descriptor is the peak rate of the connection. The admission control is very simple: it only has to check whether the sum of the required peak rates is over the total capacity. The main disadvantage of peak rate allocation is the waste of resources because statistically it is only a small fraction of the time when all the connections actually transmit traffic at the peak rate. If we design to share the bandwidth but not to share the buffer among connections, we have the rate envelope multiplexing case. This approach also called bufferless

Table 2.1 Main categories of open-loop traffic

Facility	Buffer sharing	Bandwidth sharing
Peak rate allocation	No	No
Rate envelope multiplexing	No	Yes
Rate sharing	Yes	Yes

multiplexing because in the fluid modeling framework of this method, there is no need for a buffer. Indeed, in rate envelope multiplexing the target is that the total input rate is maintained below the capacity. The events of exceeding the capacity should be preserved below a certain probability, that is, $P(\lambda_t > c) < \varepsilon$, where λ_t is the input rate process, c is the link capacity, and ϵ is the allowed probability of exceeding the capacity. In actual practice, buffers always needed to store packets that arrive simultaneously (cell scale congestion). All the excess traffic is lost, and the overall loss rate is $E[(\lambda_t - c)^+/E(\lambda_t)]$. The loss rate only depends on the stationary distribution of λ_t and not on its time-dependent properties. It is important because it means that the correlation structure has no effect on the loss rate. Therefore, the very difficult task of capturing traffic correlations (e.g., LRD) is not needed. The traffic structure can have impact on other performance measures but these can be neglected if the loss rate is small enough. For example, LRD traffic can yield to longer duration of the overloads than SRD (short-range dependence) traffic but using a small loss rate it can be neglected in practice. If we want to further increase the link utilization we have to share the buffer as well (Figure 2.1). This is the rate sharing method, also called buffered multiplexing.

The idea here is that by providing a buffer we can absorb some excess of the input rate. The excess of the queue length in the buffer at some level should be preserved below a certain probability, that is, $P(Q > q) < \varepsilon$, where q is the targeted queue length level, Q is the actual queue length, and ε is the allowed probability level of exceeding the targeted queue length. In this method much higher multiplexing gain and utilization can be achieved.

The main problem in rate sharing is that the loss rate realized with a given buffer size and link capacity depends in a complicated way on

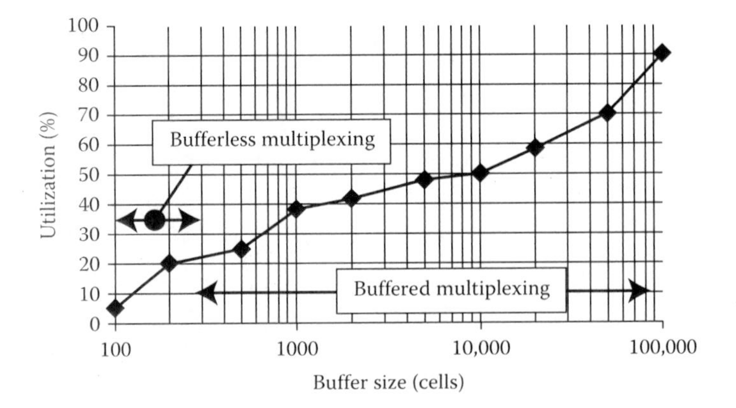

Figure 2.1 Two solutions for buffered multiplexing.

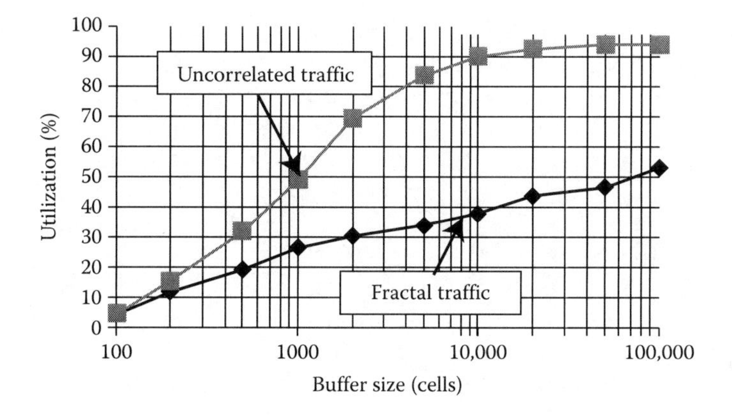

Figure 2.2 Effect of correlation structure.

the traffic characteristics including also the correlation structure. As an example the loss and delay characteristics are rather difficult to compute if the input traffic is LRD. This is the reason for the admission control methods being much more complicated for rate sharing than for rate envelope multiplexing. Moreover, the disadvantage is not only the complex traffic control but the achievable utilization is also smaller in case of fractal traffic with strong SRD and LRD properties (see Figure 2.2).

A large number of admission control strategies have been developed for both rate envelope multiplexing and rate sharing. It seems that the most powerful scheme is a kind of measurement-based admission control where the only traffic descriptor is the peak rate and the available rate is estimated in real time.

2.1.3.4 Closed-loop control of elastic traffic

Elastic traffic is generally controlled by reactive closed-loop traffic control methods. This is the principle of the TCP in the Internet and the ABR in the ATM. These protocols target at fully exploiting the available network bandwidth while keeping a fair share between contending traffic flows. Now we investigate the TCP as the general transfer protocol of the Internet. In TCP an additive increase, multiplicative decrease congestion avoidance algorithm has been implemented. If there is no packet loss the rate increases linearly but the packet transmission rate is halved whenever packet loss occurs. The algorithm tries to adjust its average rate to a value depending on the capacity and the current set of competing traffic flows on the links of its paths. The available bandwidth is shared in a roughly fair manner among the TCP flows.

A simple model of TCP, which also captures the fundamental behavior of the algorithm, is the well-known relationship between the flow throughput B and the packet loss rate p: $B(p) = c/(RTT\sqrt{p})$, where RTT is the TCP flow round-trip time and c is a constant. It should be noted that this simple formula is valid in case of a number of assumptions: RTT is constant, p is small (<1%), and the TCP source is greedy. The TCP mechanism is also assumed to be governed by the fast retransmit and recovery (no timeouts) and the slow-start phase is not modeled. More refined models were also developed but the square-root relationship between B and p seems to be the quite general rule of TCP.

We can conclude this section by stressing that the importance of choosing a good traffic model determines how successful we are in capturing the most important traffic characteristics. The basic question is the fundamental relationship among the traffic characteristics, network resources, and performance measures. Queuing models with some types of traffic models (e.g., Poisson, MMPP, MAP, etc.) are analytically tractable but others (e.g., ARIMA [autoregressive moving average], TES, FGN, etc.) are not. It remains a current research issue to develop new theoretical and applied tools to assist in solving teletraffic systems with the development of new and complex traffic models. Among them, the most promising seems to be those based on time series modeling.

2.1.4 *Internet traffic times series modeling*

The Internet traffic grows in complexity as the Internet becomes the universal communication network and conveys all kinds of information: from binary data to real-time video or interactive information. Evolving toward a multiservice network, its heterogeneity increases in terms of technologies, provisioning, and applications. The various networking functions (such as congestion control, traffic engineering, and buffer management) required for providing differentiated and guaranteed services involve a variety of timescales which, added to the natural variability of the Internet traffic, generate traffic properties that deviates from those accounted for by simple models. Therefore, a versatile model for Internet traffic that is able to capture its characteristics regardless of time, place, and aggregation level, is a step forward for monitoring and improving the Internet: for traffic management, charging, injecting realistic traffic in simulators, intrusion detection systems, etc.

Packets arrival processes are natural fine-grain models of computer network traffic. They have long been recognized as not being the Poisson (or renewal) processes, insofar as the interarrival delays are not independent. Therefore, nonstationary point processes or stationary Markov-modulated point processes have been proposed as models. However, the use of this fine granularity of modeling implies taking into account the large number of packets involved in any computer network traffic, hence huge data sets to manipulate. So a coarser description of the traffic is more convenient: the packet or byte-aggregated count processes. They are defined as the number of packets (bytes) that lives within the *k*th window of size $\Delta > 0$, that is, whose time stamps lie between $k\Delta \leq t_l < (k+1)\Delta$; it will be noted as $X_\Delta(k)$. Therefore, an objective is the modeling of $X_\Delta(k)$ with a stationary process. Marginal distributions and auto covariance functions are then the two major characteristics that affect the network performance.

Due to the point process nature of the underlying traffic, Poisson or exponential distributions could be expected at small aggregation levels Δ; but it fails at larger aggregation levels. As $X_\Delta(k)$ is by definition a positive random variable, other works proposed to describe its marginal with common positive laws such as log-normal, Weibull or gamma distributions (Scherrer et al. 2007). For highly aggregated data, Gaussian distributions are used in many cases as relevant approximations. However, none of them can satisfactorily model traffic marginals both at small and large Δs. As it will be argued in this chapter, empirical studies suggest that a Gamma distribution $\Gamma_{\alpha,\beta}$ capture best the marginals of the X_Δ for a wide range of scales, providing a unified description over a wide range scales of aggregation Δ.

In Internet monitoring projects, traffic under normal conditions was observed to present large fluctuations in its throughput at all scales. This is often described in terms of long memory, self-similarity, and multifractality that impacts the second- (or higher-order) statistics. For computer network traffic, long memory or LRD property is an important feature as it seems to be related to decreases of the QoS as well as of the performance of the network, hence need to be modeled precisely. LRD is defined from the behavior at the origin of the power spectral density $f_{X_\Delta}(\nu)$ of the process:

$$f_{X_\Delta}(\nu) \approx C\,|\,\nu\,|^{-\gamma}, |\,\nu\,| \to 0, 0 < \gamma < 1$$

Note that Poisson or Markov processes, or their declinations, are not easily suited to incorporate long memory. They may be useful in approximately modeling the LRD existing in the observation of a finite duration at the price of an increase in the number of adjustable parameters involved in the data description. But parsimony in describing data is indeed a much desired feature as it may be a necessary condition for a robust, meaningful and on-the-fly modeling. One can also incorporate long memory and short correlations directly into point processes using cluster point process models, yielding interesting description of the packet arrival processes. However, this model does not seem adjustable enough to encompass a large range of aggregation window size. An alternative relies on canonical long-range dependent processes such as fractional Brownian motion, fractional Gaussian noise, or Fractionally Autoregressive Integrated Moving Average (FARIMA). Due to the many different network mechanisms and various source characteristics, short-term dependencies also exist (superimposed to LRD) and play a central role. This leads to the idea to use the processes that have the same covariance as that of FARIMA models, as they contain both short- and long-range correlations.

A recurrent issue in traffic modeling lies in the choice of the relevant aggregation level Δ. This is an involved question whose answer mixes up the characteristics of the data themselves, the goal of the modeling as well as technical issues such as real time, buffer size, and computational cost constraints. Facing this difficulty of choosing *a priori* Δ, it is of great interest to have at disposal a statistical model that may be relevant for a large range of values of Δ. One approach is the joint modeling of the marginal distribution and the covariance structure of Internet traffic time series, in order to capture both their non-Gaussian and long-memory features. For this purpose, the best solution is to choose a non-Gaussian long-memory process whose marginal distribution is a gamma law and whose correlation structure is the one of a FARIMA process.

2.2 *Renewal traffic models*

This section presents briefly the characteristics of the renewal traffic processes, and more specific the Poisson and Bernoulli processes.

The renewal models were first used due to their mathematic simplicity. In a renewal traffic process, A_n is independent and identically distributed (i.i.d), but the distribution law can be a general one (Clark and Schimmel 2004). Unfortunately, with a few exceptions, the superposition of the independent renewal processes does not generate a new renewal process. But the mentioned exceptions have an important place in the theory and the practice of traffic description, especially those based on queues. On the other hand, the renewal processes, despite their analytic simplicity, have a main modeling drawback—the self-correlation function of $\{A_n\}$ disappears for all nonzero lags and so the important role of the self-correlation as statistically representative of the temporal dependence of the time series is lost. It has to be noted that a positive self-correlation of $\{A_n\}$ can explain the traffic variability. A variable (bursty) traffic is dominant in broadband networks and when this traffic is present in a queue system the performance (such as mean waiting times) is altered in comparison with that of the renewal traffic (which lacks temporal dependence). For these reasons, the models that capture the self-correlated nature of the traffic are essential for the prediction of the large scale networks performance.

2.2.1 *Poisson processes*

Poisson models are the oldest traffic models, dating back to the advent of telephony and the renowned pioneering telephone engineer, A. K. Erlang. In traditional queuing theory, the Poisson arrival process has been a favorite traffic model for data and voice. The traditional assumption of Poisson arrivals has been often justified by arguing that the aggregation of many independent and identically distributed renewal processes tends to a Poisson process when the number increases.

A Poisson process can be characterized as a renewal process whose interarrival times $\{A_i\}$ are exponentially distributed with rate parameter λ, that is, $P\{A_n \leq t\} = 1 - \exp(-\lambda t)$. Equivalently, it is a counting process, satisfying $P\{N(t) = n\} = \exp(-\lambda t)(\lambda t)^n/n!$, and the number of arrivals in disjoint intervals is statistically independent (a property known as independent increments). Poisson processes enjoy some elegant analytical properties. First, the superposition of independent Poisson processes results in a new Poisson process whose rate is the sum of the component rates. Second, the independent increment property renders Poisson process without memory. This, in turn, greatly simplifies queuing problems involving Poisson arrivals. Third, Poisson processes are fairly common in traffic applications that physically comprise a large number of independent traffic streams, each of which may be quite general. The theoretical basis for this phenomenon is known as Palm's theorem (Arrowsmith et al. 2005). It roughly states that under suitable but mild regularity conditions, such multiplexed streams approach a Poisson process as the number of streams grows, but the individual rates decrease so as to keep the aggregate rate constant. Thus, traffic streams on main communications arteries are commonly believed to follow a Poisson process, as opposed to traffic on upstream tributaries, which are less likely to be Poisson. However, traffic aggregation need not always result in a Poisson stream. Time-dependent Poisson processes are defined by letting the rate parameter λ depend on time. Compound Poisson processes are defined in the obvious way, by specifying the distribution of the batch size, B_n, independent of A_n.

Despite the attractiveness of the Poisson model, its validity in real-time traffic scenario has been often questioned. Barakat et al. offered evidence that flow arrivals on Internet backbone links are well matched by a Poisson process (Barabási and Bonabeau 2003). For large populations where each user is independently contributing a small portion of the overall traffic, user sessions can be assumed to follow a Poisson arrival process (Roberts 2001). Based on traces of wide-area TCP traffic, Poisson arrivals appears to be suitable for traffic at the session level when sessions are human initiated, for example, interactive FTP sessions (Paxson and Floyd 1994). However, the Poisson model does not hold for machine-initiated sessions or for any packet-level traffic.

2.2.2 *Bernoulli processes*

Bernoulli processes are the discrete-time analog of the Poisson processes (time-dependent and compound Bernoulli processes are defined in the natural way). A Bernoulli process is a finite or infinite sequence of independent random variables X_1, X_2, X_3, …, such that

- For each i, the value of X_i is either 0 or 1
- For all values of i, the probability that $X_i = 1$ is the same number p

In other words, a Bernoulli process is a sequence of independent identically distributed Bernoulli trials. Independence of the trials implies that the process is without memory. Given that the probability p is known, past outcomes provide no information about future outcomes. If the process is infinite, then from any point the future trials constitute a Bernoulli process identical to the whole process, the fresh-start property. Considering the probability of an arrival in any time slot is p, independent of any other one, it follows that for slot $\boldsymbol{k}$, the corresponding number of arrivals is binomial, and between 0 and k we have $P\{N_k = n\} = \binom{k}{n} p^n (1-p)^{k-n}$. The time between arrivals is geometric with parameter p: $P\{A_n = j\} = p(1-p)^j$ being a nonnegative integer.

2.2.3 Phase-type renewal processes

An important special case of renewal models occurs when the interarrival times are of the so-called phase type. Phase-type interarrival times can be modeled as the time to absorption in a continuous-time Markov process $C = \{C(t)\}_{t=0}^{\infty}$ with state space $\{0, 1, \ldots, m\}$; here, state 0 is absorbing, all other states are transient, and absorption is guaranteed in a finite time. To determine A_n, start the process C with some initial distribution π. When absorption occurs (i.e., when the process enters state 0), stop the process. The elapsed time is A_n which implies that it is a probabilistic mixture of sums of exponentials. Then, restart with the same initial distribution π and repeat the procedure independently to get A_{n+1}. Phase-type renewal processes give rise to relatively tractable traffic models. They also enjoy the property that any interarrival distribution can be approximated arbitrarily closely by phase-type distributions.

2.3 Markov traffic models

Unlike renewal traffic models, Markov and Markov-renewal traffic models introduce dependence into the random sequence $\{A_n\}$. Consequently, they can potentially capture traffic burstiness, because of nonzero autocorrelations in $\{A_n\}$.

Let consider a continuous-time Markov process $M = \{M(t)\}_{t=0}^{\infty}$ with a discrete-state space. In this case, $\boldsymbol{M}$ behaves as follows: it stays in a state i for an exponentially distributed holding time with parameter λ_i, which depends on i alone; it then jumps to state j with probability p_{ij}, such that the matrix $P = [p_{ij}]$ is a probability matrix. In a simple Markov traffic model, each jump of the Markov process is interpreted as signaling an arrival, so interarrival times are exponential, and their rate parameters depend on the state from with the jump occurred. This results in dependence among interarrival times as a consequence of the Markov property. Markov models in slotted time can be defined for the process $\{A_n\}$ in terms of a Markov transition matrix $P = [p_{ij}]$. Here, state i corresponds to i idle slots separating successive arrivals, and p_{ij} is the probability of a j-slot separation, given that the previous one was an i-slot separation. Arrivals may be single units, a batch of units, or a continuous quantity. Batches may themselves be described by a Markov chain, whereas continuous-state, discrete-time Markov processes can model the (random) workload arriving synchronously at the system. In all cases, the Markov property introduces dependence into interarrival separation, batch sizes and successive workloads, respectively.

Markov-renewal models are more general than discrete-state Markov processes, yet retain a measure of simplicity and analytical tractability. A Markov-renewal process $R = \{(M_n, \tau_n)\}_{n=0}^{\infty}$, is defined by a Markov chain $\{M_n\}$ and its associated jump times $\{\tau_n\}$, subject to the following constraint: the pair (M_{n+1}, τ_{n+1}) of next state and interjump time depends only on the current state M_n, but neither on previous states nor on the previous interjump times. Again, if we interpret jumps (transitions) of $\{M_n\}$ as signaling arrivals, we would have dependence on the arrival process. Also, unlike in the case of the Markov process, the interarrival times can be arbitrarily distributed, and these distributions depend on both states straddling each interarrival interval.

The Markovian arrival process (MAP) is a broad and versatile subclass of Markov-renewal traffic processes, enjoying analytical tractability. Here, the interarrival times are phase-type but with a wrinkle: traffic arrivals still occur at absorption instants of the auxiliary Markov process M, but the latter is not restarted with the same initial distribution; rather, the restart state depends on the previous transient state from which absorption had just occurred. While MAP is analytically simple, it enjoys considerable versatility. Its formulation includes Poisson processes, phase-type renewal processes, and others as special cases (Roberts 2004). It also has the property that the superposition of independent MAP traffic streams results in a MAP traffic stream governed by a Markov process whose state space is the cross-product of the component state spaces.

2.3.1 Markov-modulated traffic models

Markov-modulated models constitute an extremely important class of traffic models. The idea is to introduce an explicit notion of state into the description of a traffic stream: an auxiliary Markov process is evolving in time and its current state modulates the probability law of the traffic mechanism. Let $M = \{M(t)\}_{t=0}^{\infty}$ be a continuous-time Markov process, with state space 1,2, ..., m. Now assume that while M is in state k, the probability law of traffic arrivals is completely determined by k, and this holds for every $1 \le k \le m$. Note that when M undergoes a transition to, say, state j, then a new probability law for arrivals takes effect for the duration of state j, and so on. Thus, the probability law for arrivals is modulated by the state of M (such systems are also called doubly stochastic [Frost and Melamed 1994], but the term "Markov modulation" makes it clear that the traffic is stochastically subordinated to M). The modulating process certainly can be more complicated than a continuous-time, discrete-state Markov process (so the holding times need not be restricted to exponential random variables), but such models are far less analytically tractable. For example, Markov-renewal modulated processes constitute a natural generalization of Markov-modulated processes with generally distributed interarrival times.

2.3.2 Markov-modulated Poisson process

The most commonly used Markov-modulated model is the Markov-modulated Poisson process (MMPP) model, which combines the simplicity of the modulating Markov process with that of the modulated Poisson process. In this case, the modulation mechanism simply stipulates that in state k of M, arrivals occur according to a Poisson process at rate λ_k. As the state changes, so does the rate. MMPP models can be used in a number of ways. Consider first a single traffic source with a variable rate. A simple traffic model would quantize the rate into a finite number of rates, and each rate would give rise to a state in some Markov modulating process. It remains to be verified that exponential holding times of rates are an appropriate description, but the Markov transition matrix $Q = [Q_{kj}]$ can be easily estimated from an empirical data: simply quantize the empirical data, and then estimate Q_{kj}, by calculating the fraction of times that M switched from state k to state j.

As a simple example, consider a two-state MMPP model, where one state is an "on" state with an associated positive Poisson rate, and the other is an "off" state with associated rate zero (such models are also known as interrupted Poisson). These models have been widely used to model voice traffic sources; the "on" state corresponds to a talk spurt (when the speaker emits sound), and the "off" state corresponds to a silence (when the speaker pauses for a break). This basic MMPP model can be extended to aggregations of independent traffic sources, each of which is an MMPP, modulated by an individual Markov process M_i, as described previously.

Let $J(t) = (J_1(t), J_2(t), \ldots, J_r(t))$, where $J_i(t)$ is the number of active sources of traffic type i, and let $M(t) = (M_1(t), M_2(t), \ldots, M_r(t))$ be the corresponding vector-valued Markov process taking values on all r-dimensional vectors with nonnegative integer components. The arrival rate of class i traffic in state $(j_1, j_2, \ldots, j_r)$ of $M(t)$ is $j_i\lambda_i$.

2.3.3 Transition-modulated processes

Transition-modulated processes are a variation of the state modulation idea. Essentially, the modulating agent is a state transition rather than a state. A state transition, however, can be described simply by a pair of states, whose components are the one before transition and the one after it. The generalization of a transition-modulated traffic model to continuous time is straightforward (Adas 1997). Let $M = \{M_n\}_{n=1}^{\infty}$ be a discrete-time Markov process on the positive integers. State transitions occur on slot boundaries, and are governed by an mxm Markov transition matrix $P = [P_{ij}]$. Let B_n denote the number of arrivals in slot n, and assume that the probabilities $P\{B_n = k \mid M_n = i,\ M_{n+1} = j\} = t_{ij}(k)$, are independent of any past state information (the parameters $t_{ij}(k)$ are assumed given). Notice that these probabilities are conditioned on transitions (M_n, M_{n+1}) of M from state M_n to state M_{n+1} during slot n. Furthermore, the number of traffic arrivals during slot n is completely determined by the transition of the modulating chain through the parameters $t_{ij}(k)$.

Markov-modulated traffic models are a special case of Markovian transition-modulated ones: simply take the special case when the conditioning event is $\{M_n = i\}$. That is, $t_{ij}(t) = t_i(t)$ depends only on the state i of the modulating chain in slot n, but is independent of its state j in the next slot $n + 1$. Conversely, Markovian transition-modulated processes can be thought of as Markov-modulated ones, but on a larger state space. Indeed, if $\{M_n\}$ is Markov, so is the process $\{M_n, M_{n+1}\}$ of its transitions.

As before, multiple transition-modulated traffic models can be defined, one for each traffic class of interest. The complete traffic model is obtained as the superposition of the individual traffic models.

2.4 Fluid traffic models

The fluid traffic concept dispenses with the individual traffic units. Instead, it views traffic as a stream of fluid, characterized by a flow rate (such as bits per second), so that a traffic count is replaced by a traffic volume. Fluid models are appropriate to cases where individual units are numerous relative to a chosen timescale. In other words, an individual unit is by itself of little significance, just as one molecule more or less in a water pipeline has but an infinitesimal effect on the flow. In the B-ISDN context of ATM, all packets are fixed-size cells of relatively short length (53 bytes); in addition, the high transmission speeds (say, on the order of a gigabit per second) render the transmission impact of an individual cell negligible. The analogy of a cell to a fluid molecule is a plausible one. To further highlight this analogy, contrast an ATM cell with a much bigger transmission unit, such as a coded (compressed) high-quality video frame, which may consist of a thousand cells. A traffic arrival stream of coded frames should be modeled as a discrete stream of arrivals, because such frames are typically transmitted at the rate of 30 frames per second. A fluid model, however, is appropriate for the constituent cells. Although an important advantage of fluid models is their conceptual simplicity, important benefits will also accrue to a simulation model of fluid traffic. As an example, consider again a broadband ATM scenario. If one is to distinguish among cells, then each of them would have to count as an event. The time granularity of event processing would be quite fine, and consequently, processing cell arrivals would consume vast CPU and possibly memory resources, even on simulated timescales of minutes. A statistically meaningful simulation may often be infeasible. In contrast, a fluid simulation would assume that the incoming fluid flow remains (roughly) constant over much longer time periods. Traffic fluctuations are modeled by events signaling a change of flow rate. Because these changes can be assumed to happen far less frequently than individual cell arrivals, one can realize enormous savings in computing. In fact, infeasible simulations of cell arrival models can be replaced by feasible simulations of fluid models of comparable accuracy. In a queuing context, it is easy to manipulate fluid buffers. Furthermore, the waiting time concept simply becomes the time it takes to serve (clear) the current buffer, and loss probabilities (at a finite buffer) can be calculated in terms of overflow volumes. Because fluid models assume a deterministic service rate, these statistics can be readily computed. Typically, though, larger traffic units (such as coded frames) are of greater interest than individual cells. Modeling the larger units as discrete traffic and their transport as fluid flow would give us the best of both worlds: we can measure waiting times and loss probabilities and enjoy savings on simulation computing resources.

Typical fluid models assume that sources are bursty—of the "on–off" type. While in the "off" state, traffic is switched off, whereas in the "on" state traffic arrives deterministically at a constant rate L. For analytical tractability, the duration of "on" and "off" periods are assumed to be exponentially distributed and mutually independent (i.e., they form an alternating renewal process) (Jagerman et al. 1997). Fluid traffic models of these types can be analyzed as Markov-modulated constant rate traffic. The host of generalizations, described above for MMPP, carries over to fluid models as well, including multiple sources and multiple classes of sources.

2.5 *Autoregressive traffic models*

Autoregressive models define the next random variable in the sequence as an explicit function of previous ones within a time window stretching from the present into the past. Such models are particularly suitable for modeling processes which are large consumers of bandwidth in emerging high-speed communications networks, such as VBR-coded video. The nature of video frames is such that successive frames within a video scene vary visually very little (recall that there are 30 frames per second in a high-quality video). Only scene changes (and other visual discontinuities) can cause abrupt changes in frame bit rate. Thus, the sequence of bit rates (frame sizes) comprising a video scene may be modeled by an autoregressive scheme, while scene changes can be modeled by some modulating mechanism, such as a Markov chain.

2.5.1 *Linear autoregressive models (AR)*

The class of linear autoregressive models has the form

$$X_n = a_0 + \sum_{r=1}^{p} a_r X_{n-r} + \varepsilon_n, \quad n > 0, \tag{2.2}$$

where $(X_{-p+1}, \ldots, X_0)$ are prescribed random variables, a_r $(0 \leq r \leq p)$ are real constants and ε_n are zero mean, IID random variables, called residuals, which are independent of the $\mathbf{X}_n$. Equation 2.2 describes the simplest form of a linear autoregression scheme, called *AR*(*p*), where *p* is the order of the auto regression. In a good model, the residuals ought to be of a smaller magnitude than the X_n, in order to explain the empirical data. The recursive form in Equation 2.2 makes it clear how to randomly generate the next random element in the sequence $\{X_n\}_{n=0}^{\infty}$ from a previous one: this simplicity makes *AR* schemes popular candidates for modeling auto-correlated traffic. A simple *AR*(2) model has been used to model variable bit rate (VBR) coded video (Dobrescu et al. 2004a). More elaborate models can be constructed out of *AR*(*p*) models combined with other schemes. For example, video bit rate traffic was modeled as a sum $R_n = X_n + Y_n + K_n C_n$, where the first two terms comprise independent *AR*(1) schemes and the third term is a product of a simple Markov chain and an independent normal variate from an IID normal sequence (Ramamurthy and Sengupta 1992). The purpose of having two autoregressive schemes was to achieve a better fit to the empirical autocorrelation function; the third term was designed to capture sample path spikes due to video scene changes.

Autoregressive series are important because (1) they have a natural interpretation—the next value observed is a slight perturbation of a simple function of the most recent observations; (2) it is easy to estimate their parameters, usually with standard regression software; and (3) they are easy to forecast—again the standard regression software will do the job.

2.5.2 Moving average series (MA) models

The time series described by the model

$$X_n = \sum_{r=0}^{q} b_r \varepsilon_{n-r}, \quad n > 0, \tag{2.3}$$

is said to be a moving average process of order q – MA(q)—where b_r ($0 \le r \le q$), are real constants and ε_n are uncorrelated random variables with null average (white noise). No additional conditions are required to ensure stationarity. Note that it is easy to distinguish MA and AR series by the behavior of their autocorrelation functions (*acf*). The *acf* for MA series "cuts off" sharply while that for an AR series, it decays exponentially (with a possible sinusoidal ripple superimposed).

2.5.3 Autoregressive moving average series (ARMA) models

Modeling stationary and reversible processes using AR or MA series necessitate often the estimation of a large number of parameters, which reduces the estimation efficiency. In order to minimize this impediment, one can combine (2.2) and (2.3) in a mix model:

$$X_n = a_0 + \sum_{r=1}^{p} a_r X_{n-r} + \sum_{r=0}^{q} b_r \varepsilon_{n-r}. \tag{2.4}$$

The class of series having models of Equation 2.4 type is named ARMA(p, q) and referred to as autoregressive moving average series of order (p, q). Stationarity can be checked by examining the roots of the characteristic polynomial of the AR operator and model parametrization can be checked by examining the roots of the characteristic polynomial of the MA operator.

2.5.4 Integrated ARIMA models

An integrated autoregressive moving average (p, d, q) series, denoted as ARIMA(p, d, q), is closely related to ARMA(p, q) and can be obtain by substituting X_n in Equation 2.4 with the d-difference of the series $\{X_n\}$. ARIMA models are more general as ARMA models and include some nonstationary series. The term "integrated" denotes that the ARMA model is first approximated to the differentiate data and then added to form the target nonstationary ARIMA model. Therefore, a ARMA(p, q) process can be seen as a ARIMA(p, 0, q) process, and the random walk process can be seen as an ARIMA(0,1,0) process. It has to be noted that both ARMA and ARIMA series have autocorrelation functions with geometric decay, that is, $\rho(n) \sim r^n$ for $0 < r < 1$, when $n \to \infty$ and consequently can be used in the study of processes with short-range dependence.

2.5.5 *FARIMA models*

FARIMA processes are the natural generalizations of standard ARIMA (p, d, q) processes when the degree of differencing d is allowed to take non-integral values (Liu et al. 1999). A FARIMA(p, d, q) process $\{X_t: t = \ldots, -1, 0, 1, \ldots\}$ is defined as

$$\Phi(B)\Delta^d X_t = \Theta(B)a_t \tag{2.5}$$

where $\{a_t\}$ is a white noise and $d \in (-0.5, 0.5)$, $\Phi(\text{B}) = 1 - \varphi_1 B - \varphi_2 B^2 - \ldots \varphi_p B^p$ and $\Theta(B) = 1 - \theta_1 B - \theta_2 B^2 - \ldots - \theta_q B^q$. $\Phi(B)$, $\Theta(B)$ have no common zeroes, and also no zeroes in $|B| \leq 1$ while p and q are nonnegative integers. B is the backward-shift operator, that is, $BX_t = X_{t-1}$ $\Delta = 1 - B$ is the differencing operator and Δ^d denotes the fractional differencing operator, $\Delta^d = (1-B)^d = \Sigma_{k=0}\begin{pmatrix} a \\ k \end{pmatrix}(-B)^k$; $\begin{pmatrix} d \\ k \end{pmatrix} = \Gamma(d+1)/[\Gamma(k+1)\Gamma(d-k+1)]$; Γ denotes the Gamma function.

Clearly, if $d = 0$, FARIMA(p, d, q) processes are the usual ARMA(p, q) processes. If $d \in (0, 0.5)$, then LRD or persistence occurs in the FARIMA processes. FARIMA(0,d,0) process, that is, fractional differencing noise (FDN), is the simplest and most fundamental form of the FARIMA processes. The property of FARIMA(0,d,0) process is similar to fractional Gaussian noise (FGN) process, which can only describe LRD. The parameter d in FARIMA(0,d,0) process is the indicator for the strength of LRD, just like the Hurst parameter H in FGN process. In fact, $H = d + 0.5$. Both processes have autocorrelation functions which behave asymptotically as k^{2d-1} with different constants of proportionality.

For $d \in (0, 0.5)$, $p \neq 0$ and $q \neq 0$, a FARIMA(p, d, q) process can be regarded as an ARMA(p, q) process driven by FDN. From Equation 2.1, we obtain $X_t = \Phi^{-1}(B)\Theta(B)Y_t$, where $Y_t = \Delta^{-d}a_t$. Here, Y_t is a FDN. Consequently, compared with ARMA and FGN processes, the FARIMA(p, d, q) processes are flexible and parsimonious with regard to the simultaneous modeling of the long- and short-range dependent behavior of a time series.

2.6 TES traffic models

2.6.1 TES processes

Transform-expand-sample (TES) models provide another modeling approach geared toward capturing both marginals and autocorrelations of empirical records simultaneously, including traffic. The empirical TES methodology assumes that some stationary empirical time series (such as traffic measurements over time) is available. It aims to construct a model satisfying the following three fidelity requirements: (1) The model's marginal distribution should match its empirical counterpart (a histogram, in practice). (2) The model's leading autocorrelations should approximate their empirical counterparts up to a reasonable lag. (3) The sample path realizations generated by simulating the model should correspond to the empirical records. The first two are precise quantitative requirements, whereas the third is a heuristic qualitative one. Nevertheless, it is worth adopting this subjective requirement and keeping its interpretation at the intuitive level; after all, common sense tells us that if a model gives rise to time series which are entirely divorced in "appearance" from the observed ones, then this would weaken our confidence in the model, and vice versa.

TES processes are classified into two categories: TES^+ and TES^-. The superscript (plus or minus) is a mnemonic reminder of the fact that they give rise to TES processes with positive and negative lag-1 autocorrelations, respectively. TES models consist of two stochastic processes in lockstep, called background and foreground sequences, respectively. Background TES sequences have the form:

$$U_n^+ = \begin{cases} U_0, & n = 0 \\ \left\langle U_{n-1}^+ + V_n \right\rangle, & n > 0 \end{cases} \qquad U_n^- = \begin{cases} U_n^+, & n-\text{even} \\ 1-U_n^+, & n-\text{odd} \end{cases} \tag{2.6}$$

Here, U_0 is distributed uniformly on [0,1); $\{V_n\}_{n=1}^{\infty}$ is a sequence of IID random variables, independent of U_0, called the innovation sequence, and angular brackets denote the modulo-1 (fractional part) operator $\langle x \rangle = x - \max\{\text{integer } n{:}n \le x\}$.

Background sequences play an auxiliary role. The real targets are the foreground sequences: $X_n^+ = D(U_n^+)$, $X_n^- = D(U_n^-)$ where D is a transformation from [0,1) to the reals, called a distortion. It can be shown that all background sequences are Markovian stationary, and their marginal distribution is uniform on [0,1), regardless of the probability law of the innovations $\{V_n\}$. However, the transition structure $\{U_n^+\}$ is time invariant, while that of $\{U_n^-\}$ is time-dependent. The inversion method allows us to transform any background uniform variates to foreground ones with an arbitrary marginal distribution. To illustrate this idea, consider an empirical time series $\{Y_n\}_{n=0}^{N}$ from which one computes an empirical density $\hat{h}_Y$ and its associated distribution function $\hat{H}_Y$. Then, the random variable $X = \hat{H}_Y^{-1}(U)$ has the density $\hat{h}_Y$. Thus, *TES* foreground sequences can match any empirical distribution.

2.6.2 *Empirical TES methodology*

The empirical TES methodology actually employs a composite two-stage distortion:

$$D_{Y,\xi}(x) = \hat{H}_Y^{-1}(S_\xi(x)), \quad x \in [0,1) \tag{2.7}$$

where $\hat{H}_Y^{-1}$ is the inverse of the empirical histogram distribution based on Y, and S_ξ is a "smoothing" operation, called a stitching transformation, parameterized by $0 < \xi < 1$, and given by:

$$S_\xi(y) = \begin{cases} y/\xi, & 0 \le y < \xi \\ (1-y)/(1-\xi), & \xi \le y < 1 \end{cases} \tag{2.8}$$

For $0 < \xi < 1$, the effect of S_ξ is to render the sample paths of background TES sequences more "continuous-looking." Because stitching transformations preserve uniformity, the inversion method via $\hat{H}_Y^{-1}$ guarantees that the corresponding foreground sequence would have the prescribed marginal distribution $\hat{H}_Y$. The empirical TES modeling methodology takes advantage of this fact which effectively decouples the fitting requirements of the empirical distribution and the empirical autocorrelation function. Because the former is automatically guaranteed by TES, one can concentrate on fitting the latter. This is carried out by a heuristic search for a pair (ξ, f_V), where ξ is a stitching parameter and f_V is an innovation density; the search is declared a success on finding that the corresponding TES sequence gives rise to an autocorrelation function that adequately approximates its empirical counterpart, and whose simulated sample paths bear "adequate resemblance" to their empirical counterparts.

Stationary TES models can be combined to yield nonstationary composite ones. MPEG-coded video is such a case. It consists of three kinds of frames (called I-frames, P-frames, and B-frames), interleaved in a deterministically repeating sequence (the basic cycle starts with an I-frame and ends just short of the next 1-frame). Consequently, MPEG-coded VBR video is nonstationary, even if the corresponding I, P, and B subsequences of frames are stationary. A composite TES model can be obtained by modeling the I, B, and P subsequences separately, and then multiplexing the three streams in the correct order. The resulting multiplexed TES model obtained from the corresponding TES models of the subsequences, but with autocorrelation injected into frames within the same cycle. Although the autocorrelation functions and spectral densities were formally computed from a single sample path as if the MPEG sequences were stationary, and therefore represent averaged estimates of different correlation coefficients, they nevertheless give an indication of how well the composite TES model captured temporal dependence in the empirical data, because they were all computed from sample paths in the same way. The general good agreement between the TES model statistics and their empirical counterparts is in accord with the three fidelity requirements stipulated at the beginning of this section. These TES source models can be used to generate synthetic streams of realistic traffic to drive simulations of communications networks.

2.7 Self-similar traffic models

In the last 20 years, studies of high-quality, high-resolution traffic measurements have revealed a new phenomenon with potentially important ramifications to the modeling, design, and control of broadband networks. These works started with classical experiments, for example an analysis of hundreds of millions of observed packets over an Ethernet LAN in a R&D environment (Leland et al. 1993) or an analysis of a few millions of observed frame data generated by VBR video services (Beran 1994). In these studies, packet traffic appears to be statistically self-similar. A self-similar (or fractal) phenomenon exhibits structural similarities across all (or at least a wide range) of the timescales. In the case of packet traffic, self-similarity is manifested in the absence of a natural length of a burst: at every timescale ranging from a few milliseconds to minutes and hours, similar-looking traffic bursts are evident. Self-similar stochastic models include fractional Gaussian noise and FARIMA processes. Self-similarity manifests itself in a variety of different ways: a spectral density that diverges at the origin ($1/f^{\alpha}$ noise, $0 < \alpha < 1$), an on-summable autocorrelation function (indicating LRD), and a variance of the sample mean that decreases (as a function of the sample size n) more slowly than $1/n$. The key parameter characterizing these phenomena is the so-called Hurst parameter, H, which is designed to capture the degree of self-similarity in a given empirical record as follows.

Let $\{Y_k\}_{k=1}^{N}$ be an empirical time series with sample mean $\boldsymbol{Y(n)}$ and sample variance $S^2(n)$. The rescaled adjusted range, or R/S statistic, is given by $R(n)/S(n)$ with:

$$R(N) = \max\left\{\sum_{i=1}^{k}(Y_i - \bar{Y}(n)), 1 \le k \le n\right\} - \min\left\{\sum_{i=1}^{k}(Y_i - \bar{Y}(n)), 1 \le k \le n\right\}$$

It has been found empirically that many naturally occurring time series appear to obey the relation: $E[R(n)/S(n)] = n^H$, with n being large, the H value typically about 0.73. On the other hand, for renewal and Markovian sequences, it can be shown that the previous equation holds with $H = 0.5$, for large $\boldsymbol{n}$. This discrepancy, generally referred to as the Hurst phenomenon, is a measure of the degree of self-similarity in time series, and can be estimated from empirical data. From a mathematical point of view, self-similar traffic differs from other traffic models in the following way. Lets be a time unit representing a timescale, such as $s = 10^m$ seconds ($m = 0, \pm 1, \pm 2, \ldots$). For every timescale s, let $X^{(s)} = \{X_n^{(s)}\}$ denote the time series computed as the number of units (packets, bytes, cells, etc.) per time units in the traffic stream. Traditional traffic models have the property that, as s increases, the aggregated processes, $X^{(s)}$ end to a sequence of IID random variables (covariance stationary white noise).

On the other hand, the corresponding aggregation procedure of empirical traffic data yields time series $X^{(s)}$, which reveal two related types of behavior, when plotted against time. They either appear visually indistinguishable from one another ("exactly self-similar") but distinctively different from pure noise, or they converge to a time series with a nondegenerate autocorrelation structure ("asymptotically self-similar").

In contrast, simulations of traditional traffic models, rapidly converge to white noise after increasing the timescale by about two or three orders of magnitude. Similarly, when trying to fit traditional traffic models to self-similar traffic data, the number of parameters required typically grows, as the sample size increases. In contrast, self-similar traffic models are able to capture the observed fractal nature of packet traffic in a parsimonious manner (with about one to four parameters). Parameter estimation techniques are available for many self-similar models, as well as Monte Carlo methods for generating long traces of synthetic self-similar traffic.

Potential implications of self-similar traffic on issues related to design, control, and performance of high-speed, cell-based networks are currently under study, especially because it was shown that many of the commonly used measures for burstiness do not characterize self-similar traffic.

Contrary to commonly held beliefs that multiplexing traffic streams tends to produce smoothed out aggregate traffic with reduced burstiness, aggregating self-similar traffic streams can actually intensify burstiness rather than diminish it.

From a practical vantage point, there are also indications that traffic congestion in self-similar networks may have broadly differing characteristics from those produced by traditional traffic models. All these aspects will be detailed in Chapter 3.

chapter three

Self-similarity in traffic

3.1 Self-similar traffic and network performance

One of the most significant findings of traffic measurement studies over time has been the observed self-similarity in packet network traffic. Subsequent research has focused on the origins of such self-similarity, and the network engineering significance of this phenomenon, namely, the manner in which network dynamics (specifically, the dynamics of transmission control protocol [TCP], the predominant transport protocol used in today's Internet, which in turn exhibits mainly multimedia traffic) can affect the observed self-similarity.

Models of multimedia traffic offered to the network or to a component of the network are critical in providing high quality of service (QoS). Traffic models are used as the input to analytical or simulation studies of resource allocation strategies. We may view traffic at the application or packet level, where an application-level view may simply describe the profiled traffic as "a videoconference between three parties," while the packet-level view is based on a stochastic model that mimics the arrival process of packets associated with this application reasonably well. Clearly, in order to quantify traffic, packet-level representation of applications will be used. An important feature of multimedia traffic at the packet level having a significant impact on performance is traffic correlation. The complexity of traffic in a multimedia network is a natural consequence of integrating, over a single communication channel, a diverse range of traffic sources such as video, voice, and data that significantly differ in their traffic patterns as well as their performance requirements. Specifically, "bursty" traffic patterns generated by data sources and variable bit rate (VBR) real-time applications such as compressed video and audio tend to exhibit certain degrees of correlation between arrivals, and show LRD in time (Sahinoglu and Tekinay 1999).

The questions that arise here are how prevalent such traffic patterns are and under what conditions is performance analysis critically dependent upon taking self-similarity into account. There are different studies pointing out either the importance of self-similarity to network performance or the irrelevance of the need for capturing self-similarity in traffic modeling. To clarify this dilemma, a thorough understanding of QoS and resource allocation in a network environment is necessary. An optimal resource allocation would mean determining optimal buffer sizes, assignment of bandwidth, and other resources in order to get the desired QoS expressed in terms of parameters such as queuing delay, retransmission time, packet loss probability, and bit error rate.

3.1.1 *Quality of service and resource allocation*

The International Organization for Standardization (ISO) defines QoS as a measure for denoting how good the offered networking services are. Generally, QoS parameters are performance measures such as bit error rate, frame error rate, cell loss probability, delay, and delay variation or guarantee, which is the maximum difference between end-to-end delays experienced by any two packets, are used as the determining QoS parameters. The user and application requirements for the MCS are mapped into a communication system that tries to satisfy the requirements of the services, which are parameterized.

Parameterization of the services is defined in ISO standards through the notion of QoS. The set of chosen parameters for a particular service determines what will be measured as the QoS. Network QoS parameters describe requirements for network services. They may be specified in terms of (a) network load, characterized by average/minimal interarrival time on the network connection, packet cell size and service time in the node for the connection's packet/cell and (b) network performance, describing the requirements which the network services have to guarantee. The performance might be expressed through a source-to-destination delay bound for the connection's packet loss rate.

Services for multimedia networked applications need resources to perform their functions. Of special interest in the study of network characteristics that are shared among the application, system, and network. There are several constraints that must be satisfied during multimedia transmission: time constraints, which include delays, computing time, and signaling delay; space constraints such as system buffers; and frequency constraints like network bandwidth and system bandwidth for data transmission (Nahrstedt and Steinmetz 1994). The best utilization of resources in a network environment is only possible by first characterizing the traffic and then determining the parameters such as buffer size and bandwidth to maximize performance. The main objective in telecommunications network engineering is to have as many happy users as possible. In other words, the network engineer has to resolve the tradeoff between capacity and QoS requirements. Accurate modeling of the offered traffic load is the first step in optimizing resource allocation algorithms such that provision of services complies with the QoS constraints while maintaining maximum capacity. In recent years, as broadband multimedia services became popular, they have necessitated the development of new traffic models with self-similar characteristics.

3.1.2 Concept of self-similarity

A self-similar phenomenon displays structural similarities across a wide range of timescales. Traffic that is bursty on many or all timescales can be described statistically using the notion of self-similarity. Self-similarity is the property associated with "fractals," which are objects whose appearances remain unchanged regardless of the scale at which they are viewed. In the case of stochastic objects like time series, self-similarity is used in the distributed sense: when viewed at varying timescales, the object's relational structure remains unchanged. As a result, such a time series exhibits bursts at a wide range of timescales.

In 1993, paper titled "On the Self-Similar Nature of Ethernet Traffic" (Leland et al. 1993) was published, considered a landmark in the field of network performance modeling. Ethernet is a broadcast multiaccess system for local area networking with distributed control. The authors reported the results of a massive study of Ethernet traffic and demonstrated that it had a self-similar (i.e., fractal) characteristic. This meant the Ethernet traffic had similar statistical properties at a range of timescales: milliseconds, seconds, minutes, hours, and even days and weeks. Another consequence is that the merging of traffic streams, as in a statistical multiplexer or an asynchronous transfer mode (ATM) switch, does not result in smoothing of the traffic. Again, bursty data streams that are multiplexed tend to produce a bursty aggregate stream. This first paper sparked a surge of research around the globe. The results show the self-similarity in ATM traffic, compressed digital video streams, and Web traffic between browsers and servers. Although a number of researchers had observed over the years that network traffic did not always obey Poisson assumptions applied in queuing analysis, for the first time, this paper provided an explanation and a systematic approach to modeling realistic data traffic patterns. Following the characterization of the fractal nature of data traffic, network theorists were split into two camps, with one group advocating the rewriting of and the other disagreed to this proposal. Traditionally, networks have been described by generalized Markovian processes, which are statistical models relying on postulates framed by the Russian mathematician A. A. Markov. Markovian models of networks have limited memory of the past. They reflect SRD. In a Markovian model, smoothing of bursty data is possible. Averaging of bursty traffic over a long period of time gives rise to a smooth data stream. A network based on fractal nature will have very different parameters and congestion control techniques.

X is defined to be a wide-sense stationary random process with the mean m, variance represented by s, and autocorrelation function denoted by r. For each m, $X(m)$ defines a wide-sense stationary random process.

The process X is said to be second-order self-similar with self-similarity parameter H if the aggregated processes have the same autocorrelation structure as X. In other words, X is exactly second-order self-similar if the aggregated processes are indistinguishable from X with respect to their first- and second-order properties. The most striking feature of self-similarity is that the correlation structures of the aggregated process do not degenerate as $m \to \infty$. This is in contrast to the traditional models, all of which have the property that the correlation structure of their aggregated processes degenerates as $m \to \infty$. The Hurst parameter H is a measure of the level of self-similarity in a time series. H takes values from 0.5 to 1. In order to determine if a given series exhibits self-similarity, a method is needed to estimate H for a given series. Currently, there are three approaches to measuring H: analysis of the variances of the aggregated processes $X(m)$; analysis of the rescaled range (R/S) statistic for different block sizes; a Whittle estimator.

Since self-similarity is believed to have a significant impact on network performance, understanding the causes of self-similarity in traffic is important. In a realistic client/server network environment, the degree to which file sizes are heavy-tailed can directly determine the degree of traffic self-similarity at the link level. This causal relation is proven to be robust with respect to changes in network resources (bottleneck bandwidth and buffer capacity), network topology, the influence of cross-traffic, and the distribution of interarrival times. Specifically, for measuring self-similarity via the Hurst parameter H and the file size distribution by its power-law exponent α, it has been shown that there is a linear relationship between H and α over a wide range of network conditions.

3.1.3 Effects of self-similarity on network performance

Well-defined metrics of delay, packet loss, flow capacity, and availability are fundamental to measurement and comparison of path and network performance. In general, users are most interested in metrics that provide an indication of the likelihood that their packets will reach the destination in a timely manner. Therefore, the estimates of past and expected performance for traffic across specific Internet paths, and not simply measures of current performance, are important. Users are also increasingly concerned about path availability information, particularly as it affects the quality of multimedia applications requiring higher bandwidth and lower latency, such as Internet phone and videoconferencing. Availability of such data could help in scheduling online events such as Internet-based distance education seminars, and also influence user willingness to purchase higher service quality and associated service guarantees.

Given the ubiquity of scale-invariant burstiness observed across diverse networking contexts, finding effective traffic control algorithms capable of detecting and managing self-similar traffic has become an important problem. The control of self-similar traffic involves modulating the traffic flow in such a way that the resulting performance is optimized. Scale-invariant burstiness (i.e., self-similarity) introduces new complexities into optimization of network performance and makes the task of providing QoS together with achieving high utilization difficult. Many analytical studies have shown that self-similar network traffic can have a detrimental impact on network performance, including amplified queuing delay and packet loss rate. On the other hand, LRD is unimportant for buffer occupancy when the Hurst parameter is not very large ($H < 0.7$). One practical effect of self-similarity is that the buffers needed at switches and multiplexers must be bigger than those predicted by traditional queuing analysis and simulations. The delay-bandwidth product problem arising out of high-bandwidth networks and QoS issues stemming from the support of real-time multimedia communication have added further complexities to the problem of optimizing performance.

How much self-similarity affects network performance is modulated by the protocols acting at the transport–network layer. An exponential tradeoff relationship was observed between queuing delay and packet loss rate. It is certain that a linear increase in buffer sizes will produce a nearly exponential decrease in packet loss, and that an increase in buffer size will result in a proportional increase in the effective use of transmission capacity. With self-similar traffic, these assumptions do not hold. The decrease in packet loss with buffer size is far lower than expected and the buffer requirements begin to explode at the lower levels of utilization for higher degrees of LRD (higher values of H). Moreover, scale-invariant burstiness implies the existence of concentrated periods of high activity at a wide range of timescales, which adversely affects congestion control and is an important correlation structure, which may be exploitable for congestion control purposes. Network performance as captured by throughput, packet loss rate, and packet retransmission rate degrades gradually with increasing heavy-tailedness. The degree to which heavy-tailedness affects self-similarity is determined by how well congestion control is able to shape its source traffic into an on-average constant output stream while conserving flow.

Packet loss and retransmission rate decline smoothly as self-similarity increases in the condition of a reliable flow-controlled packet transport. The only performance indicator exhibiting a more sensitive dependence on self-similarity is mean queue length, and this concurs with the observation that queue length distribution under self-similar traffic decays more slowly than with Poisson sources. Increasing network resources such as link bandwidth and buffer space results in a linear improvement in performance. However, large buffer sizes are accompanied by long queuing delays. In the context of facilitating multimedia traffic such as video and voice in a best-effort manner while satisfying their diverse QoS requirements, low packet loss, on average, can be achieved only with a significant increase in queuing delay and vice versa. Increasing link bandwidth, given a large buffer capacity, has the effect of decreasing queuing delay much more drastically under highly self-similar traffic conditions than when traffic is less self-similar.

3.2 *Mathematics of self-similar processes*

3.2.1 *Stationary random processes*

To describe the character of the observed scaling properties of traffic data more precisely, we introduce a set of definitions on first- and second-order stationary time series.

Let consider a time series $X = X_n$, $n \in Z^+$, which represents the number of packets X_n that arrive in the nth interval of time T_i. Therefore,

$$X_n = N[nT_i] - N[(n-1)T_i] \tag{3.1}$$

The number of packets arrival till the t moment represents a counting process given by

$$N(t) = \int_0^t \mathrm{d}N(t) \tag{3.2}$$

where d$N(t)$ is a point process, which represents the arrivals of packets. For every point at time t, one find a random variable, and the complete form of X_n with variable t is a random process.

The main statistical measures of these processes are

- Mean:

$$\mu = E[X_n] \tag{3.3}$$

- Dispersion:

$$\sigma^2 = \mathrm{Var}[X_n] = E[(X_n - \mu)^2] \tag{3.4}$$

- Covariance:

$$\mathrm{Cov}(X_n, X_{n+k}) = E[(X_n - \mu)(X_{n+k} - \mu)] \tag{3.5}$$

- (Normalized) autocorrelation:

$$R(k,T_i) = \frac{\mathrm{Cov}(X_n, X_{n+k})}{\sigma^2} \tag{3.6}$$

- Dispersion index of counting:

$$F(T) = \frac{\mathrm{Var}[N(T)]}{E[N(T)]} \tag{3.7}$$

- Density of power spectrum:

$$S(\omega) = \sum_{k=-\infty}^{\infty} R(k)\mathrm{e}^{-j2k\omega} \tag{3.8}$$

  where $\omega = 2\pi f$ is the angular frequency presented in radians per second.
- The continuous component of the power spectrum:

$$S(0) = \sum_{k=-\infty}^{\infty} R(k) \tag{3.9}$$

A random process X_n is considered to be stationary if its statistical properties do not change with time. This means that the process lacks both quantitative and qualitative scale. The first-order statistics are the same on short term as well as on long term. When both first- and second-order statistics (μ and σ^2) are invariant, the random process is considered to be wide-sense stationary or second-order stationary. For such a process, the mean μ is a constant, the dispersion σ^2 is finite, and the autocorrelation function $R(k, T_i)$ depends only on the time difference k. Moreover, if $X_1, \ldots, X_n$ are noncorrelated, then $R(k, T_i) = 0$ for $n \neq n + k$, that is, for $k \neq 0$.

3.2.2 *Continuous time self-similar processes*

A stochastic process is termed fractal when a representative series of statistics presents specific scaling exponents. Because the scaling procedure is mathematically expressed by a power law, one can consider that the network traffic has fractal properties when several statistic estimations display a power-law behavior on a large range of time and frequency scales. A time-continuous stochastic process $X(t)$ is defined as statistical self-similar, with parameter $H(0.5 \leq H \leq 1.0)$ if for any a real and positive, the processes $X(t)$ and $aHX(at)$ have identical finite dimensional distributions (i.e., with the same statistic properties) for all n positive integers:

$$\{X(t_1), X(t_2), \ldots, X(t_n)\} \underset{\sim}{D} \{a^{-H}X(at_1), a^{-H}X(at_2), \ldots, a^{-H}X(at_n)\} \quad (3.10)$$

The term $\underset{\sim}{D}$ means "asymptotically equal with" from the distributions point of view. Practically, statistical self-similarity implies the validity of the following conditions:

- Mean:

$$E[X(t)] = \frac{E[X(at)]}{a^H} \quad (3.11)$$

- Dispersion:

$$\text{Var}[X(t)] = \frac{\text{Var}[X(at)]}{a^{2H}} \quad (3.12)$$

- Autocorrelation:

$$R(t, \tau) = \frac{R(at, a\tau)}{a^{2H}} \quad (3.13)$$

The Hurst parameter (H) represents a self-similarity degree, that is, the degree of persistence of the statistical phenomenon. The value $H = 0.5$ shows the lack of self-similarity, while high values of H (close to 1) shows a high self-similarity degree (or LRD) of a process. In other words, for such a process, an ascendant (or descendent) trend in the past implies, with high probability, an ascendant (or descendent) trend in the future.

3.2.3 Discrete time self-similar processes

Let consider the time series $X = X_n$, $n \in Z^+$ and the corresponding (m-aggregate) time series $X^{(m)} = X_n^{(m)}$, $n \in Z^+$ obtained by the average of over m-dimensional nonoverlapping adjacent blocks of the original (basic) time series:

$$X_n^{(m)} = \frac{1}{m} \sum_{i=nm-(m-1)}^{nm} X_i \quad (3.14)$$

$X^{(1)}$ represents the best resolution for the process in this situation. Evolutions with lower resolution of the process $X^{(m)}$ can be obtained by m-aggregation of the process X_n, such as (for $m = 4$):

$$X_n^{(4)} = \frac{X_{4n-3} + X_{4n-2} + X_{4n-1} + X_{4n}}{4} \tag{3.15}$$

The process $X^{(m)}$ represents a less detailed copy of the process $X^{(1)}$. If the statistical properties (mean, dispersion) do not change by aggregation, then the process has self-similar nature. The process $X_n^{(m)}$ can be considered also as a time average of the process X_n. For a stationary ergodic process X_n, the temporal average is equal with the mean of the all set of samples:

$$\overline{X} = E[X_n] = \frac{\sum_{i=1}^{n} X_i}{n} \tag{3.16}$$

The dispersion of the mean should be equal to the dispersion of the basic random variable divided to the number of samples:

$$\mathrm{Var}(\overline{X}) = \frac{\sigma^2}{n} \tag{3.17}$$

However, due to the persistence of the statistical properties on various timescales, the m-aggregation in a self-similar process is different from the average of all samples of length m, that is, the dispersion approaches zero slower than in the case of a stationary ergodic process:

$$\mathrm{Var}(\overline{X}) \approx \frac{\sigma^2[1 + \delta_n(R)]}{n} \tag{3.18}$$

where

$$\delta_n(R) = \frac{\sum_{i \neq j} R(i, j)}{n} \tag{3.19}$$

and the correlation function of X_i and X_j is

$$R(i, j) = \frac{\mathrm{Cov}(X_i, X_j)}{\sigma^2} \tag{3.20}$$

There are two categories of self-similar processes: exactly self-similar processes and asymptotically self-similar processes.

The process X is exactly self-similar with parameter β ($0 < \beta < 1$) if, for $m \in Z^+$, the following conditions hold:

- Dispersion

$$\mathrm{Var}[X^{(m)}] = \frac{\mathrm{Var}[X]}{m^\beta} \tag{3.21}$$

- Autocorrelation

$$R(k, X^{(m)}) = R(k, X) \tag{3.22}$$

The parameter β is related with Hurst parameter by

$$\beta = 2(1 - H) \tag{3.23}$$

One can observe that $\beta = 1$ for stationary ergodic process and the dispersion approaches zero.

The process X is asymptotically self-similar if, for big values of k the dispersion is

$$\mathrm{Var}[X^{(m)}] = \frac{\mathrm{Var}[X]}{m^\beta} \tag{3.24}$$

and the autocorrelation function is

$$R(k, X^{(m)}) \rightarrow R(k, X) \tag{3.25}$$

when $m \rightarrow \infty$.

For both categories of self-similar processes, the dispersion of $X^{(m)}$ decays slower than $1/m$ when $m \rightarrow \infty$, in comparison with the stochastic processes where the dispersion decays proportional with $1/m$ and approaches zero when $m \rightarrow \infty$ (like in the case of the white noise, which has a uniform power spectrum). Anyway, the most important characteristic of the self-similar processes is to maintain the autocorrelation when $m \rightarrow \infty$.

3.2.4 *Properties of the fractal processes*

Fractal processes are closed related to the self-similar processes, exhibiting common characteristics such as LRD, slowly decaying dispersion, heavy tail distributions, and so on. The most common self-similar process with fractal properties is the fractional Brownian motion (*fBm*).

When a self-similar process $X(t)$ has the property of stationary increments (i.e., when the finite dimensional distributions of $X[t+\tau]-X[t]$ are independent of t), it can be considered a basic fractal process with LRD, slow decaying dispersion and $1/f$ noise aspect:

$$X_n = X[nT_i] - X[(n-1)T_i] \tag{3.26}$$

3.2.4.1 *Long-range dependence*

The process X is long-range dependent if, for every fixed T_i, the autocorrelation function is not summable:

$$\sum_{k=0}^{\infty} \mathrm{Cov}(k) = \infty \tag{3.27}$$

A better definition is that a process with LRD has a covariance function with hyperbolic decay:

$$\mathrm{Cov}(X_n, X_{n+k}) \sim |k|^{-\beta} \tag{3.28}$$

when $|k| \to \infty$ and $0 < \beta < 1$ (the definition of parameter β is given by Equation 3.23.

On the contrary, for the SRD processes, the covariance function has an exponential decay:

$$\mathrm{Cov}(X_n, X_{n+k}) \sim a^{|k|} \tag{3.29}$$

when $|k| \to \infty$ and $0 < a < 1$. Therefore, a SRD process has a summable autocorrelation function

$$\sum_{k=0}^{\infty} \mathrm{Cov}(k) = \text{finite} < \infty \tag{3.30}$$

We can conclude that $\mathrm{Cov}(k)$ becomes small for big values of k, while their cumulative effect is significantly more important in the case of the LRD processes as the one shown in Figure 3.1.

3.2.4.2 *Slowly decaying dispersion*

As was already mentioned, the dispersion of the samples average decays slower in the case of self-similar processes for m big enough:

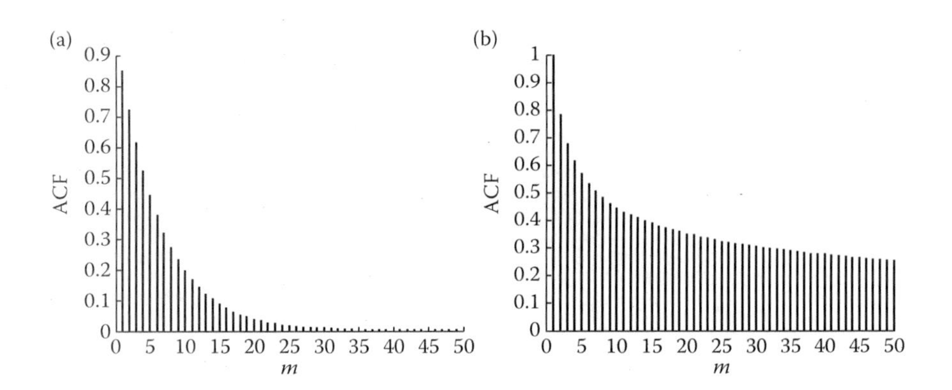

Figure 3.1 Autocorrelation structures for SRD processes (a) and for LRD processes (b).

$$\mathrm{Var}[X^{(m)}] \sim m^{-\beta} \tag{3.31}$$

On the other hand, for SRD processes $\beta = 1$ and

$$\mathrm{Var}[X^{(m)}] \sim m^{-1} \tag{3.32}$$

The property of slowly decaying dispersion can be easy shown by the graphic plot of $\mathrm{Var}[X^{(m)}]$ as function of m on a logarithmic axis (dispersion–time diagram). A straight line with negative slope <1 (on a large range of values of m) points to a slowly decaying dispersion. The same property can be expressed by the dispersion index of counting $F(T)$ (given in Equation 3.7).

The equation

$$\mathrm{Var}[X^{(m)}] = \frac{E[N(mT_i)]F[mT_i]}{m^2} \tag{3.33}$$

shows the equivalence between a dispersion–time diagram, which indicates a slowly decaying dispersion with parameter β and a plot of the dispersion index with slope $1 - \beta$.

3.2.4.3 Heavy-tailed distributions

If the abovementioned properties (covariance and autocorrelation) refer the scaling properties of time-dependent statistics, the attributes of the heavy-tailed distributions are focussed on the marginal amplitudes distribution of X_n, as defined in Equation 3.1. Packets arrivals or ON/OFF intervals are examples of such processes.

A random variable X is called heavy-tailed if the complementary cumulative distribution function (CCDF) decays as a power law:

$$\Pr P\{X \geq x\} \sim x^{-\alpha} \tag{3.34}$$

when $x \to \infty$ for $\alpha > 0$. For $0 < \alpha \leq 1$ all the moments of X are infinite and in a wide sense, the nth moment is infinite for $n \geq \alpha$. The most known type of heavy-tailed distribution is the Pareto distribution. Let us also mention that although the property of heavy-tailed distribution is not a necessary condition for self-similarity, the self-similar nature of many traffic data results from heavy-tailed distributions (the so called Joseph effect, which illustrates the idea that movements in a time series tend to be part of larger trends and cycles more often than they are completely random; the Joseph effect is quantified by the Hurst component, where movements fall between a Hurst range of 0 and 1).

3.2.4.4 1/f noise

This property is an aspect of the LRD in the frequency domain, more precisely the evidence of the fact that the density of the power spectrum (PSD—defined in Equation 3.8) of the LRD processes diverges at small frequencies, as a contrary in comparison with the SRD processes, which have a smooth PSD at small frequencies, that is,

$$S(\omega) \sim \frac{1}{|\omega|^{\gamma}} \tag{3.35}$$

when $|\omega| \to 0$ and $0 < \gamma < 1$. The relation between the parameters γ and β is

$$\gamma = 1 - \beta = 2H - 1 \tag{3.36}$$

For SRD processes, which have a finite PSD when $|\omega| \to 0$, (i.e., when $\gamma = 0$), the values of the autocorrelation function $R(k)$ decays fast for big values of k, reaching a finite sum.

3.2.4.5 Fractal dimension

The fractal dimension d of an object is defined by

$$d = \frac{\ln N}{\ln(1/\eta)} \tag{3.37}$$

where N is the number of self-similar pieces that cover an d-dimensional object, and η is the dimension of the scale unit. For SRD processes, $d = 1$, while for LRD processes, d has fractioned dimension of form $d = xx.yy\ldots$.

3.2.4.6 SRD versus LRD: Conclusion

For LRD processes, the following characteristics are representative:

- $E[X_n^m]$ is close to the second-order moment when $m \to \infty$.
- $Var[X^m]$ approaches asymptotically $Var[X]/m$ when $m \to \infty$.
- $\sum_{k=0}^{\infty} \text{Cov}(X_n, X_{n+k})$ is convergent.
- The spectrum $S(\omega)$ is finite for $\omega = 0$.

For LRD processes, the following characteristics are representative:

- $E[X_n^m]$ differs the second-order moment when $m \to \infty$.
- $Var[X^m]$ approaches asymptotically $m^{-\beta}$ when $m \to \infty$.
- $\sum_{k=0}^{\infty} \text{Cov}(X_n, X_{n+k})$ is divergent.
- The spectrum $S(\omega)$ is singular for $\omega = 0$.

3.3 Self-similar traffic modeling

We have already shown in Chapter 1 that traffic models are mainly used in traffic engineering to anticipate the performance of the traffic network and to evaluate the ways of improving it. The accuracy of a traffic model consists in the ability to model various correlation structures and marginal distributions. If a model is not able to capture the statistical characteristics of the real traffic, the network will have a poor performance, which is either underestimated or overestimated. The models presented in Chapter 1 were classified as stationary and nonstationary. In turn, stationary models are classified into two categories: short-range dependent (which includes traditional models like Markov processes and autoregressive models—AR, ARMA, ARIMA, TES, and DAR) or long-range-dependent models (fractional ARIMA, fractional Brownian motion). Traditional traffic models have an autocorrelational structure with exponential decay, so $\Sigma_k \rho_k < \infty$. In this case, the dispersion of the mean v_m decays in a similar manner as m^{-1} for large values of m, while self-similar traffic models have an autocorrelational structure with slower decay, so $\Sigma_k \rho_k \to \infty$.

In this chapter, we focus on the modeling of the self-similar characteristics, corresponding to their utilization. These models can be separated into two categories: single-source models and aggregated models. The well-known model will be analyzed in the following.

3.3.1 Single-source traffic models

Internet services are constructed usually after the client–server paradigm in order to transfer information blocks (like files or written messages) whose dimensions are better characterized by heavy-tailed distributions and eventually lead to long-range-dependent network traffic behavior. Among these distributions, the most usual in modeling sources with high variability are Pareto distribution and log-normal distribution.

3.3.1.1 Pareto distribution

Pareto distribution is a power-law function parameterized by two parameters: the shape parameter α (named also Pareto index) and the scale parameter β (named also cutoff parameter because it represents the minimum value of the random variable X). The cumulative distribution function (CDF) Pareto is given by

$$F(x) = 1 - \left(\frac{\beta}{x}\right)^{\alpha} \tag{3.38}$$

and the probability density function for $x > \beta$ and $\alpha > 0$

$$f(x) = \frac{\alpha}{\beta}\left(\frac{\beta}{x}\right)^{\alpha+1} \tag{3.39}$$

for $x \leq \beta$

$$f(x) = F(x) = 0 \tag{3.40}$$

The mean (the expected value) is

$$E[X] = \frac{\alpha\beta}{\alpha - 1} \quad \text{with } \alpha > 1 \tag{3.41}$$

The dispersion is

$$\sigma^2 = \frac{\alpha}{(\alpha-1)^2(\alpha-2)} \tag{3.42}$$

with $\alpha > 2$ and $\beta = 1$. The parameter α determines the mean and the dispersion as follows:

- For $\alpha \leq 1$, the distribution has infinite mean.
- For $1 \leq \alpha \leq 2$, the distribution has infinite mean and infinite dispersion.
- For $\alpha \leq 2$, the distribution has infinite dispersion.

The relation between parameter α and Hurst parameter is

$$H = \frac{3 - \alpha}{2} \tag{3.43}$$

3.3.1.2 *Log-normal distribution*

A log-normal distribution is a probability distribution of a random variable whose logarithm is normally distributed. If the stochastic variable $Y = \log X$ is normally distributed, then the variable X is log-normally distributed. The distribution has two parameters, the mean μ of ln(x), called also location parameter ($\mu > 0$) and the standard deviation σ of ln(x), called also the scaling parameter ($\sigma > 0$), for $0 \leq x \leq \infty$. The cumulative distribution function and the probability density function are

$$F(x) = \frac{1}{2}\left[1 + \mathrm{erf}\left(\frac{\ln x - \mu}{\sigma\sqrt{2}}\right)\right] \tag{3.44}$$

$$f(x) = \frac{1}{\sigma x \sqrt{2\pi}} \mathrm{e}^{-(\ln x - \mu)^2 / 2\sigma^2} \tag{3.45}$$

The error function erf(y) is defined as

$$\mathrm{erf}(y) = \frac{2}{\sqrt{\pi}} \int_0^y \mathrm{e}^{-t^2} \mathrm{d}t \tag{3.46}$$

The mean and the dispersion of the log-normal distribution are

$$E[X] = \mathrm{e}^{\mu + (\sigma^2/2)} \tag{3.47}$$

$$\mathrm{Var}[X] = \mathrm{e}^{2\mu + \sigma^2}(\mathrm{e}^{\sigma^2} - 1) \tag{3.48}$$

The log-normal distributions are useful in many applications such as regressive modeling and the analysis of experimental models, especially because it can model errors produced by many factors.

3.3.2 *Aggregate traffic models*

Most traffic analysis and modeling studies to date have attempted to understand aggregate traffic, in which all simultaneously active connections are lumped together into a single flow. Typical aggregate time series include the number of packets or bytes per time unit over some interval. Numerous studies have found that aggregate traffic exhibits fractal or self-similar scaling behavior, that is, the traffic "looks statistically similar" on all timescales. Self-similarity endows traffic with LRD. Numerous studies have also shown that traffic can be extremely bursty, resulting in a non-Gaussian marginal distribution. These findings are in sharp contrast to classical traffic models such as Markov or homogeneous Poisson. LRD and non-Gaussianity can lead to much higher packet losses than predicted by classical Markov/Poisson queuing analyses.

3.3.2.1 ON–OFF models

This approach was first suggested by Mandelbrot, for economical maters, based on renewal processes and later was applied by W. Willinger, M. Taqqu, R. Sherman, and D. Wilson for network communication traffic (Willinger et al. 1997).

The aggregate traffic is generated by multiplexing several independent ON–OFF sources, with each source changing these states in a given periodicity. The generated traffic is (asymptotically) self-similar due to the superposition of several renewal processes, where the renewal values are only 0 and 1, and the interval between renewal is heavy-tailed (e.g., a Pareto distribution with $1 < \alpha < 2$). This superposition presents the syndrome of the infinite dispersion (the so-called Noah effect) and leads to closing of the fractional Brownian motion (*fBm*) in self-similar aggregate traffic when the number of sources is very high (the so-called Joseph effect), with the Hurst parameter being

$$H = \frac{3-\alpha}{2} \tag{3.49}$$

The ON–OFF model is simple and therefore very attractive for the simulation of Ethernet traffic, considering each station connected in an Ethernet network as a renewal process, staying in state ON when communicating and in the state OFF when not communicating. So Ethernet traffic is the result of aggregation of several renewal processes.

3.3.2.2 Fractional Brownian motion

The concept of fractional Brownian motion (*fBm*) was introduced by Mandelbrot and Van Ness in 1968 like a generalization of the Brownian motion (*Bm*). The $Bm(t)$ process has the property that any random vector at different times has a joint distribution identical to that of a rescaled and normalized version of the random vector, which is a unique continuous Gaussian process with this property and with stationary increments. The (standard) *fBm* family of processes $B_H(t)$ is indexed by a single parameter $H \in [0, 1]$, which controls all of its properties. Larger values of H

correspond to stronger large-scale correlations, which make the sample paths appear more and more smooth. In contrast, $B_{1/2}$ is the Brownian motion without memory, which in discrete time is just a random walk. The increments of *fBm* with $H \in [0, 1]$ define the well-known fractional Gaussian noise (fGn) family of stationary processes. For Brownian motion, these increments are i.i.d., which is clearly SRD. fGn processes with $H \in [0, 1/2)$ are also SRD, though qualitatively different from white noise, whereas fGn processes with $H \in (1/2, 1]$ are LRD. Of course, this appearance can be modified by allowing for two additional parameters, namely mean and variance, which allows capturing a wide range of "smoothness" scenarios, from "highly bursty" to "very smooth." For analytical studies, the canonical process for modeling LRD traffic rate processes is fGn with $H \in (1/2, 1]$. When describing the observed self-similar behavior of measured traffic rate, we can be more flexible. The standard model is a self-similar process in the asymptotic sense or, equivalently, an LRD process.

The following definitions are specific for a *fBm* process:

The expected value of the process is zero

$$E[B(t+\tau)-B(t)]=0 \tag{3.50}$$

The dispersion is proportional to the time difference

$$\mathrm{Var}[B(t+\tau)-B(t)]=\sigma^2|\tau| \tag{3.51}$$

The fGn process $B_H(t)$ is defined as the moving average of $dB(t)$, where the past increments of $B(t)$ are mediated by the kernel $(t-\tau)^H - (1/2)$ and can be expressed by the integral fractional Weyl integral of $B(t)$:

$$B_H(t)=\frac{1}{\Gamma(H+(1/2))}\int_{-\infty}^{0}[(t-\tau)^{H-(1/2)}-(-\tau)^{H-(1/2)}]dB(\tau)+\int_{0}^{t}(t-\tau)^{H-(1/2)}dB(\tau) \tag{3.52}$$

where $\Gamma(x)$ is the Gamma function. The value of a *fBm* process at moment t is

$$B_H(t)=Xt^H \tag{3.53}$$

where X is a random variable normally distributed with zero mean and dispersion equal to 1, where due to the self-similarity

$$B_H(at)=a^H B_H(t) \tag{3.54}$$

For $H = 0.5$, one obtains the ordinary Brownian motion (*Bm*).

To conclude, let consider the main statistical properties of a *fBm* process

- Mean: $E[B_H(t)] = 0$
- Dispersion: $\mathrm{Var}[B_H(t)] = \mathrm{Var}[Xt^H] = t^{2H}$
- Autocorrelation: $R_{BH}(t, \tau) = E[B_H(t)B_H(\tau)] = 1/2(t^{2H} + \tau^{2H} - |t - \tau|^{2H})$
- Stationary increments: $\mathrm{Var}[B_H(t) - B_H(\tau)] = |t - \tau|^2$

3.3.2.3 *Fractional Gaussian noise*

The stationary process of increment X_H of *fBm* type is known as fractional Gaussian noise (fGn):

$$X_H = (X_H(t) = B_H(t+1) - B_H(t) : t \geq 0) \tag{3.55}$$

It is a process with Gaussian distribution, with mean μ and dispersion σ^2. The dispersion of the mean of the square of increments is proportional to the time difference, and the autocorrelation function is

$$R(t) = \frac{1}{2}(|t+1|^{2H} - |t|^{2H} + |t+1|^{2H}) \tag{3.56}$$

In the limit case when $t \to \infty$

$$R(t) \sim H(2H-1)|t|^{2h-2} \tag{3.57}$$

A fGn process is exactly self-similar when $0.5 < H < 1$. When $\mu_X = 0$, fGn becomes Gaussian time-continuous process, $B_H = \{B_H(t)\}_{t=0}^{\infty}$, with $0 < H < 1$ and the autocorrelation function $\mu_X(s, t) = 1/2(|s|^{2H} + |t|^{2H} - |t - s|^{2H})$

3.3.2.4 *Fractional ARIMA processes (FARIMA)*

An integrated autoregressive moving-average (p, d, q) series, denoted ARIMA (p, d, q), is closely related to ARMA (p, q). ARIMA models are more general as ARMA models and include some nonstationary series (Box and Jenkins 1976).

FARIMA processes are the natural generalizations of standard ARIMA (p, d, q) processes when the degree of differencing d is allowed to take nonintegral values. The mathematical aspects were discussed in detail in Section 2.5.5.

Let remember that for $0 < d < 0.5$, FARIMA (p, d, q) process is asymptotically self-similar with $H = d + 0.5$. An important advantage of this model is the possibility to capture both short- and long-range-dependent correlation structures in time series modeling. The main drawback is the increased complexity, but it can be reduced using the simplified FARIMA model (0, *d*, 0).

Consequently, compared with ARMA and fGn processes, the FARIMA (*p*, *d*, *q*) processes are flexible and parsimonious with regard to the simultaneous modeling of the long- and short-range-dependent behavior of a time series.

3.3.2.5 Chaotic deterministic maps

A chaotic map $f{:}V \rightarrow V$ can be defined by the following three criteria (Erramilli et al. 2002):

1. It shows sensitive dependence on initial conditions (SIC), which means that typical trajectories starting from arbitrarily close initial values nevertheless diverge at an exponential rate. SIC is the basis of the "deterministic" property of chaotic systems, because errors in the estimates of initial states become amplified and prevent accurate prediction of the trajectory's evolution.
2. It is topologically transitive or strongly mixing, which implies that it is irreducible and that it cannot be decomposed into subsets that remain disjoint under repeated action of the map.
3. Periodic points are dense in *V*.

From a modeling perspective, even low-order chaotic maps can capture complex dynamic behavior. It is possible to model traffic sources that generate fluctuations over a wide range of timescales using chaotic maps that require a small number of parameters.

If one considers a chaotic map defined as $X_{n+1} = f(X_n)$ and two trends (evolutions) with almost identical conditions X_0 and $X_0 + \varepsilon$, then SIC can be expressed by

$$|f^N(X_0+\varepsilon) - f^N(X_0)| = \varepsilon e^{N\lambda(x_0)} \tag{3.58}$$

where $f^N(X_0)$ refers to a *N* time-iterated map, and the parameter $\lambda(x_0)$ describes the exponential divergence (the so-called Liapunov exponent). For a chaotic map, $\lambda(x_0)$ should be positive for "almost all" X_0. In other words, the limits in the accuracy of the specification of the initial conditions lead to an exponential growth of the incertitude, making the traffic behavior on long-term unpredictable. Another important characteristic is the convergence of the trajectories (in the phase space) to a certain object named strange attractor, which is typical for a fractal structure.

In order to demonstrate how we can associate packet source activity with the evolution of a chaotic map, let consider the one-dimensional map in Figure 3.2 where the state variable *x* evolves according to $f_1(x)$ in the interval [0, *d*) and according to $f_2(x)$ in the interval [*d*, 1)].

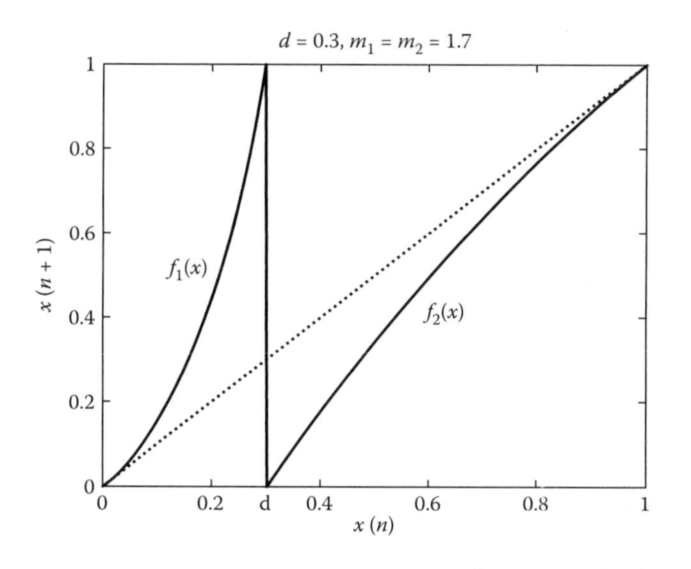

Figure 3.2 Evolutions in a chaotic map.

One can model an on–off traffic source by stipulating that the source is active and generates one or more packets whenever x is in the interval $[d, 1)]$, and is idle otherwise. By careful selection of the functions $f_1(x)$ and $f_2(x)$, one can generate a range of on–off source behavior. The traffic generated in this manner can be related to fundamental properties of the map. As the map evolves over time, the probability that is in a given neighborhood is given by its invariant density $\rho(x)$, which is the solution of the Frobenius–Perron equation:

$$\rho(x) = \int_0^1 \delta(x - f(z))\, \rho(z) dz \tag{3.59}$$

If the source is assumed to generate a single packet with every iteration in the on-state ($d \le x \le 1$), the average traffic rate is simply given by

$$\lambda = \int_0^1 \rho(x) dx \tag{3.60}$$

Another key source characteristic is the "sojourn" or run times spent in the on- and off-states. Given the deterministic nature of the mapping, the sojourn time in any given state is solely determined by the reinjection or initial point x_0 at which the map reenters that state. For example, let the map reenter the on-state at $x_0 \le d$. The sojourn time there is then the number of iterations it takes for the map to leave the interval [d, 1]:

$$f_2^{k-1}(x_0) \le d, \quad f_2^k(x_0) < d \tag{3.61}$$

One can then derive the distributions of the on- and off-periods from the reinjection density $\psi(x)$ (where the probability that the map reenters a state in the neighborhood $[x_0, x_0 + \mathrm{d}x]$ is given by $\psi[x_0]\mathrm{d}x$) and Equation 3.61. The reinjection density can be derived from the invariant density

$$\psi(x) = \rho(f^{-1}(x))\frac{\mathrm{d}f^{-1}(x)}{\mathrm{d}x} \tag{3.62}$$

One can use these relations to demonstrate the following.

1. The sojourn times in the two states are geometrically distributed if $f_1(x)$ and $f_2(x)$ are linear:

$$f_1(x) = \frac{x}{d}, \quad 0 \le x < d; \qquad f_2(x) = \frac{1-x}{1-d}, \quad d \le x \le 1 \tag{3.63}$$

   This follows from the fact that the invariant density in this case is uniform, and substituting for the linear mappings in Equation 3.60.
2. In order to match the heavy-tailed sojourn time distribution behavior observed in measurement studies, one can choose $f_1(x)$ such that as $x \to 0$, $f_1(x) \sim x + cx^m$, where $1.5 < m < 2$. Note that this function evolves very slowly in the neighborhood of 0 (a fixed point). This results in sojourn times in the off-state that are characterized by distributions with infinite variance.

One can similarly generate extended on-times by choosing $f_2(x)$ so that it behaves in a similar manner in the neighborhood of 1 (another fixed point). In this way, one can match the heavy-tailed sojourn time distribution observed in measurement studies.

Thus, chaotic maps can capture many of the features observed with actual packet traffic data, with a relatively small number of parameters. While the qualitative and quantitative modeling of actual traffic using chaotic maps is itself the subject of ongoing research, ultimately the usefulness of this approach will be determined by the feasibility of solving for the 1-D and 2-D invariant densities. We have demonstrated the potential of chaotic maps as models of packet traffic. Given that packet traffic is very irregular and bursty, it is nevertheless possible to construct simple, low-order nonlinear models that capture much of the complexity. While chaotic maps allow concise descriptions of complex packet traffic phenomena, there are considerable analytical difficulties in their application.

As we indicated earlier, a key aspect of this modeling approach is to devise chaotic mappings $f(x)$ such that the generated arrival process matches measured actual traffic characteristics, while maintaining analytical tractability. Other source interpretations relating the state variable to packet generation are possible and may be more suitable for some applications.

Performance analysis, that is, analysis of queuing systems in which the randomness in the arrival and/or service characteristics is modeled using chaotic maps, particularly systems without a conventional queuing analog requires additional work on the modeling and representation of systems of practical interest using chaotic maps, as well as investigation of numerical and analytical techniques to calculate performance measures of interest, including transient analysis.

3.3.3 *Procedures for synthetic self-similar traffic generation*

The search for accurate mathematical models of data streams in modern telecommunication networks has attracted a considerable amount of interest; several recent teletraffic studies of local and wide area networks, including the World Wide Web, have shown that commonly used teletraffic models, based on Poisson or related processes, are not able to capture the self-similar (or fractal) nature of teletraffic, especially when they are engaged in sophisticated Web services. The properties of teletraffic in such scenarios are very different from both the properties of the traditional models of data traffic generated by computers, and drawbacks can result in overly optimistic estimates of performance of telecommunication networks, insufficient allocation of communication, and data processing resources and difficulties in ensuring the quality of service expected by network.

Several methods for generating pseudo-random self-similar sequences have been proposed. They include methods based on fast fractional Gaussian noise, fractional ARIMA processes, the $M/G/1$ queue model, autoregressive processes, spatial renewal processes, etc. Some of them generate asymptotically self-similar sequences and require large amounts of CPU time. Even though exact methods of generation of self-similar sequences exist, they are usually inappropriate for generating long sequences because they require multiple passes along the generated sequences. To overcome this problem, approximate methods for generation of self-similar sequences in simulation studies of telecommunication networks have also been proposed. The most popular studies are presented in the following.

3.3.3.1 ON–OFF method

This method (also called aggregation method) was described in Section 3.3.2. It consists in the aggregation and superposition of renewal rewards process (ON/OFF), in which activity (ON) and inactivity (OFF) periods follow a heavy-tailed PDF. We also know that there are some drawbacks associated with the deployment of this technique in the network simulation procedures, such as pitfalls in choosing timescales of interest and number of samples. Despite these issues, this approach could allow an immediate use of widespread network simulation tools, such as Network Simulator 2 – ns2 or the software family from OPNET since there is no need to extend their libraries to support such analytical models.

The main advantage is the possibility to implement the algorithm for parallel processing. The main drawback of the aggregation technique is its large computational requirements per sample produced (because of the large number of aggregated sources). This is the main reason why parallelism is introduced. However, even though the computational requirements are large, its actual computational complexity is $O(n)$ for n produced samples. All the other self-similar traffic generation techniques exhibit poorer complexity than $O(n)$ and they eventually result in a slow-down of the simulation as the simulated time interval increases. As an example, Hosking's method requires $O(n^2)$ computations to generate n numbers (Jeong et al. 1998).

To illustrate the above statements, let consider the following simulation model. Cell arrivals will be represented by Run Length Encoded (RLE) tuples. An RLE tuple t_i includes two attributes, $s(t_i)$, the state of the tuple, and $d(t_i)$, the duration of the tuple. The two attributes represent the discrete time duration $d(t_i)$ over which the state $s(t_i)$ stays the same. The state is either an indication of whether the source is in the ON or OFF state (e.g., 0 for OFF and 1 for ON in a strictly alternating fashion), or the aggregate number of sources active (in the ON state) for the specified duration, that is, for N sources, $s(t_i) \in \{0, 1, \ldots, N\}$. Thus, a sequence of t_i's is sufficient for representing the arrival process from an ON/OFF source or from any arbitrary superposition of such sources. The benefit of such representation is that the activity of the source over several time slots can be encoded as a single RLE tuple. In summary, the algorithm proceeds by generating a large number, N, of individual source traces in RLE form. The utilization of each one of these sources is set to U/N, such that the aggregation of their N results in the desired link utilization to U. Each logical process (LP) of the simulation merges and generates the combined arrival trace for a separate nonoverlapping segment of time, which we call a slice. Thus, each LP is responsible for the generation of the self-similar traffic trace in the form of RLE tuples over a separate segment (slice) of time.

In logical terms, the concatenation of the slices produced by each *LP* in the proper time succession is the desired self-similar process. The *LP*s continue looping generating a different slice each time. The *LP* performs the generation of the self-similar traffic trace by going through the following three steps at each simulated time slice:

1. It generates the merge of the RLE tuple traces of the N individual sources.
2. It aggregates the merged traffic into a link speed equal to the desired access link speed.
3. It corrects the produced RLE trace by incorporating any residual cell counts.

   The generation of the individual RLE source traces is also performed in parallel. Each *LP* generates all the slices of a subset of sources that will be necessary to a number of P *LP*s during the generation of the current slices. That is, if P *LP*s are participating in the simulation, each *LP* generates the RLE tuples of P subsequent slices for the sources that it has been assigned to generate. It then sends their P –1 to the other *LP*s for each source it simulates. The individual ON/OFF sources are parameterized accordingly to fit the desired self-similar traffic. Namely:

   a. The shape value, α, of the Pareto distribution used for the ON period is set according to $H = (3 - \alpha)/2$, where H is the desired Hurst value.
   b. Since N ON/OFF sources are aggregated, the per-source utilization of U/N is determined by the ratio of the ON and OFF periods of the individual processes. That is, the average OFF period E[OFF] is set to $E[\text{OFF}] = E[\text{ON}]$.
   c. The average ON period E[ON] is set to $E[\text{ON}] = B$, the average burst length, which can be derived from traffic measurements. E[ON] does not have any impact on the self-similarity, and it can be considered a free variable.

3.3.3.2 Norros method

An exactly self-similar can be generated using a *fBm* $Z(t)$ process

$$A(t) = mt + \sqrt{am}Z(t) \quad \text{for } -\infty < t < \infty \tag{3.64}$$

$A(t)$ represents the cumulative arrivals process, that is, the total traffic that arrives until the time t. $Z(t)$ is a normalized *fBm* process with $0.5 < H < 1.0$, where m is the average rate, and a is the shape parameter (the ratio between dispersion and mean in the time unit) (Norros 1995). We call this an (m, a, H) *fBm*. When this arrival process flows into a buffer with a deterministic service, the resulting model is what we call *fBm*/D/1. Unfortunately, no simple, closed-form results have been found for this queuing model. However, Norros was able to prove the following approximating formula:

$$P_{tail}(h) \approx \exp\left(-\frac{1}{2am}\left(\frac{(C-m)}{H}\right)^{2H}\left(\frac{h}{1-H}\right)^{2-2H}\right) \tag{3.65}$$

where C is the speed at which the packets in the buffer are processed. In an infinite buffer (theoretical abstraction), $P_{tail}(h)$ represents the probability for the buffer length to exceed h bytes. (Regarded as a function of the variable h, the right-hand side is a Weibull distribution.)

There are two problems with this formula. From a theoretic standpoint, the approximations made are nonsystematic (i.e., all in the same direction, either "≤" or "≥"), so Equation 3.65 is neither an upper bound nor a lower bound of the actual $P_{tail}(h)$. The exponential could be either smaller or larger than $P_{tail}(h)$ and there is no estimation of the tightness of the approximation. From an empirical standpoint, Equation 3.65 does not fit into any of our data series, most of the time deviating by orders of magnitude. Even assuming that for an exact *fBm* process Norros' formula would be a good approximation, we still have to deal with the fact that our time series are not accurately described by *fBm* at low timescales.

3.3.3.3 *M/G/∞ queue method*

Let consider a queue M/G/∞ where the arrivals of the clients follow a Poisson distribution and the service time are obtained from a heavy-tailed distribution (with infinite dispersion). Then the process $\{X_t\}$, representing the number of clients in the system at the t moment, is asymptotically self-similar.

To exemplify this method, we present a mechanism for congestion avoidance in a packet-switched environment. If there are no resources reserved for a connection, the possibility of packets arriving at a node at a rate higher than the processing rate then appears.

Packets arriving on any of a number of input ports (A–Z) can be destined for the same output port. Although there could be, in general, many operations to be performed, they are broadly divided into two classes: serialization (i.e., the process of transmitting the bits out on the link) and everything else that happens inside the node (e.g., classification, rewriting of the TTL, and/or TOS fields in the IP header, routing table and access list look-ups, policing, shaping, etc.) called internal processing. Serialization delays are deterministic and determined by the speed of the output link. Internal processing delays are essentially random (a larger routing table requires a longer time to search) but advanced algorithms and data structures are doing, in general, a good job of maintaining overall deterministic performance. As a consequence, the only truly random part of the delay is the time a packet spends waiting for service.

The internal workings of a node are relatively well understood (by either analysis or direct measurement) and controllable. In contrast, the (incoming) traffic flows are neither completely understood nor easily controllable. The focus is therefore on accurately describing (modeling) these flows and predicting their impact on network performance.

The picture in Figure 3.3a is a simplified representation of a packet-switching node (router or switch), illustrating the mechanism of congestion for a simple M/D/1 queue model. The notation is easily explained: "M" stands for the (unique) input process, which is assumed Markovian, that is, a Poisson process; "D" stands for the service, assumed deterministic; and "1" means that only one facility is providing service (no parallel service).

It has been shown time and again that the high variability associated with LRD and SS processes can greatly deteriorate network performance, creating, for instance, queuing delays orders of magnitude higher than those predicted by traditional models. We make this point once again with the plot shown in Figure 3.3b: the average queue length created in a queuing simulation by the actual packet trace is widely divergent from the average predicted by the corresponding queuing model (M/D/1).

In view of our simple model of a node discussed earlier, hypotheses "D" and "1" seem reasonable, so we would like to identify "M" as the cause of the massive discrepancy. The question then arises: Is there an "*fBm*/D/1" queuing model? The answer is "almost," which suggests that the self-similarity is mainly caused by user/application characteristics (i.e., Poisson arrivals of sessions, highly variable session sizes or durations) and is hence likely to remain a feature of network traffic for some time to come—assuming the way humans tend to organize information will not change drastically in the future.

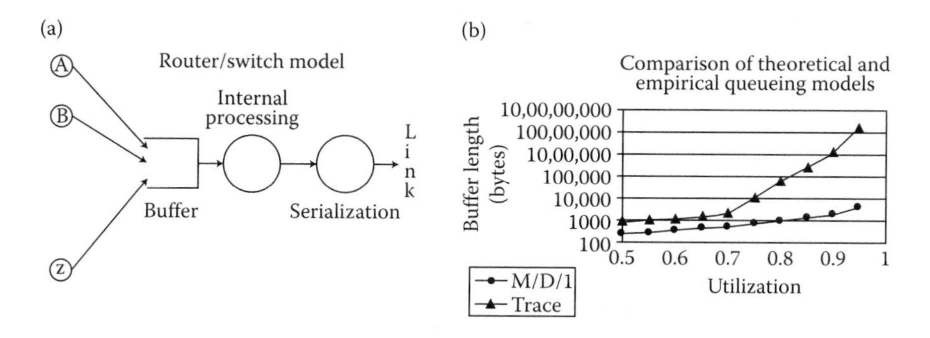

Figure 3.3 A queue model (a) and the comparison of the real traffic with the simulated traffic (b).

3.3.3.4 Random midpoint displacement (RMD) method

The basic concept of the random midpoint displacement (RMD) algorithm is to extend the generated sequence recursively, by adding new values at the midpoints from the values at the endpoints. Figure 3.4 outlines how the RMD algorithm works.

Figure 3.5 illustrates the first three steps of the method, leading to generation of the sequence ($d_{3,1}$; $d_{3,2}$; $d_{3,3}$; and $d_{3,4}$). The reason for subdividing the interval between 0 and 1 is to construct the Gaussian increments of X. Adding offsets to midpoints makes the marginal distribution of the final result normal.

Step 1. If the process $X(t)$ is to be computed for time instance t between 0 and 1, then start out by setting $X(0) = 0$ and selecting $X(1)$ as a pseudo-random number from a Gaussian distribution with mean 0 and variance $\mathrm{Var}[X(1)] = \sigma_0^2$. Then $\mathrm{Var}[X(1) - 1)(0)] = \sigma_0^2$.

Step 2. Next, $X(1/2)$ is constructed as the average of $X(0)$ and $X(1)$, that is, $X(1/2) = 1/2\,[X(0) + X(1)] + d_1$.

The offset d_1 is a Gaussian random number (GRN), which should be multiplied by a scaling factor 1/2, with mean 0 and variance S_1^2 of d_1. For $\mathrm{Var}[X(t_2) - X(t_1)] = |t_2 - t_1|^{2H}\,\sigma_0^2$ to be true, for $0 \le t_1 \le t_2 \le 1$,

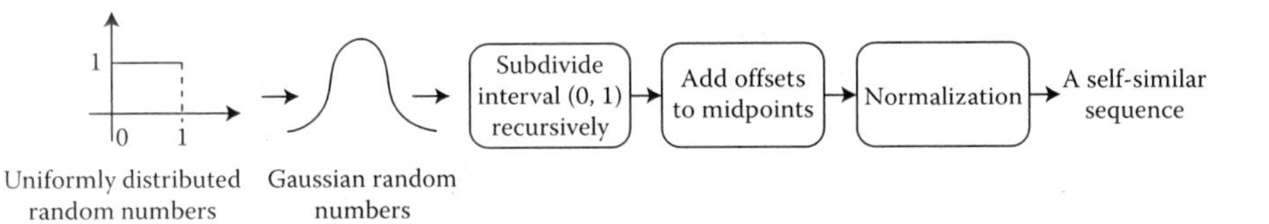

Figure 3.4 Block scheme for the RMD method.

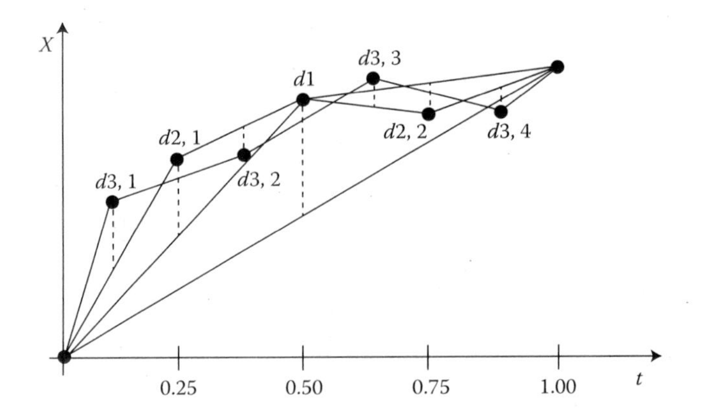

Figure 3.5 The first three steps in the RMD method.

it must be required that $\text{Var}[X(1/2) - X(0)] + S_1^2 = (1/2)^{2H}\sigma_0^2$. Thus, $S_1^2 = (1/2)^{2H}(1 - 2^{2H-2})\sigma_0^2$

Step 3. Reduce the scaling factor by $\sqrt{2}$, that is, now assume $\sqrt{8}$, and divide the two intervals from 0 and 1/2 and from 1/2 to 1 again. $X(1/4)$ is set as the average 1/2 $[X(0) + X(1/2)]$ plus an offset $d_{2,1}$, which is a GRN multiplied by the current scaling factor $\sqrt{1/8}$. The corresponding formula holds for $X(3/4)$, that is, $X(3/4) = 1/2\ [X(1/2) + X(1)] + d_{2,2}$ where $d_{2,2}$ is a random offset computed as before. So the variance S_2^2 of $d_{2,*}$ must be chosen such as follows:

$$\text{Var}\left[X\left(\frac{1}{4}\right) - X(0)\right] = \left(\frac{1}{4}\right)\text{Var}\left[X\left(\frac{1}{2}\right) - X(0)\right] + S_2^2 = \left(\frac{1}{2^2}\right)^{2H}\sigma_0^2$$

Thus, $S_2^2 = (1/2^2)^{2H}(1 - 2^{2H-2})\sigma_0^2$.

Step 4. The fourth step proceeds in the same manner: reduce the scaling factor by, that is, do $\sqrt{16}$. Then set

$$X\left(\frac{1}{8}\right) = \left(\frac{1}{2}\right)\left(X(0) + X\left(\frac{1}{4}\right)\right) + d_{3,1};\quad X\left(\frac{3}{8}\right) = \left(\frac{1}{2}\right)\left(X\left(\frac{1}{4}\right) + X\left(\frac{1}{2}\right)\right) + d_{3,2}$$

$$X\left(\frac{5}{8}\right) = \left(\frac{1}{2}\right)\left(X\left(\frac{1}{2}\right) + X\left(\frac{3}{4}\right)\right) + d_{3,3};\quad X\left(\frac{7}{8}\right) = \left(\frac{1}{2}\right)\left(X\left(\frac{3}{4}\right) + X(1)\right) + d_{3,4}$$

In the formulas above, $d_{3,*}$ is computed as a different GRN multiplied by the current scaling factor $\sqrt{16}$. The following step computes $X(t)$ at $t = 1/16, 3/16, \ldots, 5/16$ using a scaling factor again reduced by $\sqrt{2}$, and continues as indicated above. So the variance S_3^2 of $d_{3,*}$ must be chosen such that $\text{Var}[X(1/8) - X(0)] = (1/4)\text{Var}[X(1/5) - X(0)] + S_3^2 = (1/2^3)^{2H}\sigma_0^2$.

Thus, $S_3^2 = (1/2^3)^{2H}(1 - 2^{2H-2})\sigma_0^2$.

The variance S_n^2 of $d_{n,*}$ therefore, yields $S_n^2 = (1/2^n)^{2H}(1 - 2^{2H-2})\sigma_0^2$.

The main drawback of the method is the generation of a process, which is only approximate self-similar. In particular, the parameter H of the samples trajectories tend to overpass the target value for $0.5 < H < 0.75$ and to be less than the target value for $0.75 < H < 1.0$.

3.3.3.5 *Wavelet transform-based method*

This method (Abry and Veitch 1998) is based on the so-called multiresolution analysis (MRA), which consists of splitting the sequence into a (low pass) approximation and a series of details (high pass). First, the coefficients corresponding to the wavelet transform of *fBm*. Then, these are used in an inverse wavelet transform in order to obtain *fBm* samples. The disadvantage is that it is only approximate, because generally the wavelet coefficients are not correlated. Nevertheless, the problem can be compensated in the case of wavelets with a sufficient number of moments reduced to zero (i.e., the degree of the polynomial for that the internal scalar product with the wavelet is zero).

For some time after the discovery of scaling in traffic, it was debated as to whether the data was indeed consistent with self-similar scaling, or that the finding was merely an artifact of poor estimators in the face of data polluted with nonstationarities. The introduction of wavelet-based estimation techniques to traffic helped greatly to resolve this question, as they convert temporal LRD to SRD in the domain of the wavelet coefficients, and simultaneously eliminate or reduce certain kinds of nonstationarities. It is now widely accepted that scaling in traffic is real, and wavelet methods have become the method of choice for detailed traffic analysis. Another key advantage of the wavelet approach is its computational complexity (memory complexity is also in an online implementation), which is invaluable for analyzing the enormous data sets of network-related measurements. However, even wavelet methods have their problems when applied to certain real-life or simulated traffic traces. An important rule-of-thumb continues to be to use as many different methods as possible for checking and validating whether or not the data at hand is consistent with any sort of hypothesized scaling behavior, including self-similarity.

Wavelet analysis is a joint timescale analysis. It replaces $X(t)$ by a set of coefficients, such as $d_X(j, k)$, $j, k \in Z$ where 2^j denotes the scale, and the $2^j k$ instant about which the analysis is performed. In the wavelet domain, Equation 3.57 is replaced by $\text{Var}[d_X(j, k)] = c_f C 2^{(1-\beta)j}$, where the role of m is played by scale, of which j is the logarithm, where c_f is the frequency domain analog of c_γ, and C is independent of j. The analog of the variance–time diagram, which is the estimates of $\log(\text{Var}[d_X(j, \cdot)])$ against j, called the logscale diagram. It constitutes an estimate of the spectrum of the process in log–log coordinates, where the low frequencies correspond to large scales, appearing in the right in the figures below. The global behavior of data, as a function of scale, can be efficiently examined in the logscale diagram, and the exponent estimated by a weighted linear regression over the scales where the graph follows a straight line. What constitutes a straight line can be judged using the confidence intervals at each scale, which can be calculated or estimated. The important fact is that the estimation is heavily weighted toward the smaller scales, where there is much more data.

3.3.3.6 *Successive random addition (SRA)*

Another alternative method for the direct generation of fBm process is based on the successive random addition (SRA) algorithm. The SRA

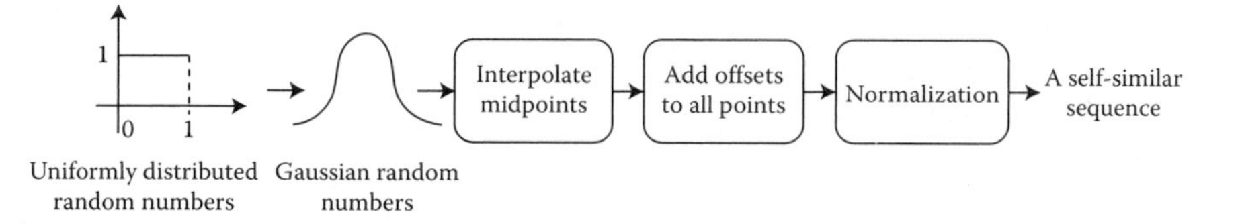

Figure 3.6 Block scheme of the SRA method.

method uses the midpoints like RMD, but adds a displacement of a suitable variance to all of the points to increase stability of the generated sequence.

Figure 3.6 shows how the SRA method generates an approximate self-similar sequence. The reason for interpolating midpoints is to construct Gaussian increments of X, which are correlated. Adding offsets to all points should make the resulted sequence self-similar and of normal distribution.

The SRA method consists of the following four steps:

Step 1. If the process $\{X_t\}$ is to be computed for time instance t between 0 and 1, then start out by setting $X_0 = 0$ and selecting X_1 as a pseudo-random number from a Gaussian distribution with mean 0 and variance $\mathrm{Var}[X_1] = \sigma_0^2$. Then $\mathrm{Var}[X_1 - X_0] = \sigma_0^2$.

Step 2. Next, $X_{1/2}$ is constructed by the interpolation of the midpoint, that is, $X_{1/2} = (1/2)\,(X_0 + X_1)$.

Step 3. Add a displacement of a suitable variance to all of the points, that is, $X_0 = X_0 + d_{1,1}$, $X_{1/2} = X_{1/2} + d_{1,2}$, and $X_1 = X_1 + d_{1,3}$.

The offsets $d_{1,*}$ are governed by fractional Gaussian noise.

For $\mathrm{Var}[X(t_2) - X(t_1)] = |t_2 - t_1|^{2H}\,\sigma_0^2$ to be true, for any t_1, t_2, $0 \le t_1 \le t_2 \le 1$, it must be required that $\mathrm{Var}[X_{1/2} - X_0] = (1/4)\,\mathrm{Var}[X_1 - X_0] + 2S_1^2 = (1/2)^{2H}\sigma_0^2$.

Thus, $S_1^2 = (1/2)(1/2^1)^{2H}(1 - 2^{2H-2})\sigma_0^2$.

Step 4. Next, Steps 2 and 3 are repeated. Therefore $S_n^2 = (1/2)(1/2^n)^{2H}(1 - 2^{2H-2})\sigma_0^2$, where σ_0^2 is an initial variance and $0 < H < 1$.

Using the above steps, the SRA method generates an approximate self-similar *fBm* process.

3.3.4 *Fast Fourier transform (FFT) method*

A faster method of generating synthetic self-similar traffic implies is the fast Fourier transform (Erdos and Renyi 1960). A complex number sequence is generated so that it corresponds to the power spectrum of a fGn process. Then, the discrete Fourier transform (DTFT) is used to obtain

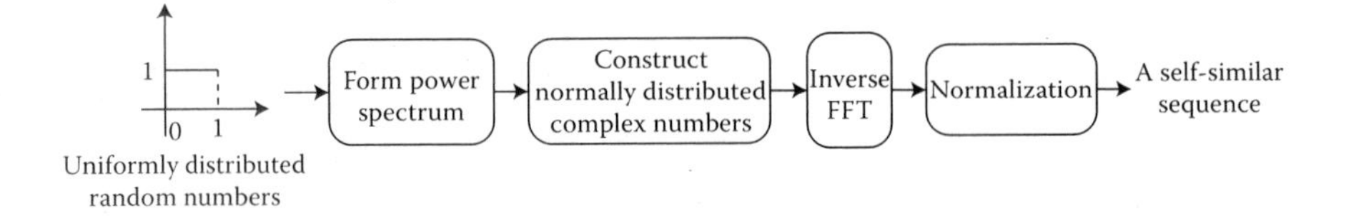

Figure 3.7 Block scheme of the FFT method.

the time correspondent of the power spectrum. Because the sequence has the power spectrum of fGn and the self-correlation and the power spectrum form a Fourier pair, it implies that this sequence has the self-correlation properties of fGn. In addition, DTFT and its inverse can be calculated using the FFT algorithm. The main advantage of this method is its swiftness.

This method generates approximate self-similar sequences based on the FFT and a process known as the fractional Gaussian noise (fGn) process (see Figure 3.7, which shows how the FFT method generates self-similar sequences). Its main difficulty is connected with calculating the power spectrum, which involves an infinite summation.

Briefly, it is based on (i) calculation of the power spectrum using the periodogram (the power spectrum at a given frequency represents an independent exponential random variable); (ii) construction of complex numbers, which are governed by the normal distribution; and (iii) execution of the inverse FFT. This leads to the following algorithm:

Step 1. Generate a sequence of the values $\{f_1, \ldots, f_{n/2}\}$ where: $\{f_i = \hat{f}(2\pi i/n), H\}$, corresponding to the power spectrum of a fGn process for frequencies from $2\pi/n$ to π, $0.5 < H < 1$. For a fGn process, the power spectrum $f(\lambda, H)$ is defined as

$$f(\lambda, H) = A(\lambda, H)[|\lambda|^{-2H-1} + B(\lambda, H)] \quad \text{for } 0 < H < 1 \quad \text{and} \quad -\pi \le \lambda \le \pi$$

where

$$A(\lambda, H) = 2\sin(\pi H)\Gamma(2H+1)(1-\cos\lambda)$$

$$B(\lambda, H) = \sum_{i=1}^{n} [(2\pi i + \lambda)^{-2H-1} + (2\pi i - \lambda)^{-2H-1}]$$

Step 2. Adjust the sequence of $\{f_1, \ldots, f_{n/2}\}$ for estimating power spectrum using periodogram.

Step 3. Generate $\{Z_1, \ldots, Z_{n/2}\}$ a sequence of complex values such that $|Z_i| = \sqrt{\hat{f}_i}$ and the phase of Z_i is uniformly distributed between 0 and 2π.

Step 4. Construct $\{Z'_0, \ldots, Z'_{n-1}$, an expanded version of $\{Z_1, \ldots, Z_{n/2}\}$

$$Z'_i = \begin{cases} 0, & \text{if } i = 0 \\ Z_{i,} & \text{if } 0 < i \leq \dfrac{n}{2} \\ \bar{Z}_{n-i}, & \text{if } \dfrac{n}{2} \leq i < n \end{cases}$$

where $\bar{Z}_{n-i}$ denotes the complex conjugate of Z_{n-i}.

$\{Z'_i\}$ retains the power spectrum used in constructing $\{Z_i\}$, but because it is symmetric about $\{Z'_{n/2}\}$, it now corresponds to the FFT of a real-valued signal.

Step 5. Calculate inverse FFT $\{Z'_i\}$ to obtain the approximate fGn sequence $\{X_i\}$.

3.4 *Evidence of self-similarity in real traffic*

As was already mentioned, the most used indicator for the level of self-similarity in time series is the Hurst parameter H. The self-similarity phenomenon is present for H in the range $0.5 \leq H \leq 1.0$, being stronger for values of H close to 1.0. But the estimation of H is a difficult task, especially because the precise value of H infinite time series should be examined, but actually we dispose of only the finite time series.

There are several methods for the estimation of the self-similarity in a time series, but the most known are statistical R/S (rescaled range) analysis, dispersion–time analysis, periodogram-based analysis, Whittle estimator, and wavelet-based analysis.

3.4.1 *Rescaled range method*

The rescaled range (R/S) is a normalized, nondimensional measure, proposed by Hurst himself to characterize the data variability. For a given set of experimental observations $X = \{X_n, n \in Z^+\}$ with the sample average $\bar{X}(n)$, sample dispersion $S^2(n)$, and sample range $R(n)$, the rescaled adjusted range (or R/S statistic) is

$$\frac{R(n)}{S(n)} = \frac{\max(0, \Delta_1, \Delta_2, \ldots, \Delta_n) - \min(0, \Delta_1, \Delta_2, \ldots, \Delta_n)}{S(n)} \tag{3.66}$$

where

$$\Delta_k = \sum_{i=1}^{k} X_i - k\overline{X} \quad \text{for } k = 1, 2, \ldots, n \tag{3.67}$$

In the case of many natural phenomena, when $n \to \infty$, then

$$E\left[\frac{R(n)}{S(n)}\right] \sim cn^H \tag{3.68}$$

with c a constant positive integer value. Applying logarithms:

$$\log\left\{E\left[\frac{R(n)}{S(n)}\right]\right\} \sim H\log(n) + \log(c) \tag{3.69}$$

So, one can estimate the value of H by the slope of a straight line that approximate the graphical plot of $\log\{E[R(n)/S(n)]\}$ as a function of $\log(n)$.

The R/S method is not very accurate and it is mainly used only to certify the existence of self-similarity in a time series.

3.4.2 Dispersion–time analysis

The dispersion–time analysis (or variance–time analysis) is based on the property of slowly decrease of the dispersion of a self-similar aggregate process:

$$\text{Var}[X^{(m)}] = \frac{\text{Var}[X]}{m^\alpha} \tag{3.70}$$

where $\alpha = 2 - 2H$.

Equation 3.8 can be written as

$$\log\{\text{Var}[X^{(m)}]\} \sim \log[\text{Var}(X)] - \alpha\log(m) \tag{3.71}$$

Because $\log\{\text{Var}[X^{(m)}]\}$ is a constant that not depends on m, one can represent Var(X) as a function of m in logarithmic axes. The line that approximates the resulting points has the slope α. The values of α between (–1; 0) represent self-similarity.

Like the R/S method, the variance–time analysis is a heuristic method. Both methods can be affected by poor statistics (few realisations of the self-similar process). They offer only a raw estimation of H.

3.4.3 Periodogram method

The estimation based on periodogram is a more accurate method than those based on aggregation, but it necessitates knowing *a priori* the mathematical parameterized model of the process. The periodogram is known also as the function of intensity $I_N(\omega)$ and represents the estimated spectral density of the stochastic process $X(t)$ (defined at discrete time intervals) and given, or the time period N by the equation:

$$I_N(\varpi) = \frac{1}{2\pi N}\left|\sum_{k=1}^{N} X_k e^{jk\varpi}\right|^2 \tag{3.72}$$

where ω is the frequency and X_k is the time series. When $\varpi \to 0$ the periodogram should be

$$I_N(\varpi) \sim |\varpi|^{1-2H} \tag{3.73}$$

The main drawback of this method is the need of a high processing capacity.

3.4.4 Whittle estimator

Whittle estimator derives from the periodogram, but uses a nongraphic representation of the probability. Considering a self-similar process with a *fBm*-type model and its spectral density $S(\omega, H)$, the value of the parameter H minimizes the so-called Whittle expression:

$$\int_{-\pi}^{\pi} \frac{I_N(\varpi)}{S(\varpi, H)} d\varpi \tag{3.74}$$

Both periodogram and Whittle estimator are derived from the maximum likelihood estimation (MLE) and offer good statistical properties, but can give errors if the model of the spectral density is false. Since the self-similarity estimation is concerned only with the LRD in the data sets, the Whittle estimator is employed as follows.

Each hourly data set is aggregated at increasing levels m, and the Whittle estimator is applied to each m-aggregated dataset using the fGn model. This approach exploits the property that any long-range-dependent process approaches fGn when aggregated to a sufficient level, and so should be coupled with a test of the marginal distribution of the aggregated observations to ensure that it has converged to the normal distribution. As m increases, SRDs are averaged out of the data set; if the value of H remains relatively constant, we can be confident that it measures a true underlying level of self-similarity. Since aggregating the series shortens it, confidence intervals will tend to grow as the aggregation level increases; however, if the estimates of H appear stable as the aggregation level increases, then we consider the confidence intervals for the unaggregated data set to be representative.

3.4.5 Wavelet-based method

This method, resembling the dispersion–time method, allows the analysis in both time and frequency domains. The wavelet estimator is constructed on the basis of the discrete wavelet transform (DWT), having the cumulative advantages of the estimators based on aggregation and of those based on MLE, but avoiding their drawbacks. The DWT uses the multiresolution analysis (MRA), which consists in the division of the sequence $x(t)$ in a low-pass approximation and several high-pass details, associated with the a_x, respectively, d_x coefficients:

$$x(t) = \text{approx}_J(t) + \sum_{j=1}^{J} \text{detail}_j(t) = \sum_{k} a_x(J,k)\varphi_{J,k}(t) + \sum_{j=1}^{J}\sum_{k} d_x(j,k)\psi_{j,k}(t) \tag{3.75}$$

The parameters $\{\varphi_{j,k}(t) = 2^{-j/2}\varphi_0\ (2^{-j}t - k)\}$ and $\{\Psi_{j,k}(t) = 2^{-j/2}\Psi_0(2^{-j}t - k)\}$ are the sets of the dilated scaling function φ_0 and wavelet function Ψ_0. The DWT consists in the sets of coefficients $\{d_x(j, k), j = 1, \ldots, J, k \in Z\}$ and $\{a_x(J, k), k \in Z\}$ defined by the vectorial products:

$$d_x(j,k) = \langle x, \Psi_{j,k} \rangle \tag{3.76}$$

$$a_x(j,k) = \langle x, \varphi_{j,k} \rangle \tag{3.77}$$

These coefficients are placed in a dyadic lattice. The details described by the wavelet coefficients grow when the resolution decrease. In the frequency domain, it corresponds to the effect of a low-pass filter, which produces a relatively constant bandwidth. On the other hand, the spectral estimators (based on periodograms) can be easily influenced, because a constant bandwidth alters the analyzed power spectrum, although it offers a perfect match. Therefore, $\text{Var}[\text{detail}_j] \sim 2^j(2^{H-2})$ and H can be estimated by the linear approximation of a plot in a diagram with logarithmic axes:

$$\log_2\left(\frac{2^j}{n_0}\sum_k |d_x(j,k)|^2\right) = (2\hat{H}-1)j + c \tag{3.78}$$

where $\hat{H}$ is a semiparametric estimator of H, n_0 is the data length, and c is finite. This estimator is efficient in Gaussian hypotheses.

3.5 Application specific models

3.5.1 Internet application-specific traffic models

3.5.1.1 Network traffic

Network traffic is obviously driven by applications. The most common applications that come to mind might include Web, e-mail, peer-to-peer file sharing, and multimedia streaming. A traffic study of Internet backbone traffic showed that Web traffic is the single largest type of traffic, on the average more than 40% of total traffic (Fraleigh et al. 2003). Although on a minority of links, peer-to-peer traffic was the heaviest type of traffic, indicating a growing trend. Streaming traffic (e.g., video and audio) constitute a smaller but stable portion of the overall traffic. E-mail is an even smaller portion of the traffic. Therefore, traffic analysts have attempted to search for models tailored to the most common Internet applications (Web, peer-to-peer, and video).

3.5.1.2 Web traffic

The World Wide Web is a familiar client–server application. Most studies have indicated that Web traffic is the single largest portion of overall Internet traffic, which is not surprising considering that the Web browser has become the preferred user-friendly interface for e-mail, file transfers, remote data processing, commercial transactions, instant messages, multimedia streaming, and other applications.

An early study focused on the measurement of self-similarity in Web traffic (Crovella and Bestavros 1997). Self-similarity had already been found in Ethernet traffic. Based on days of traffic traces from hundreds of modified Mosaic Web browser clients, it was found that Web traffic exhibited self-similarity with the estimated Hurst parameter H in the range 0.7–0.8. More interestingly, it was theorized that Web clients could be characterized as on/off sources with heavy-tailed on periods. In previous studies, it had been demonstrated that self-similar traffic can arise from the aggregation of many on/off flows that have heavy-tailed on or off periods. Heavy tails implies that very large values have nonnegligible probability. For Web clients, on periods represent active transmissions of Web files, which were found to be heavy-tailed. The off periods were either "active

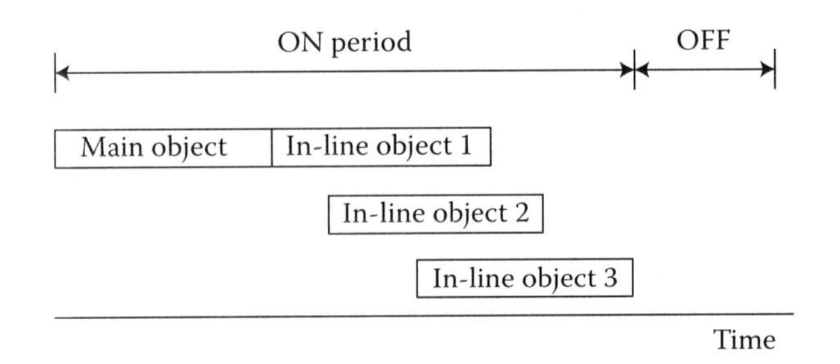

Figure 3.8 ON/OFF model with Web page.

off," when a user is inspecting the Web browser but not transmitting data, or "inactive off," when the user is not using the Web browser. Active off periods could be fit with a Weibull probability distribution, while inactive off periods could be fit by a Pareto probability distribution. The Pareto distribution is heavy-tailed. However, the self-similarity in Web traffic was attributed mainly to the heavy-tailed on periods (or essentially, the heavy-tailed distribution of Web files). For example, let consider that a Web page consists of a main object, which is an HTML document, and multiple in-line objects that are linked to the main object (see Figure 3.8). Both main object sizes and in-line object sizes are fit with (different) log-normal probability distributions. The number of in-line objects associated with the same main object follow a gamma probability distribution. In the same Web page, the intervals between the start times of consecutive in-line objects are fit with a gamma probability distribution.

TCP flow models seem to be more natural for Web traffic than "page based" consisting of a main object linked to multiple in-line objects.

In this example, the requests for TCP connections originate from a set of Web clients and go to one of multiple Web servers. A model parameter is the random time between new TCP connection requests. A TCP connection is held between a client and server for the duration of a random number of request–response transactions. Additional model parameters include sizes of request messages, server response delay sizes of response messages, and delay until the next request message. Statistics for these model parameters are collected from days of traffic traces on two high-speed Internet access links.

3.5.1.3 *Peer-to-peer traffic*

Peer-to-peer (P2P) traffic is often compared with Web traffic because P2P applications work in the opposite way from client–server applications, at least in theory. In actuality, P2P applications may use servers. P2P applications consist of a signaling scheme to query and search for file locations, and a data transfer scheme to exchange data files. While the data transfer is typically between peer and peer, sometimes the signaling is done in a client–server way. Since P2P applications have emerged only in the last decade, there are a few traffic studies to date. They have understandably focused more on collecting statistics than the harder problem of stochastic modeling. An early study examined the statistics of traffic on a university campus network (Saroiu et al. 2002).

The observed P2P applications were Kazaa and Gnutella. The bulk of P2P traffic was AVI and MPEG video, and a small portion was MP3 audio. With this mix of traffic, the average P2P file was about 4 MB, three orders of magnitude larger than the average Web document. A significant portion of P2P was dominated by a few very large objects (around 700 MB) accessed 10–20 times. In other words, a small number of P2P hosts are responsible for consuming a large portion of the total network bandwidth.

3.5.1.4 Video

The literature on video traffic modeling is vast because the problem is very challenging. The characteristics of compressed video can vary greatly depending on the type of video that determines the amount of motion (from low-motion videoconferencing to high motion broadcast video) and frequency of scene changes; and the specific intraframe and interframe video coding algorithms. Generally, interframe encoding takes the differences between consecutive frames, compensating for the estimated motion of objects in the scene. Hence, more motion means that the video traffic rate is more dynamic, and more frequent scene changes is manifested by more "spikes" in the traffic rate. While many video coding algorithms exist, the greatest interest has been in the Moving Picture Coding Experts Group (MPEG) standards. Video source rates are typically measured frame by frame (which are usually 1/30 s apart). That is, X_n represents the bit rate of the nth video frame. Video source models address at least these statistical characteristics: the probability distribution for the amount of data per frame; the correlation structure between frames; times between scene changes; and the probability distribution for the length of video files. Due to the wide range of video characteristics and available video coding techniques, it has seemed impossible to find a single universal video traffic. The best model varies from video to video file. A summary of research findings was presented by Chen (2007) and includes the following:

- The amount of data per frame has been fit to log-normal, gamma, and Pareto probability distributions or their combinations.
- The autocorrelation function for low-motion video is roughly exponential, while for broadcast video is more complicated, exhibiting both SRD at short lags and LRD at much longer lags.
- Time series, Markov modulated, and wavelet models are popular to capture the autocorrelation structure.

- The lengths of scenes usually follow Pareto, Weibull, or Gamma probability distributions.
- The amount of data in scene change frames are uncorrelated and follow Weibull, Gamma, or unknown probability distributions.
- The lengths of streaming video found on the Web appear to have a long-tailed Pareto probability distribution.

3.5.2 Models for TCP flows

Traffic sources may be limited by access control or the level of network congestion. Some protocols limit the sending rates of hosts by means of sliding windows or closed loop rate control. Clearly, traffic models must take such protocols into account in order to be realistic. Here, we focus on TCP because TCP traffic constitutes the vast majority of Internet traffic. TCP uses a congestion avoidance algorithm that infers the level of network congestion and adjusts the transmission rate of hosts accordingly. Even other protocols should be "TCP friendly," meaning that they interact fairly with TCP flows.

The transmission control protocol (TCP) has for some time carried the bulk of data in the Internet; for example, currently all Web, FTP (file transfer), and NNTP (network news) services, or some 70%–90% of all traffic, use TCP. For this reason, it is a natural choice of protocol on which to focus attention. TCP aims to provide a reliable delivery service. To ensure this, each packet has a sequence number, and its successful receipt must be signaled by a returning acknowledgment (ACK) packet. The second aim of TCP is to provide efficient delivery. An adaptive window flow control is employed where only a single window of data is allowed to be transmitted at any one time. Once one window's worth of data has been sent, the sender must wait until at least part of it has been acknowledged before sending more. This method is powerful as it is controlled by the sender, requiring no additional information from either the receiver or the network. TCP has two control windows, the sender-based *cwnd*, and the (advertised) receiver window *rwnd*. Sources use the minimum of the two. Usually we assume that *rwnd* is constant, playing the role of the maximum value of *cwnd*, the dynamical variable of interest, and all packets are of maximum segment size (MSS).

The data send rate r_s of a source with window size w, and packet acknowledgment round-trip time (RTT) is $r_s = w/\text{RTT}$. The interesting design issue in the flow control lies in how to choose *cwnd*. Ideally, the window is exactly tuned to the available bandwidth in the network. If it is too small, then the network is used inefficiently, while if too large, congestion may result. However, the Internet bandwidths can vary from kb/s to Gb/s, while RTTs can range from <1 ms to 1 s, allowing a variation in send rate of 10 orders of magnitude. The TCP flow control attempts to adaptively choose the window size using different algorithms, among them *Slow start, Congestion avoidance, Fast recovery,* and *Fast retransmission* (Erramilli et al. 2002).

3.5.2.1 TCP flows with congestion avoidance

TCP senders follow a self-limiting congestion avoidance algorithm that adjusts the size of a congestion window according to the inferred level of congestion in the network. The congestion window is adjusted by so-called additive increase multiplicative decrease (AIMD). In the absence of congestion when packets are acknowledged before their retransmission time-out, the congestion window is allowed to expand linearly. If a retransmission timer expires, TCP assumes that the packet was lost and infers that the network is entering congestion. The congestion window is collapsed by a factor of a half. Thus, TCP flows generally have the classic "saw tooth" shape. Accurate models of TCP dynamics are important because of TCP's dominant effect on Internet performance and stability. Hence, many studies have attempted to model TCP dynamics (a significant synthesis is presented in Altman et al. 2005). Simple equations for the average throughput exist, but the problem of characterizing the time-varying flow dynamics is generally difficult due to interdependence on many parameters such as the TCP connection path length, round-trip time, and probability of packet loss, link bandwidths, buffer sizes, packet sizes, and number of interacting TCP connections.

3.5.2.2 TCP flows with active queue management

The TCP congestion avoidance algorithm has an unfortunate behavior where TCP flows tend to become synchronized. When a buffer overflows, it takes some time for TCP sources to detect a packet loss. In the meantime, the buffer continues to overflow, and packets will be discarded from multiple TCP flows. Thus, many TCP sources will detect packet loss and slow down at the same time. The synchronization phenomenon causes underutilization and large queues.

Active queue management schemes, mainly random early detection (RED) and its many variants, eliminate the synchronization phenomenon and thereby try to achieve better utilization and smaller queue lengths. Instead of waiting for the buffer to overflow, RED drops a packet randomly with the goal of losing packets from multiple TCP flows at different times. Thus, TCP flows are unsynchronized, and their aggregate rate is smoother than before. From a queuing theory viewpoint, smooth traffic achieves the best utilization and shortest queue.

In conclusion, the impacts of the TCP flow control mechanisms on the traffic are very strong. TCP-type feedback appears to have the effect of modifying the self-similarity traffic behavior, but it neither generates it nor eliminates it.

chapter four

Topological models of complex networks

4.1 Topology of large-scale networks

The study of a majority of complex networks started initially from the desire to understand various natural systems, ranging from communication networks to ecologic chains or networks (Costa et al. 2003). This chapter lists the significant results in the research of complex networks, focusing on the large-scale traffic networks. Beyond the fields from where data was extracted, special emphasis will be placed on the three robust indicators for network topology: average path length, clustering coefficient, and degree distribution. To model a distributed network environment like the Internet, it is necessary to integrate the data collected from multiple points in a network in order to get a complete picture of network-wide view of the traffic. Knowledge of dynamic characteristics is essential to network management (e.g., detection of failures/congestion, provisioning, and traffic engineering like QoS routing or server selections). However, given the enormity of the huge scale and restrictions due to access rights, it is expensive (sometime impossible) to measure such characteristics directly. To solve this problem, a host of methods and tools to infer the unobservable network performance characteristics are used in large-scale networking environment. A model where inference based on self-similarity and fractal behavior is best represented is the scale-free network, and so this model will be largely discussed in this chapter.

4.1.1 World Wide Web

The World Wide Web (WWW) network is the biggest example of providing topological information. Its nodes are the documents (web pages) and its edges are the hyperlinks (URL) connecting one document to another. The interest in the WWW as a network simply exploded after discovering that the distribution of the degrees of the Web pages is given by a power law over several magnitudes (Barabási et al. 1999). Moreover, the cited work, supplemented and developed in Albert and Barabási (2002) paved the way for extensive research based on rigorous mathematical models. Because the edges of the WWW graph are directed, the network is defined by a two-degree distribution: output arcs' degree $P_{out}(k)$, which is the probability for a document to have k URL links to other documents, and the input arcs' degree $P_{in}(k)$, which is the probability that k URL link to indicate to a certain document. Several studies reported that both $P_{out}(k)$ and $P_{in}(k)$ have power law queues:

$$P_{out}(k) \sim k^{-\gamma_{out}} \quad \text{and} \quad P_{in}(k) \sim k^{-\gamma_{in}} \tag{4.1}$$

Albert, Jeong, and Barabási (2000) studied a subset of the WWW containing 325,729 nodes and found that $\gamma_{out} = 2.45$ and $\gamma_{in} = 2.1$. Other researchers used a slightly different representation of the WWW, treating every node as a domain name and considering that any two nodes are connected if a page from one refers to a page in the other. Although this method often associates thousands of pages belonging to the same domain, being a nontrivial node aggregation, the distribution of the outgoing arcs still follows a power law equal to $\gamma_{in}^{dom} = 1.94$.

The oriented nature of the WWW does not allow for measuring the clustering coefficient using Equation 4.1. A way to circumvent this problem is to transform the network in an unoriented direction, by considering each edge to be bidirectional. The original algorithms for constructing these styles of network are used for undirected graphs. Among them, small world (SW) undirected networks were first constructed by Watts and Strogatz using an algorithm described as a rewiring ring lattice (Watts and Strogatz 1998). Then, Kleinberg proposed a directed version of SW networks, which starts from a two-dimensional grid rather than a ring in his work searching for a decentralized algorithm to find shortest paths in directed SW (Kleinberg 2000). More recently, Sriram and Cliff suggested a hybrid approach closer to of Watts and Strogatz's original algorithm with the only modification that the initial condition is a ring lattice with outgoing directed edges to all k nearest neighbors, with k being the number of subscriptions in their model (Sriram and Cliff 2011). Thus, between neighboring edges, there will be two edges, one in each direction. As rewiring probability $p = 0.1$ was used, a value Watts and Strogatz showed to be large enough to exhibit short path lengths, and at the same time small enough to exhibit a high transitivity of the network.

The Web exhibits properties that identify it as a complex network. Incorporating the small world, scale-free, self-organizing properties of the Web, a theoretical framework for understanding and mapping the movement of information on such dynamic information system explores the Web using complexity and fractal theories to explain the underlying structure of the Web. In the previous years, complexity theory offers new ways to understand the evolution, topology, and relationships that comprise real networks, including the Internet and the WWW. These seemingly chaotic networks develop in ways that are self-similar, self-sustaining, and self-regulating. They have the same patterns at different scales, can replicate themselves, correct errors, and organize without central authority. Complexity theory offers a new insight to phenomena that were previously thought to be random, chaotic, or too complex to understand. Much research has been done to articulate and formalize the revolutionary ideas that explain and model the chaos, the idea that stochastic outcomes are based on the repeating relationships of simple patterns. The best models accurately and simply represent real-world phenomena dynamically. In the case of a dynamic system, however, modeling part of the system does little to facilitate our knowledge of the entire system, as the parts functioning together are frequently more revealing than the sum of the individual aspects. The ability of building this functionality, bringing together concepts from fractal geometry, complexity, and network theories provides a context for understanding and building algorithms that can describe the complex nature of the flow of information in real networks such as the Web.

4.1.2 The Internet

The Internet is a network of physical connections between computers and other telecommunication devices. The topology of the Internet is studied from two different perspectives. On the router level, the nodes are routers and the edges are the physical connections between them. On the interdomain (or autonomic system) level, each domain comprising hundreds of routers and computers is represented by only one node, and an edge connects two domains if there is at least one route connecting them. The Internet was studied considering both levels (Ravasz and Barabási 2003), and, in every case, the conclusion was that the degree distribution respects a power law. The Internet as a network also manifests phenomena like clustering and small path length (Lan and Heidemann 2003).

In the first models proposed for the simulation of traffic in large communication networks, only Poisson or renewal processes were considered. In either case, arrivals were memoryless in the Poisson case, or memoryless at renewal points, and interarrival intervals were exponentially distributed. The Poisson arrival model and exponentially distributed holding time model allowed analytically and computationally simple Markov chains to be used for modeling. A large number of works have focused on modeling Internet topologies and on developing realistic topology generators. Waxman (1988) introduced the first topology generator that became widely known. The Waxman generator was based on the classical Erdös–Rényi random graph model. After it became evident that observed networks have little in common with classical random graphs, new generators tried to mimic the perceived hierarchical network structure and were consequently called structural. Then Faloutsos et al. (1999) discovered that the degree distributions of router-level topologies of the Internet followed a power law. Structural generators failed to reproduce the observed power laws. This failure led to a number of subsequent works trying to resolve the problem.

The existing topology models capable of reproducing power laws can be roughly divided into the following two classes: causality-aware and causality-oblivious (Dimitropoulos et al. 2009). The first class includes the Barabási–Albert (BA) model, the highly optimized tolerance (HOT) model (Casal et al. 1999) and their derivatives. The models in this class employ preferential attachment mechanisms to generate synthetic Internet topologies and incremental growth of the network by adding nodes one by one and links to a graph based on a formalized network evolution process. One can show that both BA and HOT growth mechanisms produce power laws. On the other hand, the causality-oblivious approaches try to match a given (power-law) degree distribution without accounting for different forces that might have driven the evolution of a network to its currently observed state. The models in this class include random graphs with the given expected degree sequences (Chung and Lu 2002) and Markov graph rewiring models (Maslov et al. 2004). However, Internet traffic behaves very differently from such simple Markovian models. Traffic measurements made at the local area networks (LAN) and wide area networks (WAN) suggest that traffic exhibits variability (traditionally called "burstiness") over multiple time scales. The second-order properties of the counting process of the observed traffic displayed behavior that is associated with self-similarity, multifractals, and/or long-range dependence (LRD). This indicates that there is a certain level of dependence in the arrival process. Near-range and long-range dependencies often manifest themselves in a network by causing frequent and irremediable packet losses and other serious effects in the network.

Dependencies and burstiness in traffic hence brought in an enormous amount of attention from researchers. They attempted to develop mathematically based models that would help explain the nature of the systems exhibiting such phenomena and provide critical insight into the actual mechanisms that led to this behavior. Models like fractional Brownian motion and chaotic maps were suited to capture the second-order self-similar behavior of traffic. Their results were difficult to get and harder to apply, and such models did not provide insight into the actual mechanism of traffic generation. A simpler, more accurate, and analytically tractable model that provides more physical insight into why they are meaningful on physical grounds would help the network designers produce more effective and efficient designs.

The Internet is a prime example of a self-organizing complex system, having grown mostly in the absence of centralized control or direction. In this network, information is transferred in the form of packets from the sender to the receiver via routers, computers that are specialized to transfer packets to another router "closer" to the receiver. A router decides the route of the packet using only local information obtained from its interaction with neighboring routers, not by following instructions from a centralized server. A router stores packets in its finite queue and processes them sequentially. However, if the queue overflows due to excess demand, the router will discard incoming packets, a situation corresponding to congestion. A number of studies have probed the topology of the Internet and its implications for traffic dynamics.

To efficiently control and route the traffic on an exponentially expanding Internet, one must not only capture the structure of current Internet, but allow for long-term network design. After the already mentioned discovery of Faloutsos, according to which Internet is a scale-free network with a power-law degree distribution, several contributors found that the Internet flow is strongly localized: most of the traffic takes place on a spanning network connecting a small number of routers that can be classified either as "active centers," which are gathering information, or "databases," which provide information. Experimental evidence for self-similarity in various types of data network traffic is already overwhelming and continues to grow. So far, simulations and analytical studies have shown that it may have a considerable impact on network performance that could not be predicted by the traditional short-range-dependent models. The most serious consequence of self-similar traffic concerns the size of bursts. Within a wide range of time scales, the burst size is unpredictable, at least with traditional modeling methods.

This is the point from which the authors of this book assume that the traffic behavior is strongly influenced and depends on the network free-scale structure and that the self-similarity confers a fractal aspect for both traffic and topology.

4.2 Main approaches on networks topology modeling

4.2.1 Random graph theory

The random graph theory was proposed by Erdös and Rényi (1960), after Erdös had discovered the usefulness of stochastic methods when studying graph theory problems. A thorough analysis on this topic is given in the classic book of Bollobas (Bollobas 1985) complemented by the historical guide of the Erdös–Rényi approach of Karonski and Rucinski (1997). Following this publication, the most important results of random graph theory are provided, by highlighting the aspects relevant to complex networks. In their first article about random graphs, the Erdös–Rényi model is defined as a set of N nodes connected by n edges chosen randomly from the $(N(N-1))/2$ possible edges. There are a total of $C^{n}_{(N(N-1))/2}$ graphs with N nodes and n edges, which produce a probability space in which every realization is equally probable.

The random graph theory studies the probability space properties for graphs with N nodes when $N \to \infty$. Many random graph properties can be determined using probabilistic arguments. In this matter, Erdös and Rényi used a definition such that almost every graph has a Q property if the probability of having Q approaches tends to 1 for $N \to \infty$. Among the problems addressed by Erdös and Rényi are some in direct connection with understanding complex networks.

In mathematics, the construction of a random graph is often called evolution: starting from a set of N isolated vertices, the graph evolves by adding random edges. The graph obtained at different levels during this process corresponds to increasingly higher p connection probabilities, and obtaining in the end a completely connected graph (having a maximum number of edges $n = N(N - 1)/2$) for $p \to 1$. The main goal of the random graph theory is to determine the p connection probability for that a certain graph property appears. The greatest discovery of Erdös and Rényi is that many important random graph properties appear quite abruptly. Thus, for a given probability, almost every graph has a property Q (e.g., every pair of nodes is connected through a set of consecutive edges) or, on the contrary, almost no graph has it. The transition from highly improbable to highly probable for a certain property is usually quick. Thus, for random graph theory, the probability of occupation is defined as a function of the system dimension: p represents the fraction of existing edges from the total of $N(N - 1)/2$ possibilities. Larger graphs with the same p will contain more edges, and thus properties like cycles may appear for smaller values of p more easily in larger graphs than in smaller ones. This means that for many properties Q in random graphs, there is no single threshold N that is independent, but rather a threshold function must be defined depending on the system dimension and $p_c(N \to \infty) \to 0$. Also, it can be observed that the average graph degrees

$$< k >= 2n/N = p(N-1) \cong pN \tag{4.2}$$

has a critical value that is independent on the system dimension.

The component of a graph is by definition a connected, isolated subgraph, known also as a cluster in network research and percolation theory. As it was pointed out by Erdös and Rényi, there is an abrupt change in the cluster structure of a random graph when $\langle k \rangle$ tends to 1.

If $0 < \langle k \rangle < 1$, almost every cluster is either a tree either it contains exactly one cycle. Although cycles are present, almost all nodes belong to trees. The average number of clusters is of the order $N - n$, where n is the number of edges, meaning that by adding on this scale a new edge, the number of clusters decreases by one. The largest cluster is a tree and its dimension is proportional to ln N.

In a random graph with p connection probability, the k_i degree of one node i follows a binomial distribution with $N - 1$ and p parameters

$$P(k_i = k) = C_{N-1}^{k} p^{k} (1-p)^{N-1-k} \tag{4.3}$$

This probability represents the number of ways in which k edges can be drawn starting from a certain node: the probability of k edges is p^k, the probability of no additional edges is $(1 - p)^{N-1-k}$, and there are C_{N-1}^{k} equivalent ways to select the k starting points for these edges. Moreover, if i and j are distinct nodes, then $P(k_i = k)$ and $P(k_j = k)$ tend to be independent random variables. To determine the distribution of the graph degree, the number of k degree nodes, X_k, must be studied. The main goal is to determine the probability for X_k to take a given value, $P(X_k = r)$.

According to Equation 4.3, the expected value for the number of nodes with k degree is

$$E(X_k) = NP(k_i = k) = \lambda_k \tag{4.4}$$

where

$$\lambda_k = NC_{N-1}^{k} p^{k} (1-p)^{N-1-k} \tag{4.5}$$

The diameter of a graph is defined as the maximum distance between any two nodes. Strictly speaking, the diameter of an unconnected graph (i.e., containing only isolated clusters) is infinite, but it can be defined as the maximum diameter of its clusters. Random graphs tend to have small diameters when p is not too small. When $\langle k \rangle^{l}$ is equal to N the diameter is proportional to $\ln(N)/\ln(\langle k \rangle)$, thus it depends only logarithmically on the number of nodes (Albert and Barabási 2002). This means that when considering all graphs with N nodes and p connection probability, the value scale through which the diameters of these graphs can vary is very small, concentrated around

$$d = \frac{\ln(N)}{\ln(pN)} = \frac{\ln(N)}{\ln(\langle k \rangle)} \tag{4.6}$$

Let finally define the clustering coefficient of a complex network. If in a random graph, a node and its neighbors are considered, the probability for two of these neighbors to be connected is equal to the probability for two randomly chosen graphs to be connected. As a result, the clustering coefficient of a random graph is

$$C_{\text{rand}} = p = \frac{\langle k \rangle}{N} \tag{4.7}$$

The ratio between the clustering coefficient and the average degree of real networks is a function of dimension (Barabási et al. 2003).

4.2.2 *Small-world networks*

Experiments show that real networks have the small-world property just like random graphs, but they have unusually large clustering coefficients. Moreover, the clustering coefficient seems to be independent of the network size. This last property is characteristic of the ordered lattices, whose clustering coefficients are independent of the dimension and depend strictly on the coordination number. Such small size regular lattices, however, do not have small trajectories: for an n-dimensional hyper cubic lattice the average node to node distance is scaled with $N^{1/d}$, and increases with N faster than the logarithmic increase noticed for the real and random graphs. The first successful attempt to generate graphs having large clustering coefficients and small trajectory length ℓ are owed to Watts and Strogatz (1998), which proposed a model with a single parameter that interpolates between a finite-dimensional lattice and a random graph.

In order to understand the coexistence of small trajectory lengths and clustering, the behavior of the clustering coefficient $C(p)$ and the average trajectory length $\ell(p)$ will be studied as a function of the reconnection probability p. For a circular lattice $\ell(0) \cong N/2K \gg 1$ and $C(0) \cong 3/4$, so that ℓ is linearly scaled with the system dimension, and the clustering coefficient is large. Additionally, for $p \to 1$ the model converges to a random graph for that $\ell(1) \sim \ln(N)/\ln(K)$ and $C(1) \sim K/N$, so that ℓ scales logarithmically with N, and the clustering coefficient decreases with N. These limit cases may suggest that a large C coefficient is always associated with a large ℓ, while a small C with a small ℓ. Contrarily, Watts and Strogatz discovered that a large interval of p exists in which $\ell(p)$ tends to $\ell(1)$ and $C(p) \gg C(1)$. The regime starts with a quick drop of $\ell(p)$ for small values of p, while $C(p)$ remains constant, thus producing networks that are grouped, but have a small characteristic trajectory length. This coexistence of small ℓ and large C values manifests an excellent resemblance to the real networks characteristics, determining many researchers to name such systems small-world networks. In the following, the main result regarding the properties of small-world models will be presented in brief.

4.2.2.1 *Average trajectory length*

As previously stated, in the WS model there is a change in scaling the characteristic trajectory length ℓ with the increase of the fraction p for the reconnection of edges. For smaller values of p, ℓ is scaled linearly with the system dimension, while for a large p, the scaling is logarithmic. The origin for the rapid drop of ℓ is the appearance of shortcuts between nodes. Every shortcut, randomly created, is probable to connect separate parts at long distances in the graph, and thus, have a significant impact on the characteristic trajectory length of the entire graph. Even a small fraction of the shortcuts is enough to drastically decrease the average trajectory length, while on a local level the network remains very ordered. Watts demonstrated that ℓ does not start to decrease until $p \geq 2/NK$, thus guaranteeing the existence of at least one shortcut. This implies that the transition p depends on the system dimension, or, on the contrary, there is a crossover length p dependent on N^* so that if $N < N^*$ then $\ell \sim N$, but if $N > N^*$ then $\ell \sim \ln(N)$. Today, it is widely accepted that the characteristic trajectory length follows the general scaling form

$$\ell(N,p) \sim \frac{N}{K} f(pKN^d) \tag{4.8}$$

where $f(u)$ is an universal scaling function that satisfies

$$f(u) = \begin{cases} \text{constant} & \text{if } u \ll 1 \\ \ln(u)/u & \text{if } u \gg 1 \end{cases} \tag{4.9}$$

Several attempts have been made to calculate the exact distribution of the trajectory length and the average trajectory length ℓ. Dorogovtsev and Mendes (2002) have studied a simpler model that contained a circular lattice with oriented edges of length ℓ and a central node connected with the probability p to the nodes of the lattice through nonoriented edges of length 0.5. They calculated precisely the distribution of the trajectory length through this model, showing that ℓ/N depends only on the scaling variable pN, and the functional shape of this dependence is similar to $\ell(p)$ obtained numerically in the WS model.

4.2.2.2 *Clustering coefficient*

Besides the small average trajectory length, the small-world networks have a relative large clustering coefficient. The WS model manifests this duality for a wide range of the reconnection probability p. In a regular lattice ($p = 0$), the clustering coefficient does not depend on the lattice size but only on its topology. While the edges of the lattice are randomly reconnected, the clustering coefficient remains close to $C(0)$ until larger values for p. According to this, $C'(p)$ is the ratio of the average number of edges between the neighbors of a node to the average number of possible edges between those neighbors.

4.2.2.3 *Degree of distribution*

In the WS model, for $p = 0$, every node has the same degree K, so that the degree distribution is a delta function centered around K. A nonnull

p introduces disorder in the network, widening the degree distribution but still keeping the average degree K. Because only one end of each edge is reconnected (a total of $pNK/2$ edges), each node has at least $K/2$ edges after the reconnection process. As a consequence, for $K > 2$, there are no isolated nodes, and the network is normally connected for a wide range of connection probabilities, unlike a random graph formed of isolated groups.

4.2.2.4 Spectral properties

The spectral density $\rho(\lambda)$ of a graph highlights important information about its topology. To be specific, it was observed that for large random graphs $\rho(\lambda)$ converges to a semicircle. Thus, it comes as no surprise that the spectrum of the WS model depends on the reconnection probability p. For $p = 0$, the network is regular and periodic, so that $\rho(\lambda)$ contains numerous singularities. For intermediate values of p these singularities $\rho(\lambda)$ retains a strong inclination. Finally, when $p \to 1$, then $\rho(\lambda)$ tends to the semicircle law that describes random graphs. Thus, the e high regularity of small-world models for a variety of values for p is highlighted by the results regarding the spectral properties of the Laplace operator, which provides information about the time evolution of a diffuse graph field.

4.2.3 The scale-free (SF) model

Empirical results demonstrate that many large networks are scale-free, meaning that the degree of distribution follows a power law for large values of k. Moreover, even for those networks for that $P(k)$ has an exponential tail, the degree of distribution deviates from the Poisson distribution (Li et al. 2005). It was seen in Sections 4.2.1 and 4.2.2 that the random graph theory and the WS model cannot reproduce this characteristic. While random graphs can be directly constructed to have degree distributions of power law type, they can only postpone an important question: what is the mechanism responsible for generating scale-free networks? In this section, it will be shown that in order to answer this question a deviation from topology modeling to constitution and evolution of the network is necessary. At this point, the two approaches do not appear to be distinct, but it will become clear that a fundamental difference exists between random graph and small-world modeling, and the modeling of power law graphs. While the purpose of the previous models was to build a graph with correct topological characteristics, the modeling of scale-free networks will focus on capturing the dynamics of the network. Thus, the main assumption behind evolutionary or dynamic networks is that if the processes that assembled the present networks are correctly captured, then their topology can also be obtained.

4.2.3.1 Definition of the scale-free network (SFN) model

The origin of the power law degree distribution in networks was first addressed by Albert and Barabási (2002), who argued that the free-scale nature of real networks starts from two generic mechanisms common to many real networks. The network models discussed so far imply starting from a fixed number of vertices *N*, which are afterward connected or reconnected randomly, without altering *N*. By contrast, the majority of real networks describe open systems that grow during their life span by further adding new nodes. For example, the WWW network grows exponentially over time by adding new Web pages, or the research literature grows constantly by the publishing of new documents.

Additionally, the network models discussed until now assume that the probability for two nodes to be connected (or their connections to be modified) is independent of the node degree, meaning that the new edges are placed randomly. The majority of the real networks display a preferential attachment, and thus the connection probability of one node depends on its degree. For example, a Web page will rather include links to popular documents with an already large degree, because this kind of documents are easy to find and well known; a new manuscript is more likely to cite a well-known publication and so a largely cited one, than the less cited and less known ones.

These two ingredients, growth and preferential attachment, inspired the introduction of the scale-free model (SF) that has a power law degree distribution. The SF model algorithm is

1. *Growth:* starting from a small number m_0 of nodes, at each step, a new node is added with $m(\leq m_0)$ edges that link the new node to m different already existing ones.
2. *Preferential attachment:* When node to which the new node will link to are chosen, the probability Π that the new node to be connected to node i depends on the degree k_i of node i, so that

$$\Pi(k_i) = \frac{k_i}{\sum_j k_j} \tag{4.10}$$

After t steps this algorithm generates a network with $N = t + m_0$ nodes and mt edges.

4.2.3.2 General SFN properties

Barabási and Albert suggested that scale-free networks emerge in the context of growing network in which new nodes connect preferentially to the most connected nodes already in the network. Specifically,

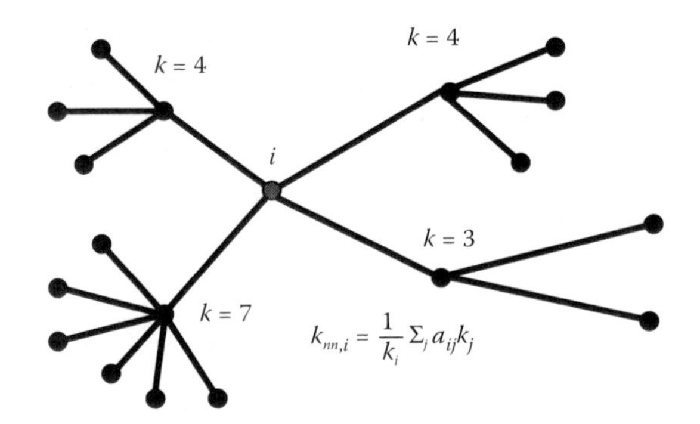

Figure 4.1 A scale-free network graph.

$$p_i(n+1) = \frac{k_i(n)}{\sum_{1-n_0}^{n} k_i(n)} \tag{4.11}$$

where n is the time and number of nodes added to the network, n_0 is the number of initial nodes in the network at time zero, k_i is the degree of node i and $p_i(n+1)$ is the probability of a new node, added at time $n+1$ linking to node i.

As illustrated in Figure 4.1, as time ticks by, the degree distribution of the nodes becomes more and more heterogeneous since the nodes with higher degree are the most likely to be the ones new nodes link to. Significantly, scale-free networks provide extremely efficient communication and navigability as one can easily reach any other node in the network by sending information through the "hubs," the highly connected nodes. The efficiency of the scale-free topology and the existence of a simple mechanism, leading to the emergence of this topology led many researchers to believe in the complete ubiquity of scale-free network. Note that scale-free networks are a subset of all small-world networks because (i) the mean distance between the nodes in the network increases extremely slowly with the size of the network and (ii) the clustering coefficient is larger than for random networks.

4.2.3.3 Diameter of scale-free networks

It was shown that scale-free networks with degree exponent $2 < \lambda < 3$ possess a diameter $D \sim \ln N$, smaller even than that of random and small-world networks (Dorogovtsev and Mendes 2001). If the network is fragmented, we will only be interested in the diameter of the largest cluster (assuming there is one). In this study, we consider the diameter of a scale-free network defined as the average distance between any two sites on the graph. Actually, it easier still to focus on the radius of a graph, $L \equiv \langle l \rangle$ as the average distance of all sites from the site of highest degree in the network. The diameter of the graph D is restricted to $L \leq D \leq 2L$ and thus scales like L.

4.2.3.4 Average trajectory length

The average trajectory length is smaller in SFNs than in random graphs for every N, thus indicating that the heterogeneous scale-free topology is more efficient in closing in the nodes than the homogeneous topology of random graphs. Note that average trajectory length in the SF model increases almost logarithmically with N. One can observe that the prediction of Equation 4.6, although gives a good approximation for the random graph, systematically underestimates the average trajectory length of the SF network, and also the average trajectory length of real networks.

4.2.3.5 Node degree correlation

In the random graph models with random average degree distributions, the node degrees are not correlated. Krapivsky and Redner demonstrated that for SF models, the correlations develop spontaneous between the degrees of already connected nodes (Krapivsky and Redner 2001). It is considered that all the pair of nodes with degree k and l are connect by an edge. Without losing generality, it can be assumed that the node with degree k was added later to the system, implying from that moment on that $k < l$, such that older nodes have a higher degree than the newer ones, and, for simplicity, $m = 1$ is used.

4.2.3.6 Clustering coefficient

Even if the clustering coefficient has been previously studied for the WS model, there is no analytical prediction for the SF model. It was discovered that the clustering coefficient of a scale-free network is almost five times larger than the one of a random graph, and this factor increases slightly with the number of nodes. Nevertheless, the clustering coefficient of the SF model decreases with the network dimension following approximately the power law $C \sim N^{-0.75}$, which, although being a slower decrease than $C = \langle k \rangle N^{-1}$ noted for random graphs, it is still different from the small-world behavior, where C is independent of N.

4.3 Internet traffic simulation with scale-free network models

4.3.1 Choosing a model

To model a distributed network environment like the Internet, it is necessary to integrate data collected from multiple points in a network in order to get a complete picture of network-wide view of the traffic. Knowledge of dynamic characteristics is essential to network management (e.g., detection of failures/congestion, provisioning, and traffic engineering like QoS routing or server selections). However, because of a huge scale and access rights, it is expensive (sometime impossible) to measure such characteristics directly. To solve this, methods and tools for inferencing of unobservable network performance characteristics are used in large-scale networking environment. A model where inference based on self-similarity and fractal behavior can be applied is the scale-free network (SFN) model.

The simulation of scale-free networks is necessary in order to study their characteristics like fault-tolerance and resistance to random attacks. However, large-scale networks are difficult to simulate due to the hefty requirements imposed on CPU and memory. Thus, a distributed approach to simulation can be useful, particularly for large-scale network simulations, where a single processor is not enough.

From the different variants that have been suggested for the SFN model, one of them known as the "Molloy–Reed construction" (Molloy and Reed 1998) will be considered in the following. This model ignores the evolution and assumes only the degree of distribution and no correlations between nodes. Thus, the site reached by following a link is independent of the origin. This means that a new node will more probably attach to those nodes that are already very well connected, that is, they have a large number of connections with other nodes from the network. Poor connected nodes, on the other hand, have smaller chances of getting new connections (see Figure 4.2.).

Besides following the repartition law mentioned above, some other restrictions (for example, those related to cycles and long chains) had to be applied in order to make the generated model more realistic and similar to the Internet. Another obvious restriction is the lack of isolated components (see Figure 4.3).

A more subtle restriction is related to the time-to-living (TTL), which is a way to avoid routing loops in a real Internet. This translates in a restriction for our topology—there can be no more than 30 nodes to get from any node to any other node. Another subtle restriction is that the generated network will also have redundant paths, that is, multiple possible routes between nodes. In other words, the Internet model topology should not "look" like a tree, but should rather have numerous cycles.

The algorithm used for the generation of the scale-free network topology is generating networks with a cyclical degree that can be controlled, in our case, approximately 4% of the added nodes form a cycle. One more restriction is that we try to avoid long-line type of scale-free networks—a succession of several interconnected nodes—structure that does not have a real-life Internet equivalent, so our algorithm makes sure such a model is not generated.

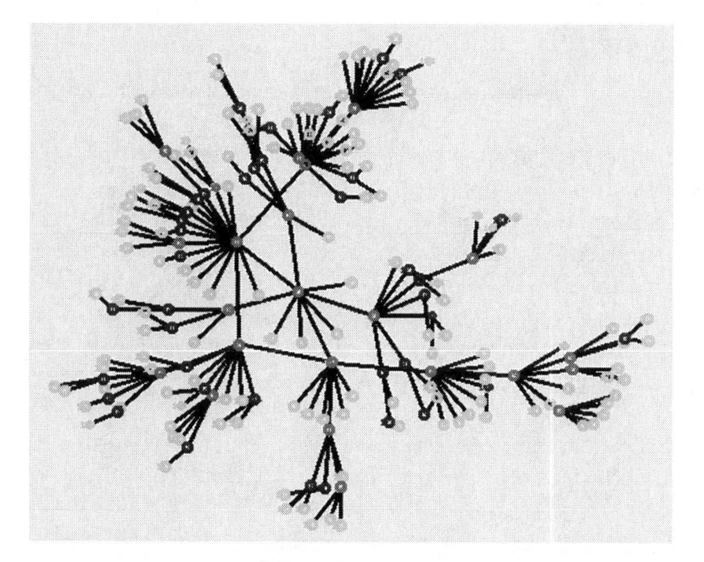

Figure 4.2 Graphic representation of the generated network for 200 network nodes and $\lambda = 2.35$.

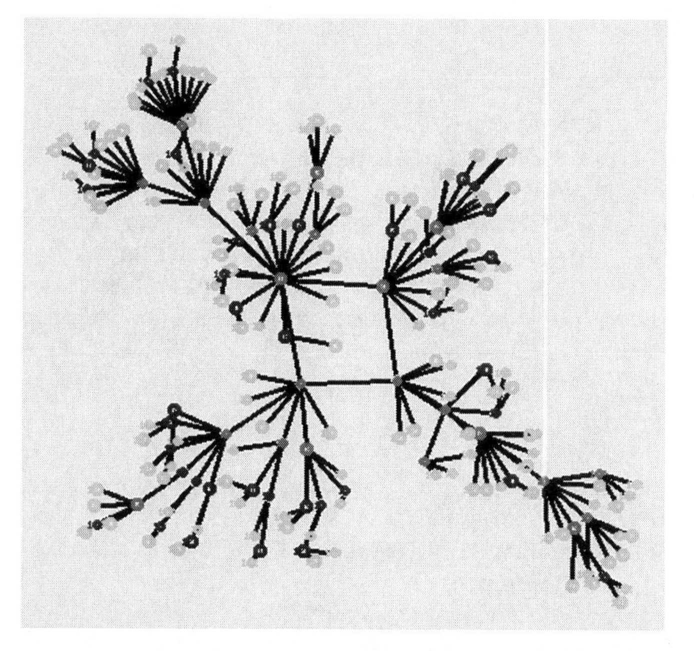

Figure 4.3 Graphic representation of the generated network for 200 network nodes and $\lambda = 2.85$.

4.3.2 Proposed Internet model

The generated topology consists of three types of nodes: routers, servers, and customers.

Routers are defined as nodes with one or several links that do not initiate traffic and do not accept connections. Routers can be one of the following types: routers that connect primarily customers, routers that connect primarily servers and routers that connect primarily other routers. Routers that connect primarily customers have hundreds or thousands of type one connections (leaf nodes) and a reduced number of connections to other routers. Routers that connect primarily servers have a reduced number of connections to servers in the order of tenths and reduced number (2 or 3) connections to other routers. Routers that connect primarily routers have a number in the order of tenths of connections to other routers and do not have connections to neither servers nor customers.

Servers are defined as nodes with one connection but sometimes could have two or even three connections. Servers only accept traffic connections but do not initiate traffic.

Customers (end-users) are defined as nodes that have only one connection, very seldom two connections. Customers initiate traffic connections toward servers at random moments but usually in a time succession. For our proposed model, we chose a 20:80 customer-to-server ratio.

4.3.2.1 Scale-free network design algorithm

We designed and implemented an algorithm that generates those subsets of the scale-free networks that are close to a real computer network such as the Internet. Our application is able to handle very large collections of nodes, to control the generation of network cycles, and the number of isolated nodes. The application was written in Python being, as such, portable. It runs very fast on a decent machine (less than 5 min for 100.000 nodes model).

Network generation algorithm:

1. Set node_count and λ
2. Compute the optimal number of nodes per degree
3. Create manually a small network of three nodes
4. For each node from four to node_count
    a. Call add_node procedure
    b. While adding was not successful
    c. Call recompute procedure
    d. Call add_node procedure
5. Save network description file

add_node procedure

1. According to the preferential attachment, compute the degree of the parent node
2. If degree could be chosen then exit procedure
3. Compute the number of links that the new node shall establish with descendants of its future parent, according to copy model
4. Chose randomly a parent from the nodes having the degree as computed above
5. Compute the descendant_list, the list of descendants of the newly chosen parent
6. Create the new node and links
7. For each descendant of the descendant_list and create the corresponding links
8. Exit procedure with success code

recompute procedure

1. For each degree category
   a. Calculate the factor needed to increase the optimal count of nodes per degree
   b. If necessary increase the optimal number of nodes per degree
2. Exit procedure

The algorithm starts with a manually created network of several nodes, and then using preferential attachment and growth algorithms, new nodes are added. We introduced an original component, the computation in advance of the number of nodes on each degree level. The preferential attachment rule is followed by obeying to the restriction of having the optimal number of nodes per degree.

We noticed that the power law is difficult to follow while the network size is growing, as a result we calculate again the optimal number of nodes per degree level at given points in the algorithm. This is necessary because the bigger the network the higher the chance that a new node will be attached only to some specific very-connected nodes. In a real network, such as the Internet this will not happen. If only the preferential and growth algorithms are followed, then the graph will have no cycles, which is not realistic, therefore we introduced a component from the "copy model" for graph generation in order to make the network graph include cyclical components. This component ensures that each new node is also attached to some of its parent-node descendants using a calibration method. The calibration method computes the number of additional links that a new node must have with the descendants of its parent. This number depends on how well-connected is the parent and it also includes a random component.

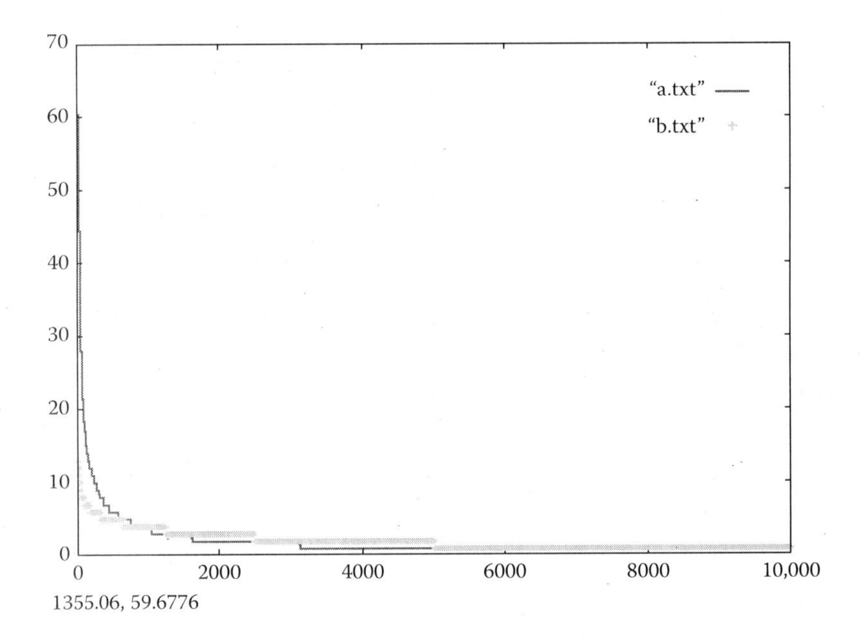

Figure 4.4 Graphic representation of the distribution law for a scale-free network model and for a randomized network with 10,000 nodes.

The output of the application is a network description file that can be used by several tools like, for instance, a tool to display the power law. This file is stored using a special format needed in order to reduce the amount of disk writes.

In Figure 4.4, we compare an almost random network distribution law and a free-scale distribution law. On the *Y*-axis, we represent the number of connections and on the *X*-axis, the number of nodes having this number of connections. It was impossible to obtain a completely random network given the limitations imposed by the Internet model, so further only the scale-free network model will be discussed.

4.3.2.2 Traffic generation

Traffic generation is an essential part of the simulation as such, and we decided to initiate randomly between 1 and 3 simultaneous traffic connections from "customer" nodes and for the sake of simplicity, we used ftp sessions to randomly chosen destination servers. We also decided that the links connecting routers should have higher speeds than lines connecting customers to routers, for example—server-router 1 Gbps, client-router 10 Mbps, router-router 10, 100 Mbps or 1 Gbps depending on the type of router. The code generated respecting these two conditions is added to the network description file, being ready to be processed by the simulator.

4.3.2.3 *Single-CPU simulation*

We used a modular approach that allows us to later reuse components for different parts of the simulation. For example, the same network model generated by the initial script can be used for both single-CPU and distributed simulations, allowing a comparison between the two types of simulation. Standalone simulations were run under University of California Berkeley's NS2 network simulator. NS2 (The Network Simulator) (Chung 2011). We noticed that on a single machine, as network size increases, very soon we hit the limit of the network sizes that can be simulated due to resource limitations, mostly on memory, but also high CPU load. In case of small size network models, such as with a few nodes, simulations can be run on a single machine. The results provided by NS2 were visualized using the nam (network animator) software package. The topology generator gives different colors to different type of nodes: server, client, and router. Details about the networking traffic through each network node are parsed from the simulator output.

4.3.2.4 *Multi-CPU simulations*

Unfortunately NS was not designed to run on parallel machines. The main obstacle in running NS in a distributed or parallel environment is related to the description of objects in the simulation. As such, we ran our distributed simulations under Georgia Tech's extension to NS2, *pdns* [PDNS], which uses a syntax close to that of NS2, the main differences being a number of extensions needed for the parallelization so that different instances of *pdns* can communicate with each other and create the global image of the network to be simulated. Each simulator running on different nodes needs to know the status of other simulators. Furthermore, if we try to split the network description file into separate files and run each of these in separate simulation contexts, we need to find a way to communicate parameters between the simulation nodes. All simulations are running 40 s of simulated traffic scenarios.

4.3.2.5 *Cluster description*

In order to create a parallel or distributed environment, we have built a cluster using commodity hardware and running Linux as operating system (Mocanu and Ţarălungă 2007). The cluster can run applications using a parallelization environment. We have written and tested applications using parallel virtual machine (PVM), which is a framework consisting of a number of software packages that accomplish the task of creating a single machine that spans across multiple CPUs, by using the network interconnection and a specific library (Grama et al. 2003). Our cluster consists of a "head" machine and a number of six cluster nodes. The "head" provides all services for the cluster nodes—IP allocation, booting services, file system (NFS) for storage of data, facilities for updating, and managing and controlling the images used by the cluster nodes as well as access to the cluster. The "head" computer provides an image for the operating system that is loaded by each of the cluster nodes since the cluster nodes do not have their own storage media. The application partition is mounted read-only while the partition where data is stored is mounted read-write and accessible to the users on all machines in a similar manner providing transparent access to user data. In order to access the cluster, users must connect to a virtual server located on a head machine. This virtual server can also act as a node in the cluster when extra computation power is needed.

4.3.2.6 Network splitting

In order to use PDNS simulation, we needed to split the network into several quasi-independent subnetworks. Each instance of PDNS handles a specific subnetwork, thus the dependencies between them need to be minimal, that is, there shall be as few as possible links between nodes located in different subnetworks. In order to have a federated simulation approach, we designed and implemented a federalization algorithm in order to split the original generated network into several small ones. The algorithm that generates n federative components chooses the most n linked nodes, assigned them to an empty federation and starts a procedure similar to the breadth-first search algorithm. Each node is marked as being owned by a federation. The *pdns* script generator takes as input the generated network description and the generated federations, respectively. Depending on the connectivity of nodes, they are assigned the role of routers, servers, end-users, and the corresponding traffic scenario is associated with them.

4.3.3 Simulation results and discussion

We have decided to run simulations for 40 s of traffic for a scale-free network model with 10000 nodes. At such a scale, a one-node processing is impossible because the cluster node runs out of memory. Still, to get valid results, we had run the simulation on a much more powerful machine with plenty of memory and virtual memory. We chose two different scenarios, one with a moderate network traffic and another scenario with a heavy network traffic. Each scenario was simulated five times under similar load conditions, using two to six CPUs and we noted the time used for the actual simulation (in seconds). The results are presented in Tables 4.1 and 4.2.

For the first scenario, we noted that there is a point where adding more nodes in the simulation does not help but rather increases the simulation time. In this scenario, the optimum number of nodes is 5.

Table 4.1 Scale-free network model with 10,000 nodes and moderate network traffic (40 seconds)

	Number of cluster nodes used					
	1	2	3	4	5	6
Run 1	Failed	68	46	32	29	40
Run 2	Failed	68	41	31	30	37
Run 3	Failed	67	43	32	33	31
Run 4	Failed	68	45	30	29	43
Run 5	Failed	67	45	32	31	40
Average	135	67.6	44	31.4	30.4	38.2

Table 4.2 Scale-free network model with 10,000 nodes and heavy network traffic (40 seconds)

	Number of cluster nodes used					
	1	2	3	4	5	6
Run 1	Failed	319	338	135	173	165
Run 2	Failed	343	357	140	176	171
Run 3	Failed	347	351	134	177	166
Run 4	Failed	316	347	139	177	165
Run 5	Failed	308	320	138	178	163
Average	1139	326.6	342.6	137.2	176.2	166

The second scenario requires much more resources as can be seen from the single-processor simulation, which again failed on the cluster nodes but was successful on a more powerful machine, although it takes a longer time. Also in this simulation, we see that adding more nodes (in our case more than 4), the simulation process is slower.

One can observe that the 2-CPU simulation is actually faster than the 3-CPU simulation, although the optimal number of nodes is not 2. Running *pdns* is more efficient than running NS2 especially on large-size network models where sometimes *pdns* is the only solution. However, there are limitations in the number of cluster nodes that could process a given network model since more nodes are used, more traffic links between different cluster nodes are to be simulated and therefore more time is spent on interprocessor communication. Anyway, it is very important to split the network model correctly into smaller subnetworks (federations) since there is a tradeoff between the degree of separation and federation balancing—the more separated the subnetworks are, the more unbalanced they become.

4.4 Improvement of Internet traffic QoS performance

4.4.1 Quality of service (QoS) requirements in high-speed networks

The new broadband high-speed networks are designed to integrate a wide range of audio, video and data traffic within the same network. In these networks, the stringent quality of service (QoS) requirements of real-time multimedia applications are met by the mean of special services providing guarantees. One distinguishes between two main kinds of such services: deterministic and statistical (Chao and Guo 2001). The deterministic service, also called guaranteed service within the Internet framework, is aimed to provide strict guarantees on the QoS requirements. The mathematical basis of the deterministic service is a network calculus that allows obtaining deterministic bounds on delay and buffering requirements in a communication network. This model provides bounds only on the maximal delay that traffic can undergo through the different elements in the network. The statistical service is intended to provide only statistical (probabilistic) guarantees. For services offering statistical guarantees, one is rather interested in bounding the mean delay or the probability that the delay exceeds a certain value. The main advantage of the statistical service over the deterministic service is that it can achieve higher network utilization, at the expense of some minor quality degradation.

Two notable models of statistical services are used to meet the demand for QoS: integrated services (IntServ) and differentiated services (DiffServ). The IntServ model is characterized by resource reservation; before data is transmitted, applications must set up paths and reserve resources along the path. IntServ aims to support applications with different levels of QoS within the TCP/IP (transport control protocol/Internet protocol) architecture. IntServ, however, requires the core routers to remember the state of a large number of connections, giving rise to scalability issues in the core of the network. It is therefore suitable at the edge network where the number of connections is limited. The DiffServ model is currently being standardized to provide service guarantees to aggregate traffic instead of individual connections. The model does not require significant changes to the existing Internet infrastructure or protocol. DiffServ does not suffer from scalability issues, and hence is suitable at the core of the network. It is therefore believed that a significant part of the next generation high-speed networks will consist of IntServ at the edge and DiffServ at the core of the network.

Unfortunately, all of these algorithms have poor burst loss performances, even at low traffic loads, because the real traffic is bursty. To solve this problem, we discuss in the following a novel burst assembly algorithm with traffic-smoothing functions (the traffic is smoothed by restricting the burst length to a threshold). Compared with existing algorithms, our scheme can improve network performance in terms of the burst loss ratio. The simulation results show that our proposed scheme can lighten burst loss ratio for TCP flows of a high-speed network.

4.4.2 Technologies for high-speed networks

In high-speed networks, all communications are limited by the electronic processing capabilities of the system. Although hardware-based high-speed electronic IP routers with capacity up to a few hundred gigabits per second are available now, there is still a serious mismatch between the transmission capacity of wavelength division multiplexing (WDM) optical fibers and the switching capacity of electronic IP routers. With IP traffic as the dominant traffic in the networks, the traditional layered network architecture is no longer adapted to the evolution of the Internet. In the multilayered architecture, each layer may limit the scalability of the entire network, as well as adding the cost of the entire network. As the capabilities of both routers and optical cross-connects (OXCs) grow rapidly, the high data rates of optical transport suggest bypassing the SONET/SDH and ATM layers and moving their necessary functions to other layers. This results in a simpler, more cost-efficient network that can transport very large volumes of traffic. On the other hand, the processing of IP packets in the optical domain is still not practical yet, and the optical router control system is implemented electronically. To realize an IP-over-DWDM architecture, several approaches, such as wavelength routing (WR) (Zang et al. 2004), optical packet switching (OPS) (O'Mahony et al. 2001), and optical burst switching (OBS) (Xiong et al. 2004), have been proposed. Of all these approaches, optical burst switching (OBS) can achieve a good balance between the coarse-gained wavelength routing and fine-gained optical packet switching, thereby combining others' benefits while avoiding their shortcomings.

4.4.2.1 Characteristics of OBS networks

In an OBS network (shown in Figure 4.5), the edge routers and core routers connect with each other with WDM links. The edge nodes are responsible for assembling packets into bursts and deassembling bursts into packets. The core nodes are responsible for routing and scheduling based on the burst header packets.

The architecture of edge router aims to eliminate optical to electronic to optical (O/E/O) conversions and electronic processing loads, which are the bottlenecks of an OBS network. The ingress part of an edge node assembles multiple IP packets with the same egress address into a switching granularity called a burst. A burst consists of a burst header packet (BHP) and a data burst (DB). The BHP is delivered on a control channel;

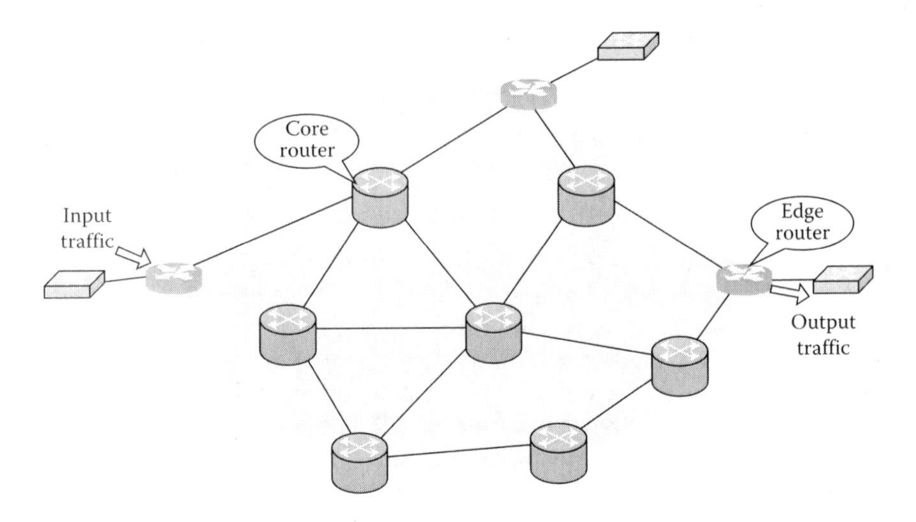

Figure 4.5 OBS network model.

its corresponding DB is delivered on a data channel without waiting for a confirmation of a successful reservation. A channel may consist of one wavelength or a portion of a wavelength, in case of time-division or code-division multiplexing. When a BHP arrives at a core node, the core node converts it into an electronic signal, performs routing and configures the optical switching according to the information carried by the BHP. The DB remains in the optical domain without O/E/O conversion when it cuts through the core node.

4.4.2.2 Burst assembly schemes

Burst assembly at the edge router is an important issue for OBS networks. Basically, there are two assembly schemes: threshold-based and timer-based. In a threshold-based scheme, a burst is created and sent into the optical network when the total size of the packets in the queue reaches a threshold value L_b. The shortcoming of the threshold-based scheme is that it does not provide any guarantee on the assembly delay that packets will experience. In a timer-based scheme, a timer is started at the initialization of the assembly. A burst containing all the packets in the buffer is generated when the timer reaches the burst assembly period T_b. A large time-out value T_b results in a large packet buffering delay at the edge node. On the other hand, a too small T_b results in too many small bursts and a high electronic processing load.

The choice of burst assembly algorithms depends on the type of traffic being transmitted. Timer-based algorithms are suitable for time-constrained traffic such as real-time applications because the upper bound of the burst assembly delay is limited. For a time-insensitive application

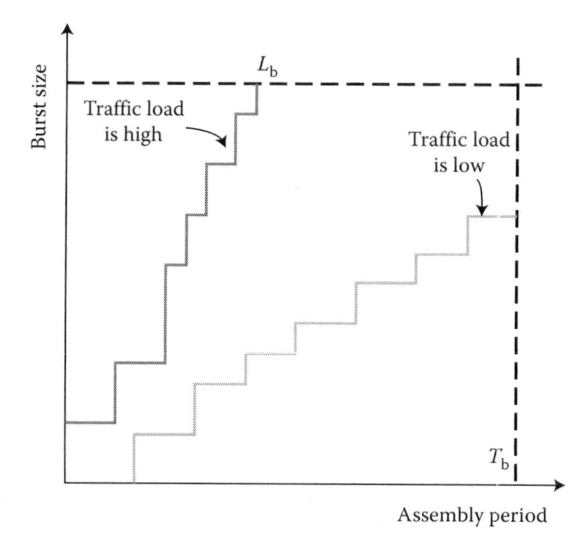

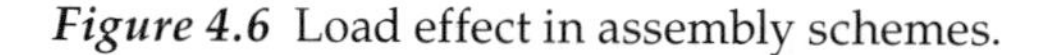
Figure 4.6 Load effect in assembly schemes.

such as file transmission, to reduce the overhead of control packets and increase OBS transmission efficiency, a threshold-based scheme may be more appropriate. Figure 4.6 shows the effect of load on timer-based and threshold-based assembly schemes.

How to choose the appropriate time-out or threshold value for creating a burst is still an open issue. A smaller assembly granularity leads to a higher number of bursts and a higher number of contentions, but the average number of packets lost per contention is less. Also, it will increase the number of control packets. On the other hand, a higher assembly granularity will lead to a higher burst assembly delay and the average number of packets lost per contention is larger.

4.4.3 Traffic-smoothing using burst assembly schemes

4.4.3.1 Analysis of assembled traffic

A challenging issue in OBS is how to assemble IP packets into bursts at ingress nodes. As we assume there is no buffer in OBS networks, a burst loss event will occur if multiple bursts from different input ports are destined for the same output port at the same time. The burst arrival process is determined by the traffic characteristics such as the burst interarrival time and the burst length distributions, which are dependent on the burst assembly strategy. For the timer-based assembly algorithm, the burst size is unlimited when too many packets arrive suddenly. A larger burst is more likely to be blocked or to block other bursts during transmission. For the threshold-based and hybrid assembly algorithms, a large number of bursts will be generated and injected into the network in a short period when many packets arrive.

The network traffic characteristics have attracted considerable research attention because the traffic properties greatly influence the network's performance. Let consider in a simple traffic model for a burst assembler $\{A(t), t > 0\}$ be the cumulative amount of the input traffic in bits of the burst assembly function arriving during time interval $(0, t]$, and $\{A^{OBS}(t), t > 0\}$ be the cumulative amount of output traffic of the burst assembly function during time interval $(0, t]$, that is, assembled traffic.

We define $X_n (n \in N)$ as a sampled process of input traffic with the unit time of τ. Therefore, $X_n = \{A(n\tau) - A((n-1)\tau), n \in N\}$. We assume X_n to be a wide-sense stationary discrete stochastic process, with constant mean $\mu = E[X_n]$. Let $S_{n,m}$ be the sum of m consecutive numbers from X_n; then $S_{n,m} \equiv \sum_{j=(n-1)m}^{nm} X_j$.

We define $X_n^{(m)} \equiv S_{n,m}/m$ and use the variance–time function $\mathrm{Var}[X_n^{(m)}] = E[(X_n^{(m)} - \mu)^2]$ to represent the variance of input traffic. Here, $E[X_n^{(m)}] = E[X_n] = \mu$.

Similarly, we also define $Y_n (n \in N)$ as a sampled process of assembled traffic with the unit time interval of τ and assume it to be a wide-sense stationary discrete stochastic process. Here, $Y_n = \{A^{OBS}(n\tau) - A^{OBS}((n-1)\tau), n \in N\}$. Then, the variance of assembled traffic can be measured by the variance–time function

$$\mathrm{Var}[Y_n^{(m)}] = E[(Y_n^{(m)} - \mu^{OBS})^2] \tag{4.12}$$

where $Y_n^{(m)}$ is the mean value of m consecutive numbers from Y_n, and μ^{OBS} is the mean of $Y_n^{(m)}$ and can be expressed by $E[Y_n^{(m)}] = E[Y_n] = \mu^{OBS}$.

Here, we analyze the assembled traffic through a model based on fractional Brownian motion (FBM) proposed by the authors in a previous work (Dobrescu et al. 2004b). The FBM model is defined by

$$A(t) = \lambda t + \sqrt{\lambda a} Z_H(t); T \in R \tag{4.13}$$

where λ is the arrival rate for the packets, a is the variance of the coefficient for the arrival packet, and $Z_H(t)$ is the normalized FBM with zero mean and variance of $\mathrm{Var}[Z_H(t)] = |t|^{2H}$, where H is the Hurst parameter and satisfies $H \in [0{:}5; 1)$.

In our simulations, we have considered for large timescales ($m > 200$) that the variance–time curve is approximated using the FBM model for the parameter set $\lambda = 564$ Mb/s, $a = 2.5 \times 10^6$, and $H = 0.85$. In small timescales ($m < 200$), the traffic can be approximated by an FBM model with the parameter set = 564 Mb/s, $a = 9 \times 10^5$, and $H = 0.75$.

Since the variance of the aggregate of uncorrelated traffic will equal the sum of the individual source's variance, we only have analyzed the variance of the assembled traffic of one burst source, using a timer-based burst assembly scheme. With a timer-based burst assembly algorithm, all packets in the assembly buffer will be sent out as a burst when the timer reaches the burst assembly period T_b. After a burst is generated, the burst is buffered at the edge node for an offset time before being transmitted to give its BHP enough time to reserve wavelengths along its route. During this offset time period, packets belonging to that queue will continue to arrive. Because the BHP that contains the burst length information has already been sent out, these arriving packets could not be included in the generated burst. When a burst is generated at one buffer, the future arriving packets will be stored at another buffer until the next assembly cycle. We assume that the interarrival time of the bursts from the same burst source is fixed as T_b. Accordingly, there will be no packets left in the assembly buffer at time $kT_b (k \in N)$. Therefore, $A^{OBS}kT_b = A(kT_b)$. We denote $Q(t)$ as the number of bits that are buffered at the edge router. So $A^{OBS}(t) = A(t) - Q(t)$.

For the timer-based assembly algorithm, the $Q(t)$ bits are at most the packets that arrive during $[0, s]$, where $s \in [0,T_b]$. For simplification, we assume s is uniformly distributed in $[0, T_b]$ and $Q(s)$ is a Gaussian process with mean λ_s and variance $\lambda a s^{2H}$. We denote $\overline{E_Q}$ as the mean value of $Q(t)$ observed at any sample point. Then, $\overline{E_Q} = \lim_{T \to \infty} \frac{1}{T} \int_0^T Q(t) dt$.

The difference between the original traffic and the assembled traffic can be denoted by

$$\Delta V(t) = \mathrm{Var}[A^{\mathrm{OBS}}(t)] - \mathrm{Var}[A(t)] \tag{4.14}$$

To describe the variance of assembled traffic $A^{OBS}(t)$, the variance–time function becomes

$$\mathrm{Var}[Y_n^{(m)}] = E\left[\frac{A(m\tau)}{m}\right] + \frac{1}{m^2} \Delta V(m\tau) \tag{4.15}$$

From Equation 4.15, we can see there is an increase in the variance in short timescales. This indicates that the timer-based assembly algorithm will increase the variance of traffic. As the timescale increases, the difference between the original and assembled traffic becomes negligible because the traffic does not change significantly for large timescales.

4.4.3.2 Description of the proposed burst assembly algorithm

The simulation results on the discussed model show that the timer-based assembly algorithm could not reduce the variance of the real traffic, but do increase it in small timescales. Defining the traffic burstiness as the variance of the bit rate in small timescales, one can observe that a larger burstiness indicates that the traffic is burstier and more likely to exceed the capacity of the network and it results in burst loss events. So, a larger burstiness implies a higher burst loss ratio in bufferless OBS networks. One way to reduce the burst loss ratio is to control the burst sources at the edge nodes and thereby inject the bursts more smoothly into the network. This section discusses a novel burst assembly algorithm with traffic-smoothing functions, to reduce the burstiness of the assembled traffic.

A simple way to reduce the burstiness is a peak rate restriction. Conceptually, the number of bursts simultaneously arriving at an input port is most likely to reach a maximum value when the traffic is at a peak rate. Reducing the number of overlapping bursts on a link is for each ingress node to restrict the assembled traffic to a specified rate. In a timer-based assembly scheme, because the bursts are generated periodically, the traffic rate can be restricted by restricting the burst length to a threshold. Based on this idea, we propose a scheme, called adaptive timer-based, to reduce the burstiness of traffic.

For this scheme, we suppose each edge router has G queues to sort the arriving packets. Let the timer of queue $Q[i]$ be denoted by $T[i]$ and the length of $Q[i]$ be denoted by $L[i]$. The threshold for generating a burst is $L_{th}[i]$. When the value of the queue length $L[i]$ is smaller than $L_{th}[i]$, all packets in $Q[i]$ will be assembled into a burst. Otherwise, a burst is generated with the length of $L_{th}[i]$ and the other packets are left in $Q[i]$.

The scheme is thus implemented using the following algorithm:

- Step 1. When a packet with a length of b arrives at $Q[i]$, then if $Q[i]$ is empty, start timer $T[i]$, $L[i] = b$; else push packet into $Q[i]$, with $L[i] = L[i] + b$.
- Step 2. When $T[i] = T_b$ if $(L[i] > L_{th}[i])$, $L_b = L_{th}[i]$, $L[i] = L[i] - L_{th}[i]$, $T[i] = 0$ and restart timer $T[i]$; else $L_b = L[i]$, $L[i] = 0$, $T[i] = 0$ and stop timer $T[i]$.
- Step 3. Generate a burst with length L_b and send it into the OBS network.

Figure 4.7 shows a comparison of the timer-based (a) and adaptive timer-based (b) assembly algorithms. For the timer-based assembly algorithm, all packets in the assembly buffer will be multiplexed to a burst every assembly period. This makes for various burst sizes because the number of packets that arrive in each assembly period varies. On the other hand, the burst size in the adaptive timer-based assembly algorithm is restricted by the threshold $L_{th}[i]$. After an $L_{th}[i]$ length of packets are assembled into a burst, the other packets will be left in the assembly queue for a future assembly process. So, the adaptive timer-based scheme can avoid a sudden increase in the burst size and makes the burst sent out more smooth than the timer-based assembly algorithm does.

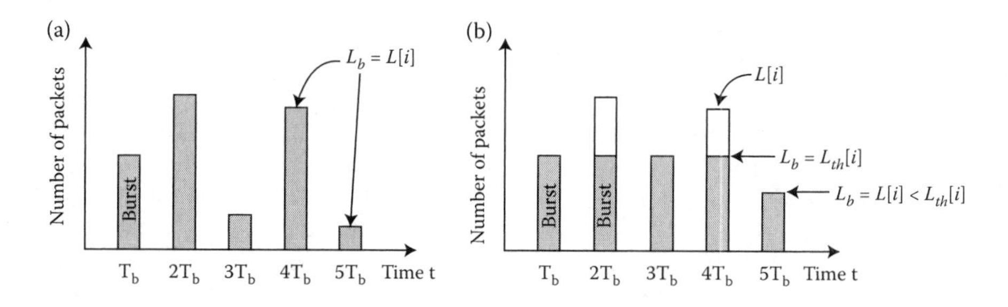

Figure 4.7 Comparison of timer-based (a) and adaptive timer-based (b) algorithms.

Take into consideration the choice of $L_{th}[i]$ to take advantage of the effects of the peak rate restriction. It is clear that the restriction of on the peak rate should be bigger than the average rate and the threshold $L_{th}[i]$ must exceed the average burst length $\overline{L_M}(i)$, otherwise the traffic would be blocked at the edge nodes. As $L_{th}[i]$ is close to $\overline{L_M}(i)$ ($\alpha \rightarrow 0$), the transmission is almost the same as in a CBR transmission. However, this will result in enormous backlogs at the edge routers. The choice of α is a tradeoff between the effects of the peak rate restriction and the edge buffering delay. When too many packets suddenly arrive, if we still assemble packets into bursts using a small threshold $L_{th}[i]$, the packets will suffer a large edge buffering delay.

4.4.4 Simulation results

To check the efficiency of our schemes, we have simulated a multiple hop network with a ring topology on a dedicated platform. We use OPNET as a simulation tool to study the performance of our schemes and compare them with existing dropping schemes, especially when the offset times are varied during transmission. The shortest-path-first routing method is used to establish a route between each pair of edge nodes $E_i(i = 1$ to $16)$, and the maximum hop distance is 10. Bursts are generated at each edge node E_i. We assume that the burst interarrival time follows an exponential distribution and the burst size follows a normal distribution. Note that these assumptions are the same as the ones in (Yu et al. 2002). Therefore, in our simulations, we have use the same ring topology and environment to test the performance of the integrated scheme when performing joint QoS with absolute constraints. The average burst size is 50 kbytes. All bursts are assumed to have the same initial offset time (the default value is 5 ms, which is small enough even for real-time applications). For a core node $C_i(i = 1$ to $16)$, we assume that each output link consists of 16 wavelengths with a transmission rate of 1 Gbps per wavelength. The basic processing time for BHP at each core node is set to be 0.1 ms. To investigate the service differentiation, we consider four classes, a load distribution of $\lambda_1 = \lambda_2 = \lambda_3 = \lambda_4$, and proportional parameters of $s_1 = 1$, $s_2 = 2$, $s_3 = 4$, and $s_4 = 8$.

For simplicity, in the simulations, the traffic load refers to the average traffic load on the links connecting two core nodes. Regarding the performance of the advanced timer-based assembly algorithm, we set $M = 50$. Figure 4.8 shows the impact of parameter α on the burst loss ratios for different traffic loads. It shows that the burst loss ratio decreases as α decreases. On the other hand, the average edge buffering delay increases as α decreases, as shown in Figure 4.9.

How to choose α is a tradeoff between the burst loss ratio and the average edge buffering delay. A suitable α should be able to achieve a low burst loss ratio while not increasing the edge buffering delay too much. From Figures 4.8 and 4.9, we can see that when α is smaller than 0.1, the burst loss ratio decreases slowly and is almost the same as the one for $\alpha = 0.1$ while the average edge buffering delay obviously increases.

Regarding the performance of the DBA scheme, Figures 4.10 and 4.11 show the impact of N_r and T_{wait} on the average burst loss ratios for a low traffic load (0.3) and for a high traffic load (0.89), respectively. In these figures, the curve "*aL*" denotes the burst loss ratio when T_{wait} is set to *a* times the average burst length (*L*). The initial offset time is set to 30 ms.

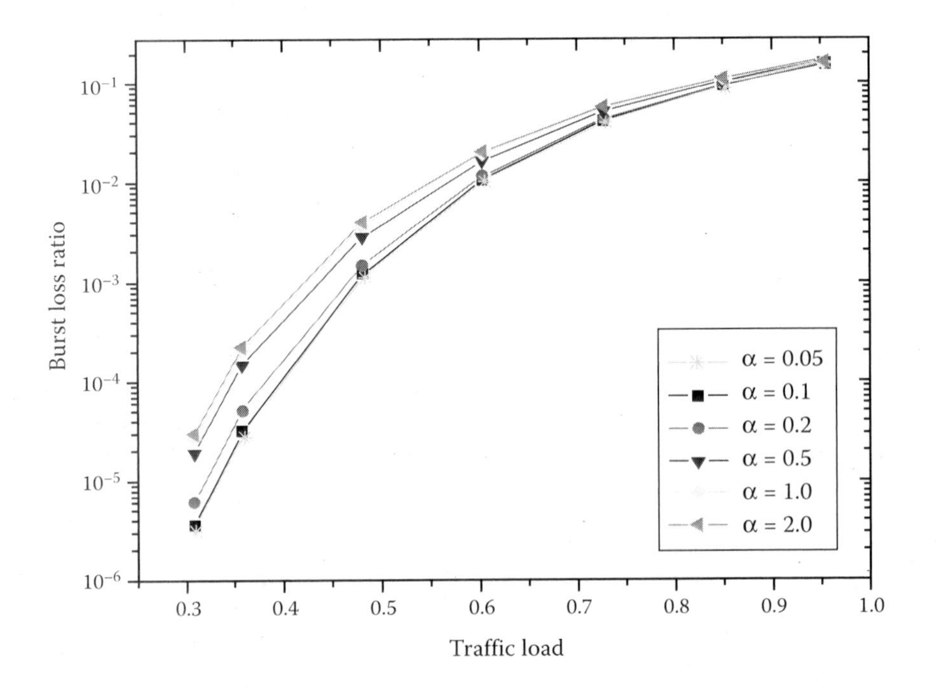

Figure 4.8 Impact of α on burst loss ratio.

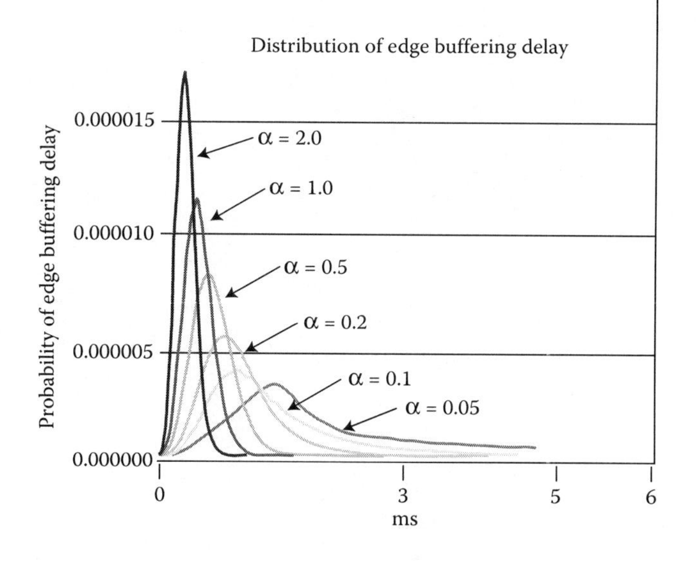

Figure 4.9 Impact of α on edge buffering delay.

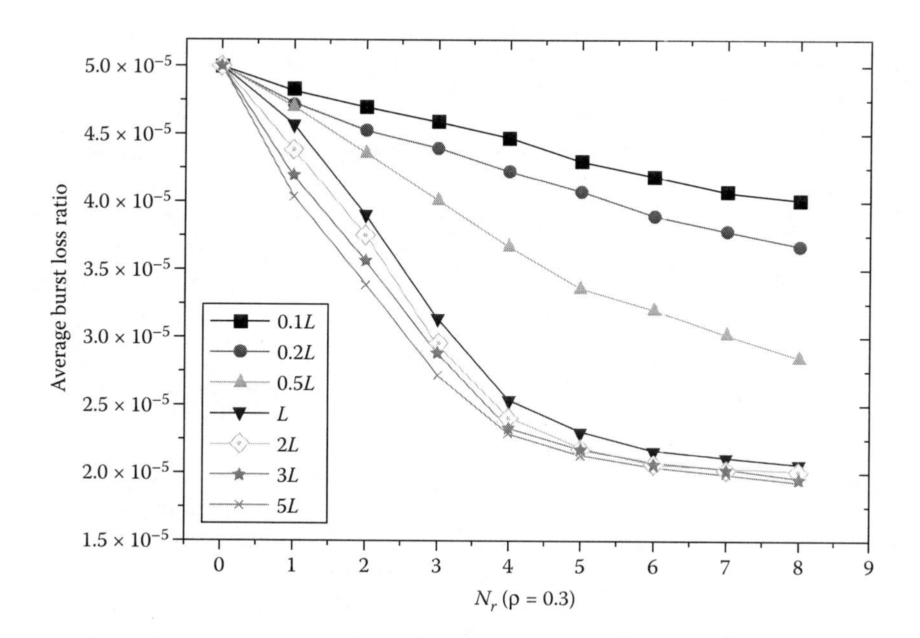

Figure 4.10 Average burst loss ratios versus N_r with traffic load 0.3.

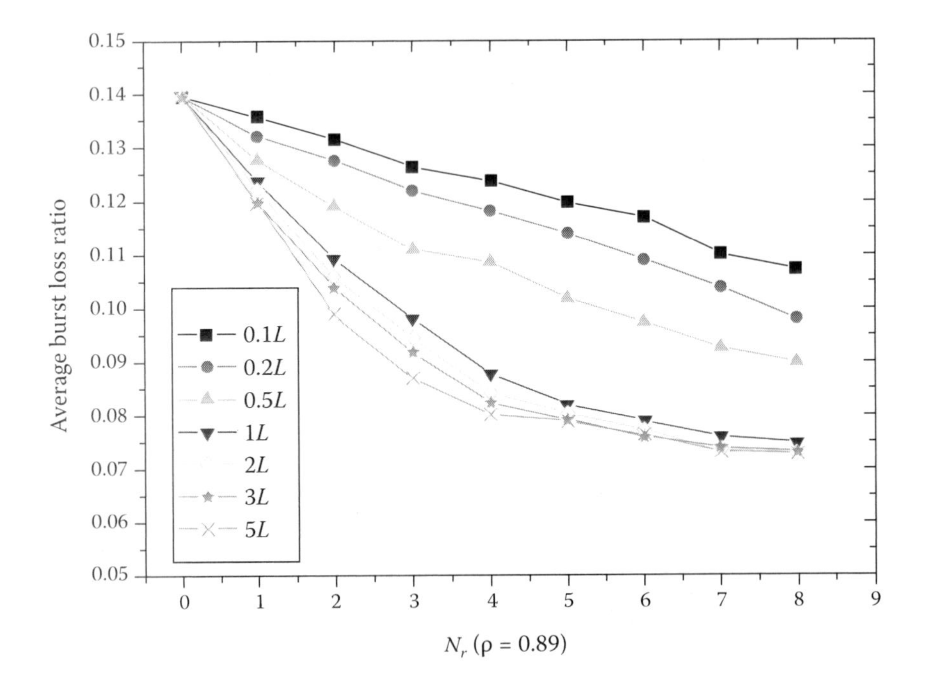

Figure 4.11 Average burst loss ratios versus N_r with traffic load 0.89.

The results indicate that as N_r and T_{wait} increase, the burst loss ratio decreases. However, when T_{wait} exceeds the average burst length duration and N_r exceeds 4, the burst loss ratios decrease slowly and eventually become almost the same. Thus, in the simulations we set 0.4 ms (average burst length duration = average burst length (50 Kbytes)/bandwidth (1 Gbps)) and 4 as the default values of T_{wait} and N_r.

Regarding the performance of the integrated scheme presented in this section, the results obtained by simulation show that the proportions of different classes are close to the ratios of predefined parameters and are independent of the traffic load for the integrated scheme. Therefore, the integrated scheme achieves proportional differentiated services for multiclass traffic. Figure 4.12 shows the end-to-end BHP queueing delay normalized to the average burst length duration for each class during transmission.

We can see that the lower priority bursts have larger queueing delay than do the higher priority bursts. For example, the queueing delay is several microseconds for class 2, tens of microseconds for class 3, and hundreds of microseconds for class 4. The simulation results show that although the integrated scheme improves the burst loss performance at the expense of increasing extra offset time, the increase of end-to-end delay is very small (at most hundreds of microseconds) and would be negligible for real applications. As an advantage over the existing priority schemes,

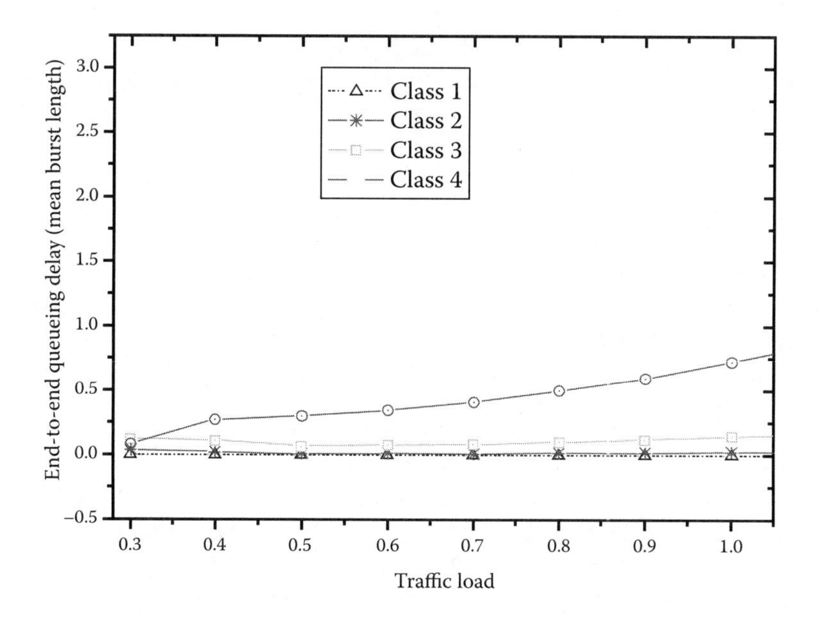

Figure 4.12 End-to-end BHP queueing delay.

our integrated scheme does not need any complex burst segmentation or wavelength preemption support, so it has a simple implementation. Moreover, it provides controllable and predictable proportional differentiated services for each class.

4.5 *Fractal approaches in Internet traffic simulation*

4.5.1 *Using parallel discrete-event simulations (PDES) for scalability evidence in large-scale networks modeling*

Large-scale networks are a class of complex systems made of a large number of interacting individual elements with a finite number of local state variables each assuming a finite number of values. The dynamics of the local state variables are discrete events occurring in continuous time. We assume that between two consecutive updates, the local variables stay unchanged. Another important assumption is that the interactions in the underlying system to be simulated have finite range. Often the dynamics of such systems is inherently stochastic and asynchronous. The simulation of such systems is rather nontrivial and in most cases the complexity of the problem requires simulations on distributed architectures, defining the field of parallel discrete-event simulations (PDES) (Toroczkai et al. 2006). Conceptually, the computational task is divided among N

processing elements (PE-s), where each processor evolves the dynamics of the allocated piece. Due to the interactions among the individual elements of the simulated system, the PEs must coordinate with a subset of other PEs during the simulation. In the PDES schemes, update attempts are self-initiated and are independent of the configuration of the underlying system, thus their performance also becomes independent. The update dynamics, together with the information sharing among PEs, make the parallel discrete-event simulation process a complex dynamical system itself. In fact, it perfectly fits the type of complex systems we are considering here: the individual elements are the PEs, and their states (local simulated time) evolve according to update events, which are dependent on the states of the neighboring PEs.

The design of efficient parallel update schemes is a challenging problem, due to the fact that the dynamics of the simulation scheme itself is a complex system, whose properties (efficiency and scalability) are hard to deduce using classical methods of algorithm analysis. Since one is interested in the dynamics of the underlying complex system, the PDES scheme must simulate the physical time variable of the complex system. When the simulations are done on a single-processor machine, a single time stream is sufficient to time-stamp the updates of the local configurations, regardless whether the dynamics of the underlying system is synchronous or asynchronous. When simulating asynchronous dynamics on distributed architectures, however, each PE generates its own virtual time, which is the physical time variable of the particular computational domain handled by that PE. In conservative PDESs schemes, a PE will only perform its next update if it can obtain the correct information to evolve the local configuration (local state) of the underlying physical system it simulates. Specifically, when the underlying system has nearest-neighbor interactions, each PE must check the interaction topology of the underlying system to see if those are progressed at least up to the point in virtual time where the PE itself did. Based on the fundamental notion of discrete-event systems that the state of a local state variable remains unchanged between two successive update attempts, we have developed a framework for simulation of jobs scheduling on a parallel processing platform.

4.5.1.1 *Parallel platform model*

The proposed model for the platform consists of P processor units. Each processor p_i has capacity $c_i > 0$, $i = 1,2,\ldots, P$. The capacity of a processor is defined as its speed relative to a reference processor with unit-capacity. We assume for the general case that $c_1 \leq c_2 \leq \cdots \leq c_P$. The total capacity C of the system is defined as $C = \sum_{i=1}^{P} c_i$. A system is called homogeneous when $c_1 = c_2 \cdots = c_P$. The platform is conceived as a distributed system. The distributed-system model, as shown in Figure 4.13, consists of P machines connected by a network.

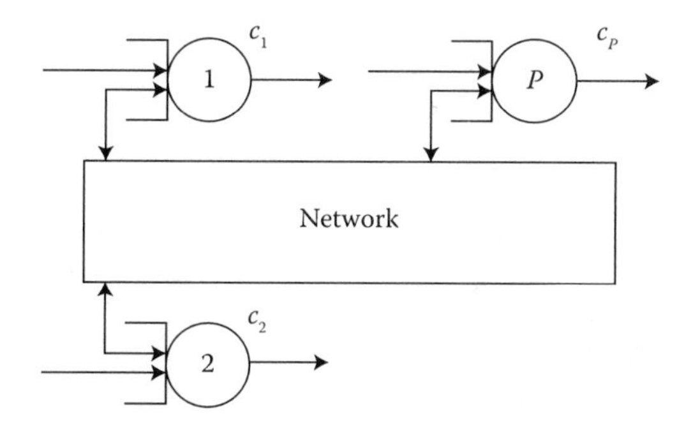

Figure 4.13 The model of the distributed platform.

Each machine is equipped with a single processor. In other words, we do not consider interconnections of multiprocessors. The main difference with multiprocessor systems is that in a distributed system, information about the system state is spread across the different processors. In many cases, migrating a job from one processor to another is very costly in terms of network bandwidth and service delay, and that the reason that we have considered for the beginning only the case of data parallelism for a homogenous system. The intention was to develop a scheduling policy with two components. The global scheduling policy decides to which processor an arriving job must be sent, and when to migrate some jobs. At each processor, the local scheduling policy decides when the processor serves which of the jobs present in its queue.

Jobs arrive at the system according to one or more interarrival time processes. These processes determine the time between the arrivals of two consecutive jobs. The arrival time of a job j is denoted by A_j. Once a job j is completed, it leaves the system at its departure time D_j. The response time R_j of job j is defined as $R_j = D_j - A_j$. The service time S_j of job j is its response time on a unit-capacity processor serving no other jobs; by definition, the response time of a job with service time s on a processor with capacity c' is s/c'. We define the job set $J(t)$ at time t as the set of jobs present in the system at time t:

$$J(t) = \{j \mid A_j \le t < D_j\}$$

For each job $j \in J(t)$, we define the remaining work $W_j^r(t)$ at time t as the time it would take to serve the job to completion on a unit-capacity processor. The service rate $\sigma_j^r(t)$ of job j at time t $(A_j \le t < D_j)$ is defined as $\sigma_j^r(t) = \lim_{\tau \to t} dW_j^r(\tau)/d\tau$. The obtained share $\omega_j^s(t)$ of job j at time t $(A_j \le t < D_j)$ is defined as $\omega_j^s(t) = \sigma_j^r(t)/C$. So, $\omega_j^s(t)$ is the fraction of the total system capacity C used to serve job j, but only if we assume that $W_j^r(t)$ is always a piecewise-linear, continuous function of t. Considering $W_j^r(A_j) = S_j$ and $W_j^r(D_j) = 0$ we have $\int_{A_j}^{D_j} \omega_j^s(t)\mathrm{d}t = \int_{A_j}^{D_j} \sigma_j^r(t)\mathrm{d}t = S_j / C$.

One can define an upper bound on the sum of the obtained job shares of any set of jobs $\{1,\ldots,J\}$ as

$$\omega_{\max}(t) = C^{-1} \sum_{i=1}^{\min(J,P)} c_i \tag{4.16}$$

Since Equation 4.1 imposes upper bonds on the total share, the maximum obtainable total share at time t is defined as

$$m^T(t) = c^{-1} \sum_{p=1}^{\min(|J(t)|,P)} c_p \tag{4.17}$$

In a similar manner, for a group g of processors ($g = 1, 2, \ldots, G$), the maximum obtainable group share at time t is defined as

$$m^G(t) = c^{-1} \sum_{p=1}^{\min(|J_g(t)|,P)} c_p \tag{4.18}$$

Let consider as an example a system with $P = 3$, $c_1 = 2$ şi $c_2 = c_3 = 1$. For simplicity, we assume that there is no job migration and that jobs are only served by processor 1, or wait in its queue. At time $t = 0$, job 1 with service time $S_1 = 4$ enters the system, job 0 is already present, $W_0^J(0) = 2$. There are no other jobs present in the system, nor do any other jobs arrive. We consider the first-come first-served (FCFS) policy. Figure 4.14 shows the Gantt chart for the execution of the two jobs (0 and 1) and Figure 4.15 shows the remaining work $W_1^J(t)$, the service rate $\sigma_1^J(t)$, and the obtained share of job $o_1^J(t)$.

4.5.1.2 *Share-scheduling policy for parallel processing*

Let consider the above mentioned model of a parallel processing platform with P processors, and assume that each job j has a weight w_{jp} on processor p, representing the fraction of time job j spends on processor p. We consider also that job j can switch between processors such that it appears to be served by more than one processor at the same time, at total service rate of $\sum_{p=1}^{P} w_{jp}c_p$ and the following conditions are respected:

$$0 \le w_{jp} \le 1, \text{ for all j, p}; \quad \sum_j w_{jp} \le 1 \text{ for all p}; \quad \sum_p w_{jp} \le 1 \text{ for all j}$$

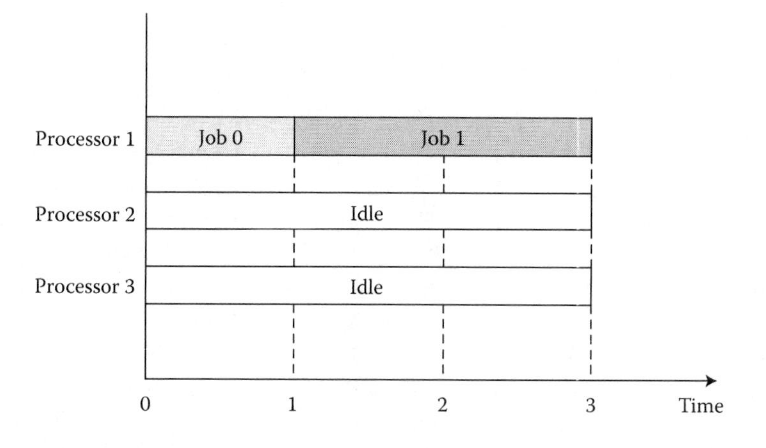

Figure 4.14 The Gantt chart for the FCFS policy.

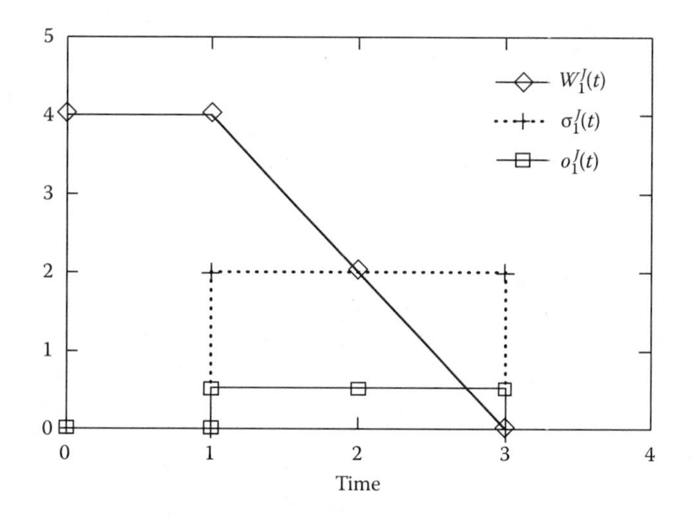

Figure 4.15 Job's scheduling parameters for the FCFS policy.

Under these assumptions, the proposed policy defined as job-priority multiprocessor sharing (JPMPS) can be applied without difference for both multiprocessors and distributed systems with free job migration. Under JPMPS, jobs $1, \ldots, J$ ($J \geq 1$) can be given service rates $\sigma_1^J \geq \sigma_2^J \geq \ldots \geq \sigma_J^J, J > 0$, if and only if, for $j = 1, \ldots, J$,

$$\sum_{k=1}^{j} \sigma_k^J \leq \sum_{p=1}^{\min(k,P)} c_p \tag{4.19}$$

When we compare this result to the definition of $m^T(t)$ in Equation 4.2 and $m^G(t)$ in Equation 4.3 it is obvious that, under JPMPS, the set of jobs present in the system can always be provided with a share of $m^T(t)$ and also all jobs of group g can be provided with $m^G(t)$.

The objective of share scheduling is to provide groups with their feasible group shares. For each group one can define a constant share, the required group share r_g^G, as the fraction of the total system capacity c that group g is entitled to, with $\sum r_g^G = 1$. The required group shares are assumed to be constant over time and to be known in advance to the system. When the required group shares of two groups are the same, the system should treat them equally, and when one exceeds the other, the system should give preferential treatment to the group with the r_g^G. The required group share plays an important role in the definition of the feasible group share, because the feasible group share depends only on the required group share, the number of jobs present of the group and the average processor capacity.

We state the following three requirements for the definition of the feasible group share $f_g^G(t), g = 1, \ldots, G$:

1. $0 \leq f_g^G(t) \leq m_g^G(t)$
2. $f_g^G(t)$ depends only on $|J_g(t)|, r_g^G, P, c_i (i = 1, \ldots, P)$
3. if $|J_g(t)| \geq P$ then $f_g^G(t) = r_g^G$

The first requirement means that we should not promise to a group more than the maximum we can provide. The second requirement means that the feasible group share only depends on the number of jobs of the group and not on the coincidental presence of jobs of other groups (i.e., groups are promised a share of the system that does not depend on the activity of other groups). The third requirement states that when the number of jobs in a group exceeds a threshold, the feasible group share equals the required group share. This threshold may be different for different groups, provided they have different required group shares.

From requirements 2 and 3, it follows that:

$$f_g^G(t) \leq r_g^G \quad \text{and} \quad \sum_g f_g^G(t) \leq 1 \tag{4.20}$$

This means one can never promise more than the total system capacity to all groups together.

Because Internet is a spontaneously grown collection of connected computers, we have utilized our PDES model for solving current problems due to the distributed computation on the Internet, when often the computed tasks have little or no connection to each other. Therefore, complex problems can be solved in real time on the Internet if the task allocation problem, which is rather complex by itself, is solved. However, if we assume that task allocation is resolved and the PE communication topology on the Internet is a scale-free network, the scalability properties of a PDESs scheme on such networks still remain to be emphasized.

With this aim, we have studied the fundamental scalability problem of conservative PDES schemes where events are self-initiated and have identical distribution on each PE using as model of scale-free networks the Barabási–Albert model. The scalability properties have been studied for a causally constrained PDES scheme hosted by a network of computers where the network is scale-free following a "preferential attachment" construction. The simulation exhibited slow logarithmic decay as a function of the number of PEs. The numerical results confirm the validity of any scalable parallel simulation for systems with asynchronous dynamics and short-range interactions.

4.5.2 *Evidence of attractors in large-scale networks models*

One of the most interesting theoretic and applied scientific result the quantification of how can be evidenced the attractors in dynamic systems which can be modeled by complex networks. In this aim, for the main three models (EA, Erdös–Rényi; BA, Barabási–Albert; and WA, Watts–Strogatz) analyzed in this chapter, we tried to observe specific events, where each state variable (associated to each node) is used to represent a specific prototype pattern (attractor). By assuming that the attractors spread their influence among its neighboring nodes through a diffusive process, it is possible to overlook the specific details of specific dynamics and focus attention on the separability among such attractors. Costa developed a study where this property is defined in terms of two separation indices (one individual to each prototype and the other considering also the immediate neighborhood) reflecting the balance and proximity to attractors revealed by the activation of the network after a diffusive process (Costa 2007). The separation index also considering the neighborhood was found to be much more informative, while the best separability was observed for the Watts–Strogatz model. The effects of the involved parameters on the separability were investigated by correlation and path analyses. The obtained results suggest the special importance of some measurements in underlying the relationship between topology and dynamics.

In order to infer the influence of the topologic features on the dynamics properties (i.e., attractor separation) correlation analysis and path analysis were applied. A series of interesting results were obtained regarding not only the separation indices, but also the importance of several local and global topologic features on the overall attractor separation. The simulation results confirm the correlation and path analyses of the effects of the topologic features of the network on the respective attractor separability.

In the same direction is oriented the study of the influence of the geometrical properties of the complex networks on the dynamics of the processes supported by the network infrastructure (for example, routing problems, inference, and data mining). In real growing networks, topological, structural, and geometrical properties emerge spontaneously from their dynamical rules. Nevertheless we still miss a model in which networks develop an emergent complex geometry, but a fractal analysis can confirm that a single two parameter of a network model, the growing geometrical network, can generate complex network geometries, combining exponential growth and small-world properties with finite spectral dimensionality. In one limit, the nonequilibrium dynamic rules of these networks can generate scale-free networks with clustering and communities.

In particular, growing complex networks evolving by the preferential attachment mechanism have been widely used to explain the emergence of the scale-free degree distributions which are ubiquitous in complex networks. In the next chapter, the interdependence between the fractal properties of the network topology and the self-similar aspect of the information flow will be highlighted and explained.

chapter five

Topology and traffic simulations in complex networks

5.1 Example of building and simulating a network

NS2 (Chung 2011) is a network simulator built using discrete events, developed at Berkeley University of California, and it simulates a variety of IP networks (Fall and Varadhan 2008). The simulator benefits from implementations of the TCP and UDP protocols, typical behavior of traffic sources such as FTP, Telnet, Web, CBR, and VBR, router queue management mechanisms such as DropTail, RED, and CBQ, routing algorithms such as Dijkstra, and many more. NS2 also implements multicasting and some MAC level protocols for LAN simulations. The current version of NS (version 2) is written in C++ and OTcl (Tcl scripting language with object-oriented extensions developed by MIT).

As it can be seen in Figure 5.1, from a simplified user perspective, NS is an object-oriented Tcl script interpreter (OTcl) that has an event timer for simulation, network object libraries, and network initialization libraries. In other words, in order to use NS, a program must be written in the OTcl scripting language. In order to set up and run a network simulation, the user must write an OTcl script that sets an event timer, defines network topology by using network objects and corresponding library functions, and instructs traffic sources when to start and stop packet transmission through the event timer. If the user wants to create a new network object, this can be accomplished by writing an altogether new object or by building a composite one using the existing object library.

Besides the network objects, another major component of NS is the event timer. An NS event is a unique ID for a packet, with a predefined time step and a pointer to the object that uses the event. In NS, a timer keeps track of the simulation time and raises all events from the queue that are scheduled for the current time step by invoking the corresponding network components. Network components communicate among them by exchanging packets.

NS is written not only in OTcl but also in C++. For a better performance, NS separates the data engine implementation from the processing and control engine. In order to minimize the event and package processing time, the timer and the main network objects are written and compiled

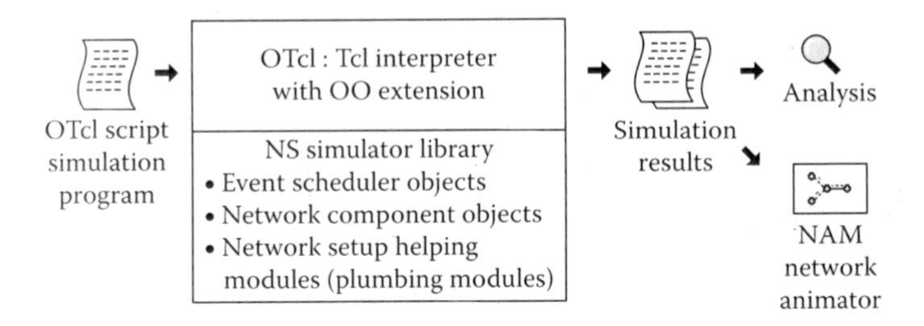

Figure 5.1 Simplified user perspectives on NS.

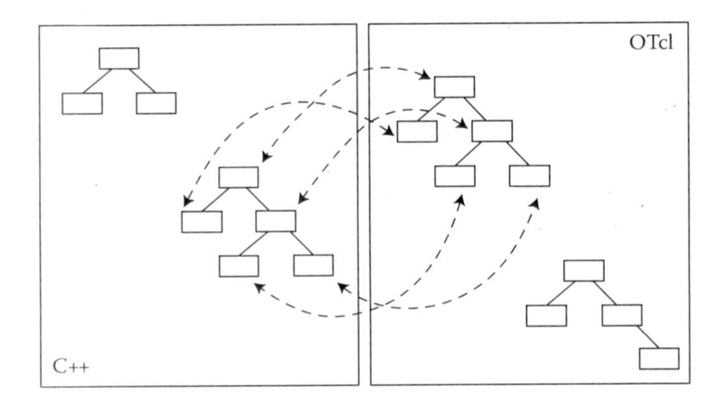

Figure 5.2 C++ and OTcl duality.

using C++. These compiled objects are available to the OTcl interpreter through a linking mechanism that creates an OTcl object corresponding to each C++ object. Figure 5.2 illustrates an example of C++ and OTcl object hierarchy.

5.1.1 *Simple simulation example*

This section describes an NS simulation script and explains the role of each line. The example is an OTcl script that creates a trivial network configuration and runs the simulation scenario in Figure 5.3. In order to run a simulation, the code is saved in a file "ns-simple.tcl" and the command "ns ns-simple.tcl" is given at the console.

The network consists of four nodes (n0, n1, n2, and n3) shown in Figure 5.3. The duplex links n0–n2 and n1–n2 have a bandwidth of 2 Mbps with a 10 ms delay. The duplex link n2–n3 has a bandwidth of 1.7 Mbps with a 20 ms delay. Every node uses a DropTail queue with a maximum size of 10. A TCP agent is attached to n0, and a connection is established to a sink TCP agent attached to n3. Implicitly, the maximum size of a packet generated by a TCP agent is 1 KB. A TCP sink agent generates and transmits

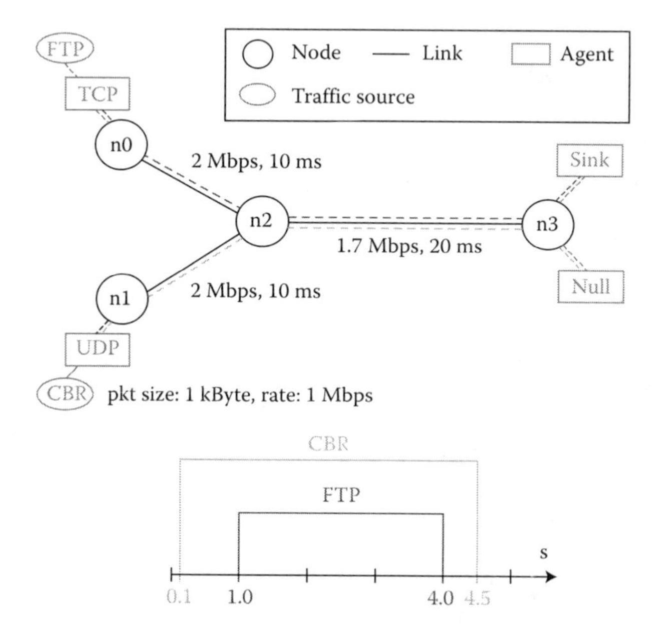

Figure 5.3 Trivial network topology and test scenario.

ACK to the sender and frees the received packets. The UDP agent attached to n1 is connected to the null agent attached to n3. A null agent frees the received packets. Two FTP and CBR traffic generators are attached to the TCP and UDP agents, and the CBR is configured to generate 1 kb packets with a rate of 1 Mbps. The CBR is set up to start at 0.1 s and stop at 4.5 s, while the FTP is setup to start at 1.0 s and stop at 4.0 s.

```
#Create a simulator object
set ns [new Simulator]

#Define different colours for data flows (for NAM)
$ns color 1 Blue
$ns color 2 Red

#Open the NAM trace file
set nf [open out.nam w]
$ns namtrace-all $nf

#Define a 'finish' procedure
proc finish {} {
 global ns nf
 $ns flush-trace
 #Close the NAM trace file
 close $nf
```

```
 #Execute NAM on the trace file
 exec nam out.nam &
 exit 0
}

#Create four nodes
set n0 [$ns node]
set n1 [$ns node]
set n2 [$ns node]
set n3 [$ns node]

#Create links between the nodes
$ns duplex-link $n0 $n2 2Mb 10ms DropTail
$ns duplex-link $n1 $n2 2Mb 10ms DropTail
$ns duplex-link $n2 $n3 1.7Mb 20ms DropTail

#Set Queue Size of link (n2-n3) to 10
$ns queue-limit $n2 $n3 10

#Give node position (for NAM)
$ns duplex-link-op $n0 $n2 orient right-down
$ns duplex-link-op $n1 $n2 orient right-up
$ns duplex-link-op $n2 $n3 orient right

#Monitor the queue for link (n2-n3). (for NAM)
$ns duplex-link-op $n2 $n3 queuePos 0.5


#Setup a TCP connection
set tcp [new Agent/TCP]
$tcp set class_ 2
$ns attach-agent $n0 $tcp
set sink [new Agent/TCPSink]
$ns attach-agent $n3 $sink
$ns connect $tcp $sink
$tcp set fid_ 1

#Setup a FTP over TCP connection
set ftp [new Application/FTP]
$ftp attach-agent $tcp
$ftp set type_ FTP

#Setup a UDP connection
set udp [new Agent/UDP]
$ns attach-agent $n1 $udp
set null [new Agent/Null]
$ns attach-agent $n3 $null
$ns connect $udp $null
$udp set fid_ 2
```

```
#Setup a CBR over UDP connection
set cbr [new Application/Traffic/CBR]
$cbr attach-agent $udp
$cbr set type_ CBR
$cbr set packet_size_ 1000
$cbr set rate_ 1mb
$cbr set random_ false

#Schedule events for the CBR and FTP agents
$ns at 0.1 "$cbr start"
$ns at 1.0 "$ftp start"
$ns at 4.0 "$ftp stop"
$ns at 4.5 "$cbr stop"

#Detach tcp and sink agents (not really necessary)
$ns at 4.5 "$ns detach-agent $n0 $tcp ; $ns detach-agent $n3
$sink"

#Call the finish procedure after 5 seconds of simulation
time
$ns at 5.0 "finish"

#Print CBR packet size and interval
puts "CBR packet size= [$cbr set packet_size_]"
puts "CBR interval= [$cbr set interval_]"

#Run the simulation
$ns run
```

The previous script is explained below. In general, an NS script starts with instantiating a simulator object.

`set` *`ns`* `[new Simulator]`: Generates an NS simulator instance, at attributes it to the *ns* variable (italics are used for variable names). This line does the following:

- Initializes package formats
- Creates a timer
- Selects the implicit address format

The simulator object has methods for the following actions:

- Creates composite objects such as nodes and links (explained later)
- Connects the network components
- Sets component parameters
- Creates connections between agents
- Specifies Nam display options

Most methods are used for initializing the simulation and some are used for displaying with Nam.

`$ns color fid color`: Is used for specifying the color of the packets of a flux given by its flux identifier (fid). This method of the simulator object is used only for Nam display and has no other effect upon the simulation.

`$ns namtrace-all file-descriptor`: This method determines the simulator to record the simulation in Nam input format. It also provides the filename where are later to be written data generated with the command `$ns flush-trace`. Similarly, the `trace-all` function is used for following the simulation in a general format.

`proc finish {}`: Is called after the simulation has ended by the command `$ns at 5.0 "finish."` This function specifies postsimulation processes.

`set n0 [$ns node]`: The `node` method creates a node. An NS node is a composite object containing address and port classifiers. Users can create nodes by separately creating address and port classifiers and connecting them afterwards.

`$ns duplex-link node1 node2 bandwidth delay queue-type`: Creates two simplex links with specified bandwidth and delay and connects the two given nodes. In NS, the output queue of a node is implemented as part of a link, and thus the users must specify the queue type when creating links. In the above simulation script, a DropTail queue has been used. If a RED queue is preferred, it is necessary only to replace the word "DropTail" with "RED."

`$ns queue-limit node1 node2 number`: This line sets the queue limit for the two simplex links that connect node 1 and node 2 at the specified number.

`$ns duplex-link-op node1 node2 ...`: The following lines are used for Nam display. To see the effect of these lines, they can be commented out and the simulation rerun.

Now the basic network setup is done. The next step is to set up the traffic agents such as TCP and UDP, the traffic sources FTP and CBR, and to attach them to the respective agents.

`set tcp [new Agent/TCP]`: This line creates a TCP agent. Generally, the user can create this way any agent of traffic source. Agents and traffic sources are in fact basic objects (noncomposite), the majority of which are implemented in C++ and linked to OTcl. Thus, there are no object-specific methods. The simulator can create object instances. To create agents or traffic sources, a user must only now be the names of these classes (Agent/TCP, Agent/TCPSink, Application/FTP, and so on).

`$ns attach-agent node agent`: The `attach-agent` function attaches an agent object already created to a node object. In fact, all that this function does is to call the method `attach` of the specified node, which attaches the specified agent to itself. Thus, the user can accomplish the same results by writing, for example, `$n0 attach $tcp`. Similarly, each agent object has a method `attach-agent` that attaches a traffic source to itself.

`$ns` `connect` *`agent1 agent2`*: After two communicating agents are connected, the next step is to establish a logical connection between them. This line sets a network connection by specifying the destination address for each address port pair in the network.

Assuming that the network configuration is ready, the next step is writing a simulation scenario (in other words, planning the simulation). The simulator object has many scheduling functions. The most used of them is the following:

`$ns` `at` *`time "string"`*: This method of the simulator object instructs the timer to schedule the specified command at the given moment in simulation time (*scheduler_* is the variable that refers to the timer object created with the command [new Scheduler] in the beginning of the script). For example, *`$ns`* `at` *`0.1 "$cbr start"`* determines the timer to call the function `start` of the CBR traffic source object, that prompts CBR to start transmitting data. In NS, usually a traffic source does not directly transmit data, but notifies the associated agent that it has to send a certain data quantity and the agent creates and then sends the packages.

After all network configurations, scheduling and postsimulation procedures are done, the next and last step is running the simulation. This is accomplished through the *$ns* run command.

5.2 Construction of complex network topologies

The three types of complex networks of interest for simulating are random graphs, small-world networks, and scale-free networks. Each of them has specific topological properties, and this can be graphically illustrated with the help of the Nam animator. Because of readability concerns, the networks analyzed with Nam will have small to medium dimension (on the order of tens or hundreds of nodes). Large dimension networks (thousands, tens of thousands, or hundreds of thousands) cannot be analyzed visually using Nam not because the animator is somehow limited but because the nodes become indiscernible and the links between them form a thick weaved strip of lines (Ţarălungă et al. 2007). Thus, for the analysis of large dimension networks, the only accessible tools are those of statistics, which can highlight degree of distributions, average trajectory lengths, and clustering coefficients. These statistical analyses together with other indicators (like those for simulation performance) are presented in detail in the following paragraphs. Also, note that for simulations of networks with a large number of nodes and long computation time, the most suitable machines are those with parallel architectures (Wilkinson and Pearson 2005).

In Mocanu and Ţarălungă (2007), the authors, together with a research group, constructed a model of the Internet network, based upon a growth algorithm with a modified preferential attachment. Given the difficulties of modeling such a high dimension network (Floyd and Paxson 2001) with specific topological properties, only original algorithm variants have been simulated in this thesis: Erdös–Rényi, Watts–Strogatz, and Albert–Barabási models.

5.2.1 Construction of a random network

Using the Erdös–Rényi model described in Section 4.2, an algorithm for generating a random graph was implemented in a .NET application using the C# language. The graphical user interface is shown in Figure 5.4. The application generates an OTcl script, which can be directly fed into NS for simulating. This script can be manually altered if necessary. Through the interface, various parameters for traffic generating can be configured: simulation time, percent of nodes that fail at given moments, the percent of nodes that initiate traffic, the maximum number of applications that generate traffic on a certain node, and so on. Of them, the most important

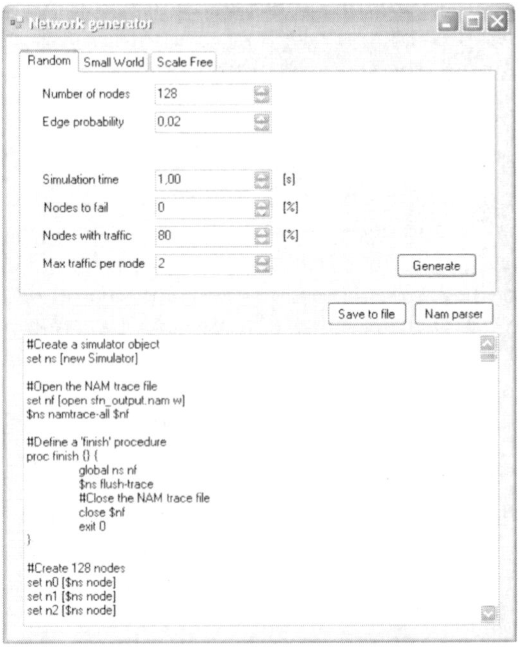

Figure 5.4 Graphical user interface for the complex network generator. The screenshot depicts the creation of a random network with 128 nodes and 0.02 connection probability.

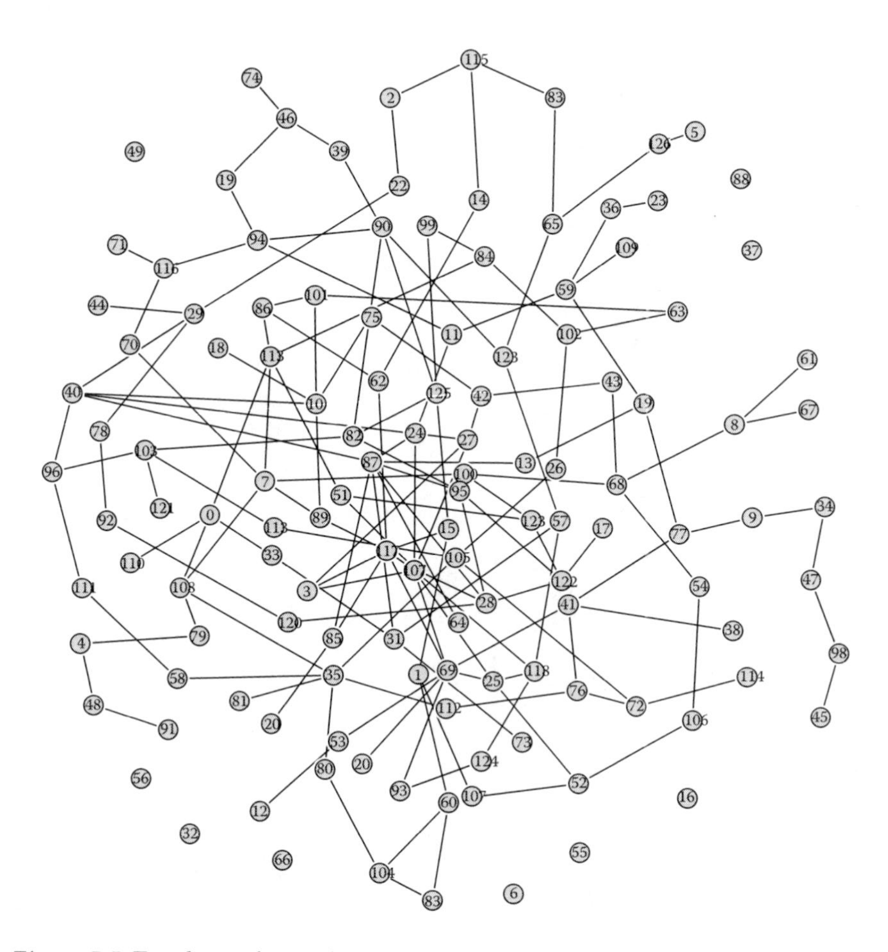

Figure 5.5 Topology of a random network with 128 nodes and a 0.02 connection probability. Because of the small connection probability, isolated nodes can be observed.

concerning speed is the simulation time. If it is large, the actual simulation takes a lot of time. Thus, it is advisable that for topological analysis simulations, the simulation time be short because the traffic is of no interest.

For illustrating the topology of a random graph, a 128-node network has been generated with a connection probability of 0.02. Figure 5.5 depicts this topology animated with Nam and rearranged for visual comfort.

5.2.2 Construction of a small-world network

For constructing a small-world network the Watts–Strogatz model (Watts and Strogatz 1998) was employed. The model construction parameters are the number of nodes (network dimension), starting degree, and node

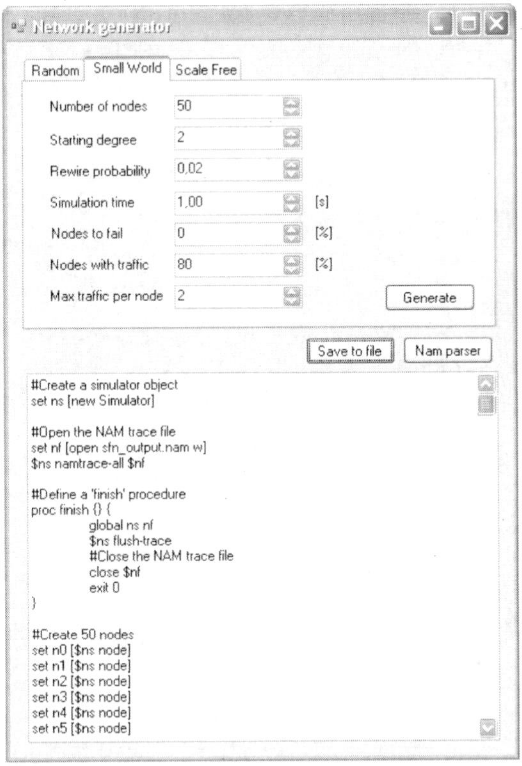

Figure 5.6 Graphical user interface for the complex network generator. The screenshot depicts the creation of a small-world network with 128 nodes, starting degree 2 and 0.02 reconnection probability.

reconnection probability. Figure 5.6 depicts the GUI for generating the OTcl script for such a small-world network model.

As a result of network generation, the NS simulation script was also generated. After simulating, Nam was used for visualizing the small-world topology created. Evidently, traffic was also generated, but this is of no interest at this point. In Figure 5.7, the positioning of the 50 nodes can be observed, as a circular lattice, as well as their respective connections.

5.2.3 *Construction of a scale-free network*

For the construction of a scale-free network, the Albert–Barabási model (Albert and Barabási 2002b) was used with the following parameters: $N = 128$ nodes, $m_0 = 5$ initial number of nodes and $m = 1$ edges per iteration (Figure 5.8).

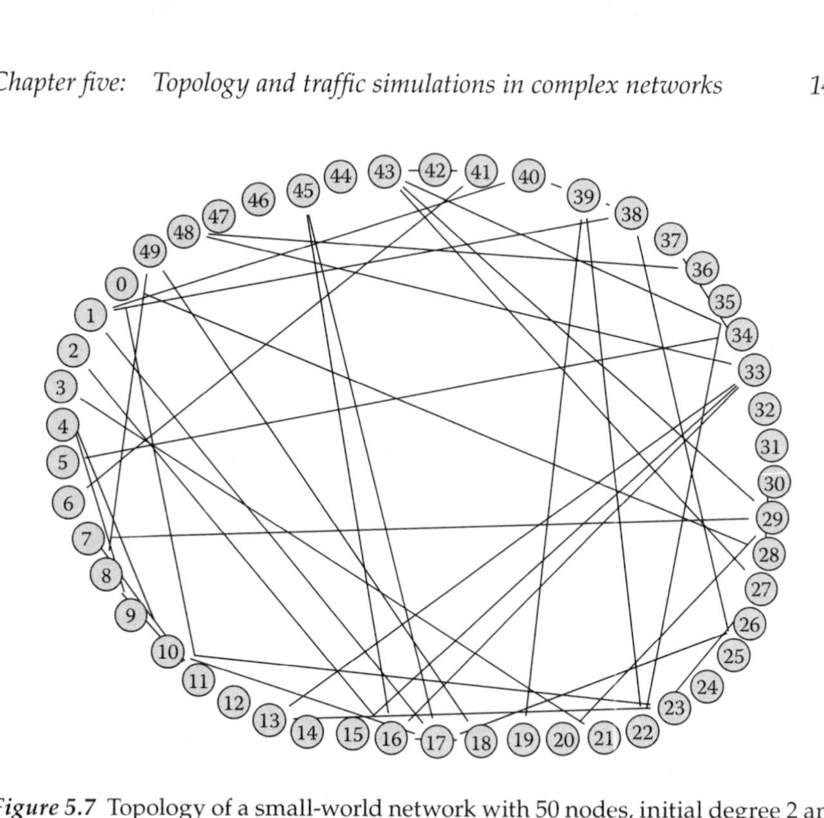

Figure 5.7 Topology of a small-world network with 50 nodes, initial degree 2 and reconnection probability 0.02.

The scale-free network topology is presented in Figure 5.9. Note the tree-like structure and the strongly connected nodes, as well as the lack of cycles (sequences of nodes that form a closed chain). Lack of cycles and clusters is due to the fact that for $m = 1$, a new node that is to be attached to the network will connect through a single edge, thus forming an arborescent structure.

5.3 *Analyses and topological comparisons of complex networks*

With the help of the algorithms described in Section 4.2, several numerical simulations were done for three types of complex networks: random, small-world, and scale-free. Networks with dimensions covering four orders of magnitude have been simulated, with $N = 100$, 1000, 10,000, and 100,000 nodes. For random networks, the connection probability was $p = 0.01$; for the small-world networks, the reconnection probability $p = 0.01$ and the starting degree $k = 2$; and for scale-free networks, the initial number of nodes $m_0 = 5$ and $m = 1$ edges added per step were considered.

For every simulated network, the clustering coefficients and average trajectory length were calculated. There were cases when these indicators could not be calculated due to the extremely high numerical processing

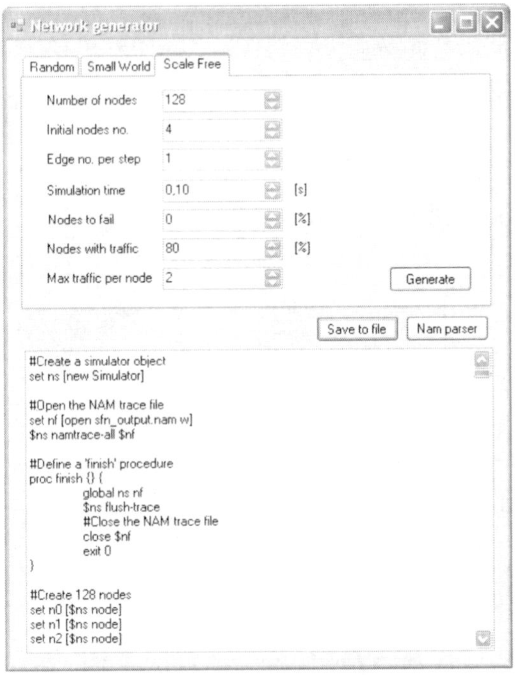

Figure 5.8 Graphical user interface for the complex network generator. The screenshot depicts the creation of a scale-free network with 128 nodes, 4 initial nodes and 1 edge added per node.

time (Ţarălungă et al. 2007). In any case, at least two degrees of magnitude were covered.

In the following, various diagrams and tables for the simulated networks are presented with the node degree diagrams dominating.

Below, the node degree diagram of a random network is presented, ordered from the most connected node to the least connected one. Note the quasi-linear characteristic of degree evolution (Figure 5.10).

To better illustrate the probability distribution characteristic, the node degree distribution was graphically presented, calculated as the number of nodes with degree k, X_k, as a function of k. Note that the distribution is very well approximated by a binomial $C_{N-1}^{k}p^{k}(1-p)^{N-1-k}$ distribution, with $N-1$ and p (Figure 5.11).

For comparison, given in Figure 5.12 the decreasing node degree evolutions for the four random networks of different dimensions, on logarithmic axes. The quasi-linear characteristic of the degree distribution is retained for all orders of magnitude.

Entering the field of small-world networks, Figure 5.13 is a decreasing diagram of nodes for such a network with reconnection probability

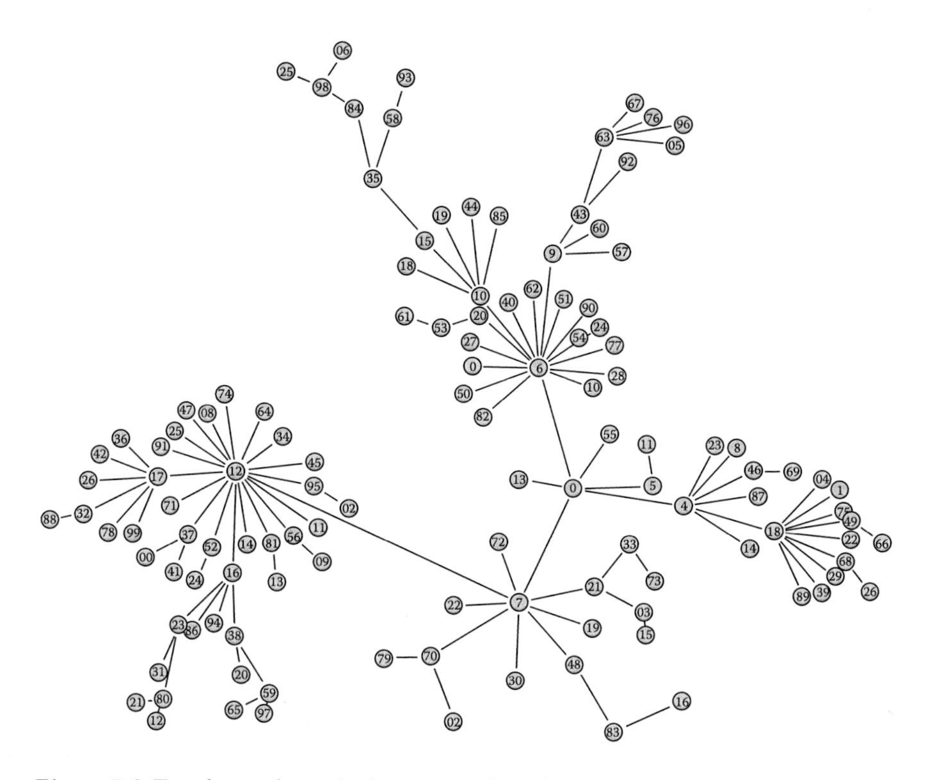

Figure 5.9 Topology of a scale-free network with 128 nodes.

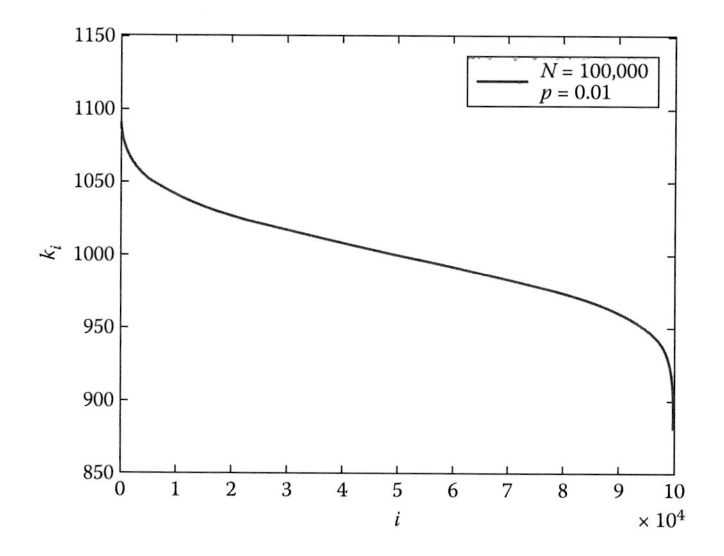

Figure 5.10 Node degree diagram (descending) for a random network with $N = 100{,}000$ and $p = 0.01$.

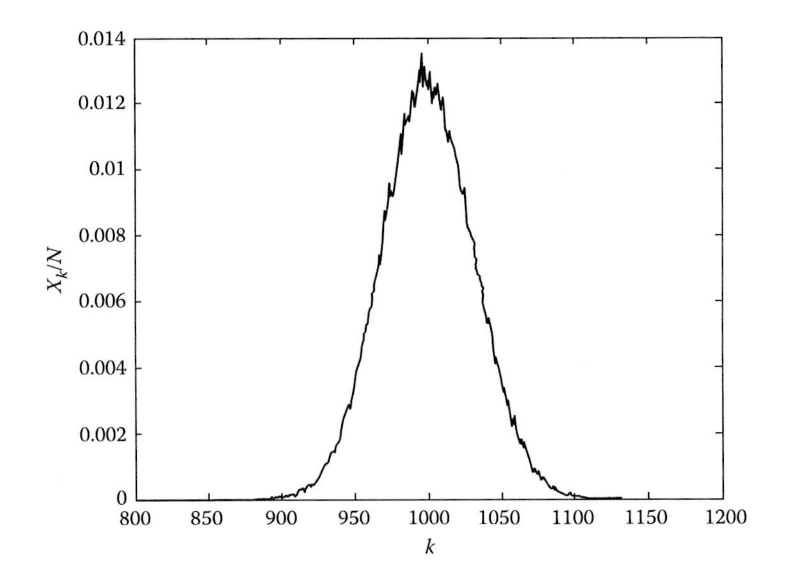

Figure 5.11 Node degree distribution, obtained from numerically simulating a random network with $N = 100{,}000$ and $p = 0.01$. X_k represents the number of nodes in the network that have the degree equal to k.

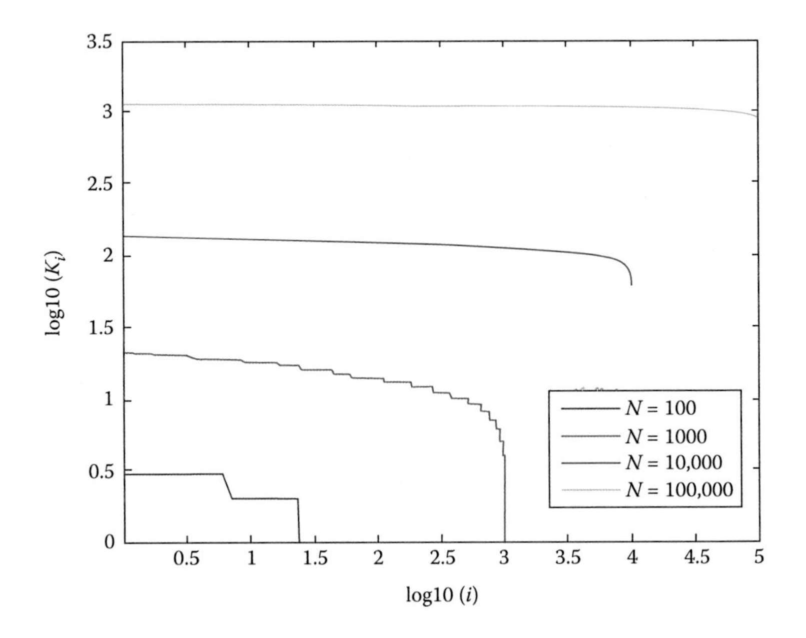

Figure 5.12 Comparison between node degree evolutions (descending) for four random networks with dimensions spanning over four orders of magnitude. Logarithmic axes.

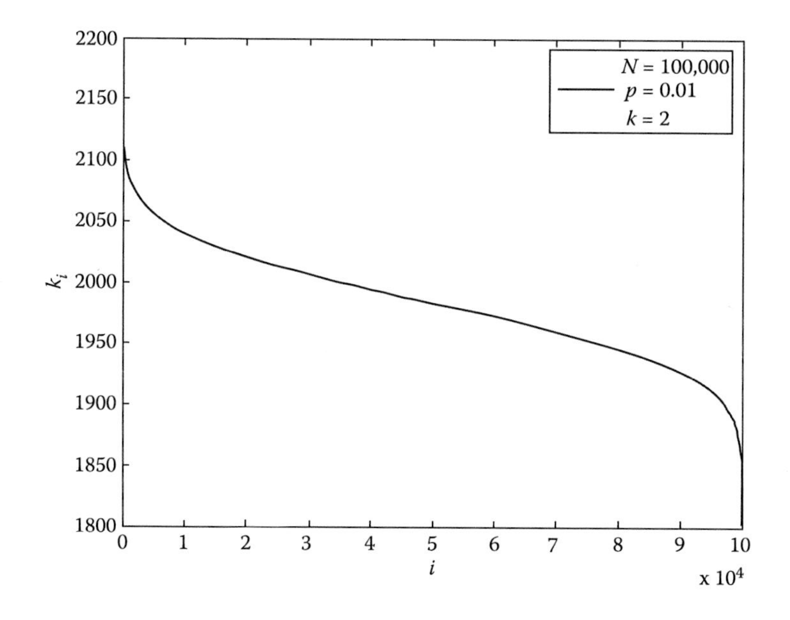

Figure 5.13 Node degree diagram (descending) in a small-world network with $N = 100{,}000$, $k = 2$, and $p = 0.01$.

$p = 0.01$ and starting degree $k = 2$. Opting for a higher than one value for the starting degree, it can be seen that node degrees have significantly larger values (almost double) than the node degree of the random network of the same dimension presented previously.

Figure 5.14 is a graphical representation of the node degree distribution calculated as the number of nodes with k degree, X_k, as a function of k. The shape of the degree distribution is similar to that of the random graphs: a prominent peak at $\langle k \rangle = K$ and decreases exponentially for a large k. Thus, the network topology is relatively homogeneous, all nodes having approximately the same number of edges.

As in the case of random graphs, in Figure 5.15, the descending degree evolution of the four small-world network types are presented for different dimension on logarithmic axes. The quasi-linear characteristic is retained for all the four orders of magnitude.

For scale-free networks, the node degree distribution has a significantly different distribution than previous networks. Figure 5.16 illustrated, on logarithmic axes, the node distribution calculated from a scale-free network with $N = 100{,}000$ nodes, $m_0 = 5$ initial number of nodes, and $m = 1$ edges added per step. The power law characteristic $P(k) \sim k^{-\gamma}$ with negative slope $-\gamma$ can be observed.

As in the other cases, Figure 5.17 presents the node degree evolution for four scale-free networks with dimensions spanning over four orders of

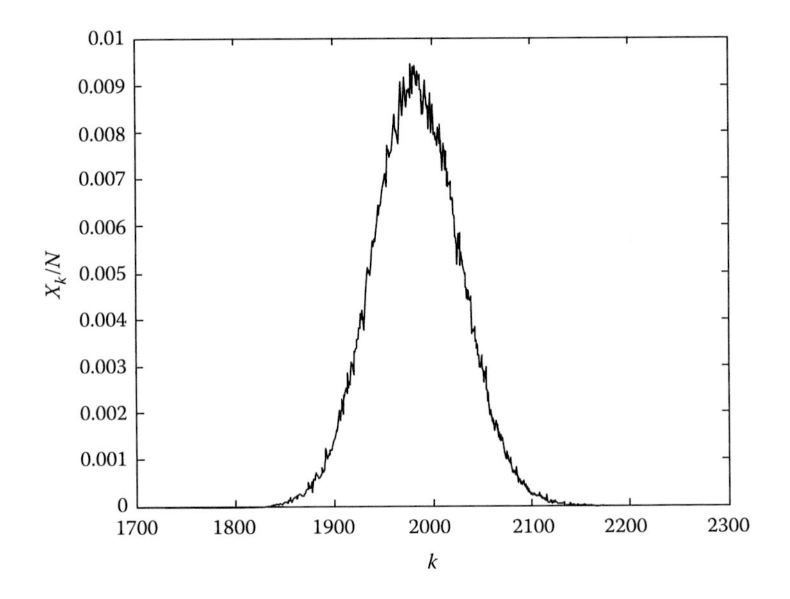

Figure 5.14 Node degree distribution obtained from numerical simulation of a small-world network with $N = 100{,}000$, $k = 2$, and $p = 0.01$. X_k represents the number of nodes in the network having the degree equal to k.

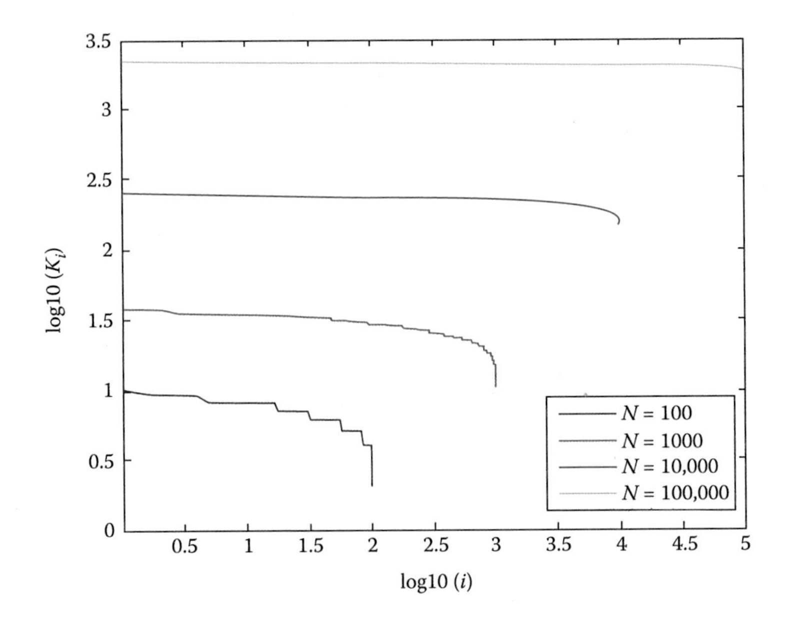

Figure 5.15 Comparison between node degree evolution (descending) for four small-world networks with dimensions spanning over four orders of magnitude. Logarithmic axes.

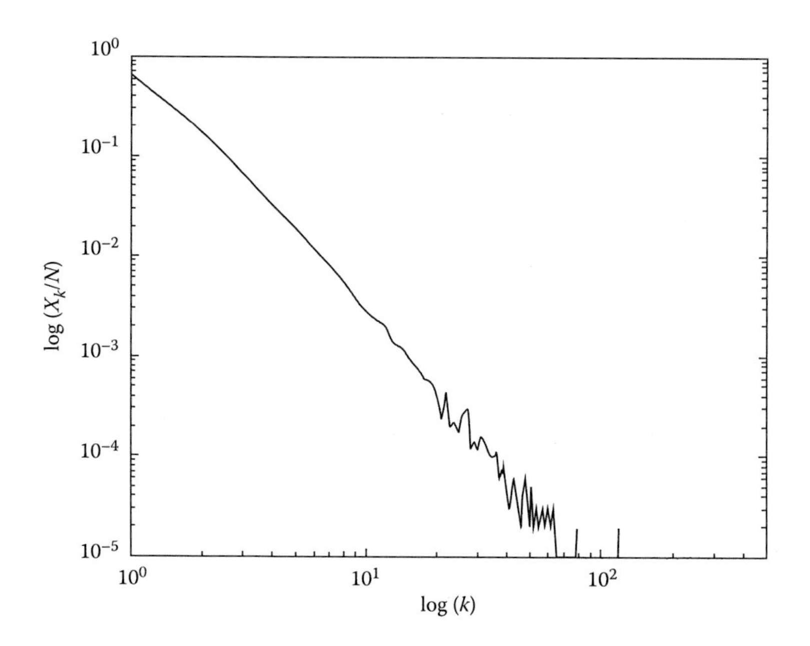

Figure 5.16 Node degree distributions obtained from numerical simulation of a scale-free network, with $N = 100{,}000$, $m_0 = 5$, and $m = 1$. X_k represents the number of nodes in the network with the degree equal to k.

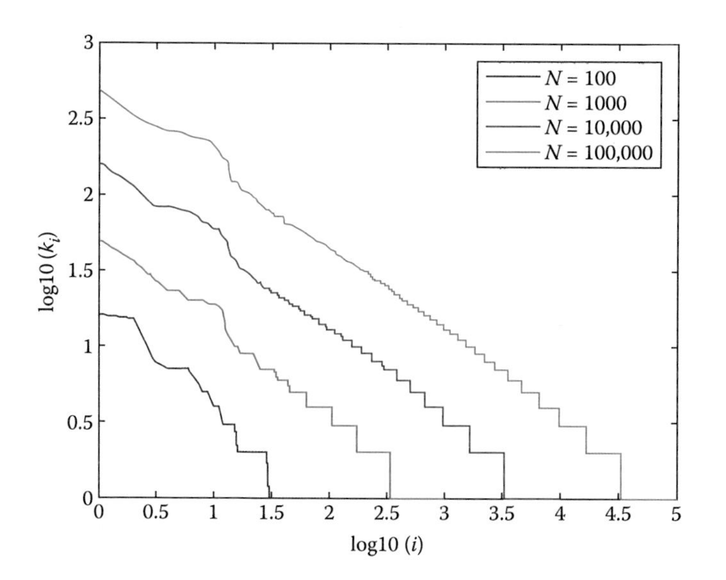

Figure 5.17 Comparison between node degree evolution (descending) for four scale-free networks with dimensions spanning over four orders of magnitude. Logarithmic axes.

magnitude. The characteristic scaling can be easily noticed for all orders of magnitude. For this reason, for further comparison between different networks, dimensions of 100,000 nodes will be used. Results could then be extended to networks of any dimension, smaller or larger (Figure 5.17).

To illustrate the difference between the three types of degree evolutions in the analyzed complex networks, the comparative linear diagram is presented for node degree-ordered descending. The similarity between random and small-world network characteristics is obvious, as well as the huge difference to the scale-free one. All evolutions are drawn for networks of 100,000 nodes (Figure 5.18).

Using a logarithmic diagram for both axes, the difference between the laws describing the three network types can be observed more precisely. For the scale-free network, the negative slope $-\gamma$ is clear, while for the other two networks, the evolution is almost constant (Figure 5.19).

The average trajectory lengths have also been calculated for the three network types. They are centralized in Table 5.1. Note that for random networks, this length tends to increase with N, while for small-world networks, the length seems to decrease. For scale-free networks simulated using the unmodified Albert–Barabási model, with no cycles, the average trajectory lengths increase with the network dimension.

Due to numerical processing complexity for determining the shortest path in a graph (using Dijkstra), the average trajectory length could not be determined for large networks (i.e., 10,000 or 100,000). The average

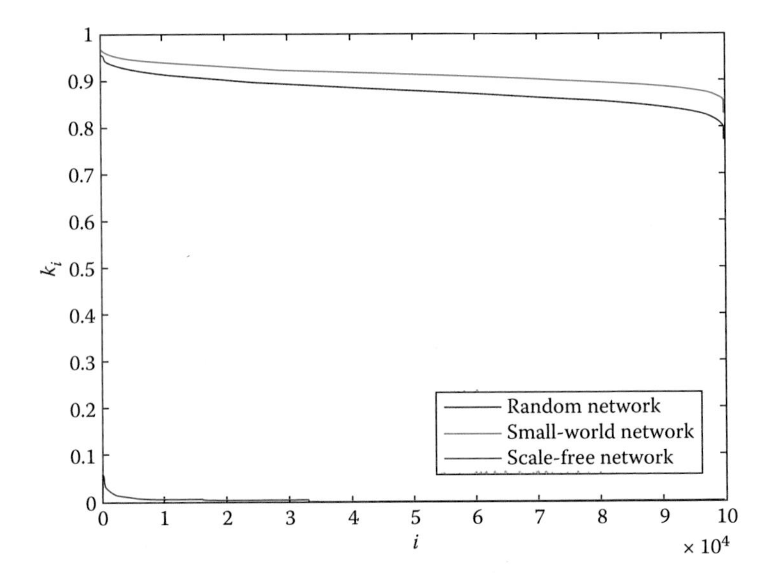

Figure 5.18 Comparison between node degree evolutions (descending) for random, small-world and scale-free networks of the same dimension ($N = 100{,}000$).

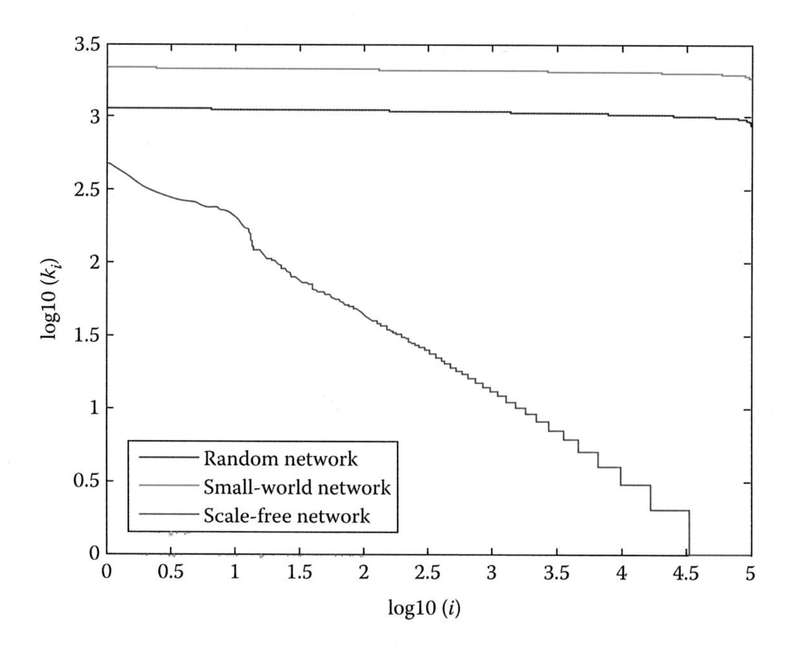

Figure 5.19 Comparison between node degree evolution (descending) for random, small-world and scale-free networks of the same dimension ($N = 100{,}000$). Logarithmic axes.

Table 5.1 Average trajectory length for three complex network types

	Random network	Small-world network	Scale-free network
$N = 100$	0.0657	4.0073	4.5224
$N = 1000$	3.2872	2.6631	7.4441

trajectory length for a network was calculated as an average of the length of every node. Thus, $\ell = (1/N)\sum_{i=1}^{N} \ell_i$, where ℓ_i is the average minimal trajectory between node i and all other nodes (Figure 5.20).

By graphically representing the dependences of the average trajectory length as functions of the network dimension, on semilogarithmic scales, it can be seen that they follow the theoretical estimations given by

$$\ell_{\text{random}} \sim \frac{\ln(N)}{\ln(\langle k \rangle)}$$

$$\ell_{\text{small-world}} \sim \frac{N}{K} f(pKN^d)$$

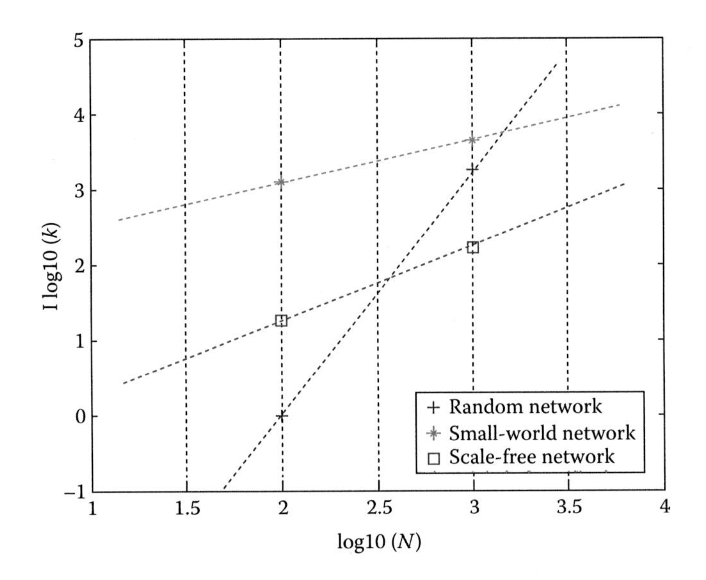

Figure 5.20 Comparison between the average trajectory lengths for random networks (plus) with $p = 0.01$, small-world networks (star) with $p = 0.01$ and $k = 2$ and scale-free networks (squares) with $m_0 = 5$ and $m = 1$.

where

$$f(u) = \begin{cases} \text{constant} & \text{if } u \ll 1 \\ \dfrac{\ln(u)}{u} & \text{if } u \gg 1 \end{cases}$$

$$\ell_{\text{scale-free}} \sim \frac{\ln(N/z_1)}{\ln(z_2/z_1)}$$

Clustering coefficients for each networks have been calculated using the average of the individual coefficients for each node. Thus, the clustering coefficient for the whole network is $C = (1/N)\sum_{i=1}^{N} C_i$, where $C_i = (2E_i/k_i(k_i - 1))$, E_i being the total number of edges between all the neighbors of node i, and k_i being the degree of node i.

In Table 5.2, the clustering coefficients are grouped for the three network types. For scale-free networks, simulated using the Albert–Barabási model in order for them to have cycles, a value of $m = 2$ has been used. Otherwise, the clustering coefficients would have been zero.

As in the case of average trajectory lengths, the numerical processing effort for computing the clustering coefficients is high, thus for larger networks (i.e., 10,000 and 100,000 nodes) they could not be determined (Ţarălungă et al. 2007).

Table 5.2 Clustering coefficients for three complex network types

	Random network	Small-world network	Scale-free network
$N = 100$	0	0.9222	0.1401
$N = 1000$	0.0124	0.3412	0.0264
$N = 10{,}000$	0.0101	–	0.0050

5.4 *Self-similar traffic simulation*

For simulating self-similar traffic, the method of superimposed ON–OFF sources was employed, considering a Pareto distribution. According to 3.3.2, the aggregated traffic is generated by multiplexing a series of independent ON–OFF sources (renewal processes), where each source alternates these states at certain time intervals. The generated traffic is asymptotically self-similar if many renewal processes superimpose one another, renewal values are restricted to 0 and 1, and renewal times are with heavy tail (Popescu 2001). For this, a Pareto distribution with $1 < \alpha < 2$ can be used.

The Pareto distribution is a power curve with two parameters, that is, the shape parameter α and the lower cutoff parameter β (or minimal value for random variable X). The Pareto cumulative distribution function (CDF) is given by $F(x) = 1 - (\beta/x)^{\alpha}$, and the probability density function $f(x) = (\alpha/\beta)(\beta/x)^{\alpha+1}$ for $x > \beta$ and $\alpha > 0$. Moreover, the α parameter is dependent on the Hurst parameter through $H = (3 - \alpha/2)$.

For illustrating the self-similar traffic simulation, a scale-free topology network with 40 nodes was used. In order to have a nonzero clustering coefficient, two edges were added per iteration. Thus, cycles have been obtained, which is a positive thing for the routing mechanisms in traffic simulation. The network topology is given in Figure 5.21. A small-sized network was used so as to be able to graphically illustrate the topology and some result, and also to keep the simulation time under control (Estan et al. 2003).

For traffic simulation, TCP agents with Pareto traffic applications objects have been used. These application objects are provided in the NS2 library (Fall and Varadhan 2008). For the shape coefficient α, a value of 1.4 was used, thus coming to an expected H value of 0.8. Below, the Tcl script is reproduced for defining a Pareto application and associating it to a TCP agent.

```
set p0 [new Application/Traffic/Pareto]
$p0 set packetSize_ 200
$p0 set burst_time_ 500ms
$p0 set idle_time_ 1s
$p0 set rate_ 1000k
$p0 set shape_ 1.4
$p0 attach-agent $tcp0
```

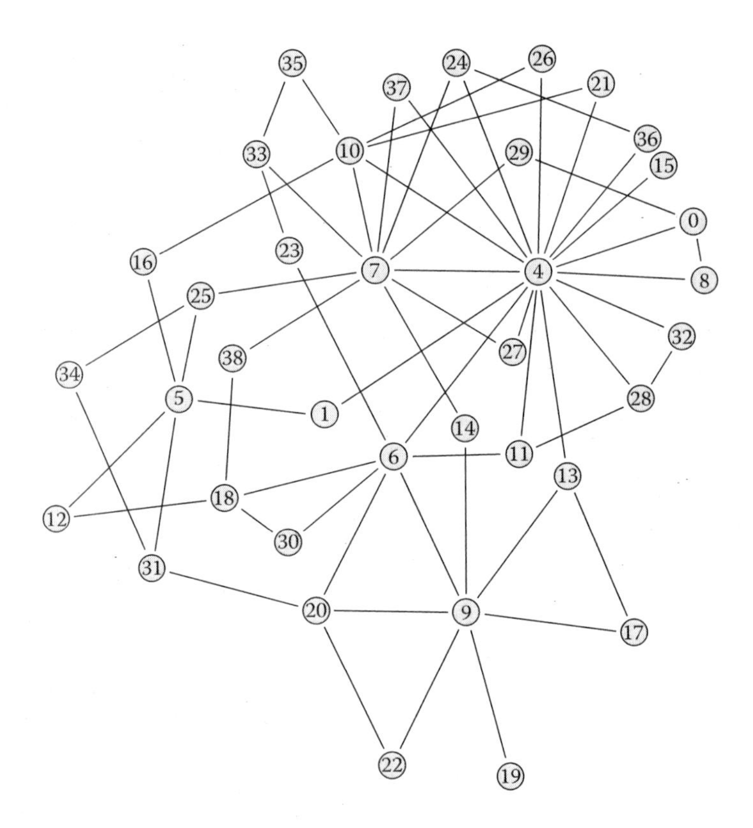

Figure 5.21 Scale-free network topology used for self-similar traffic. For generating the network the Albert–Barabási model was employed with: $N = 40$, $m_0 = 5$, and $m = 2$.

A number of 32 sources have been used, randomly associated to network nodes. The simulation lasted for 200 s so as to capture enough packets, and to be able to represent the aggregated packet number over as many time units as possible. The size of the generated Nam file was over 1.9 GB, which is more than enough to provide consistent results. Figure 5.22 shows the diagrams of the cumulated packet number over three time steps covering three degrees of magnitude: 1 s, 100 ms, and 10 ms. One can easily observe the strong variation of the packets number on all three timescales.

For a clearer graphical evidence, after the model of Leland et al. (1994), the diagram of the aggregated number of packets is zoomed over intervals of 20 time units (Figure 5.23). The red section of the histogram is the enlarged area. The strong traffic variation can be clearly seen for all three diagrams, thus visually confirming the incidence of the self-similarity phenomenon.

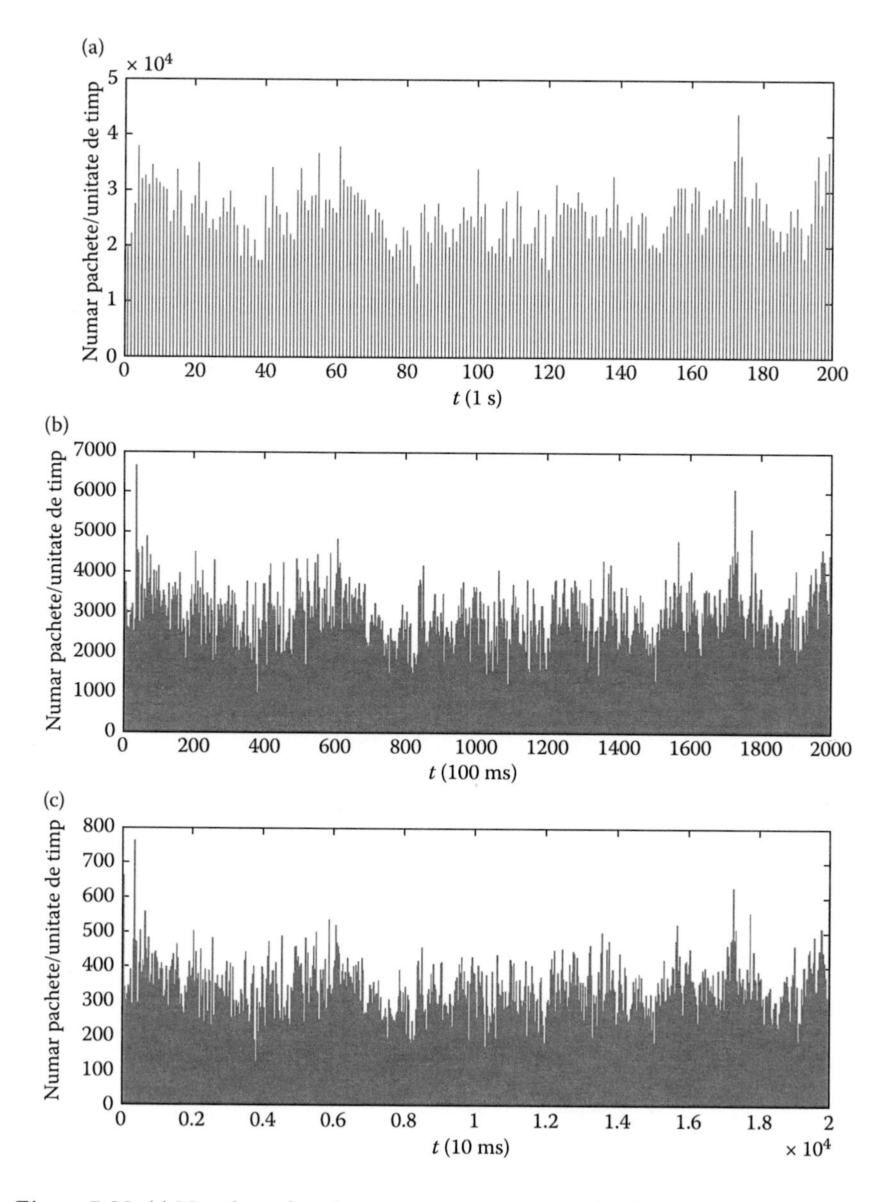

Figure 5.22 (a) Number of packets over a 1 s time unit. (b) Number of packets over a 100 ms time unit. (c) Number of packets over a 10 ms time unit.

For a statistical estimation of the Hurst parameter, three estimation methods have been employed: domain rescaling diagram (Figure 5.24), dispersion–time diagram (Figure 5.25), and the periodogram (Figure 5.26). The estimation is based on the time series of the aggregated number of packets per time unit of 10 ms (20,000 samples). Theoretically, the expected

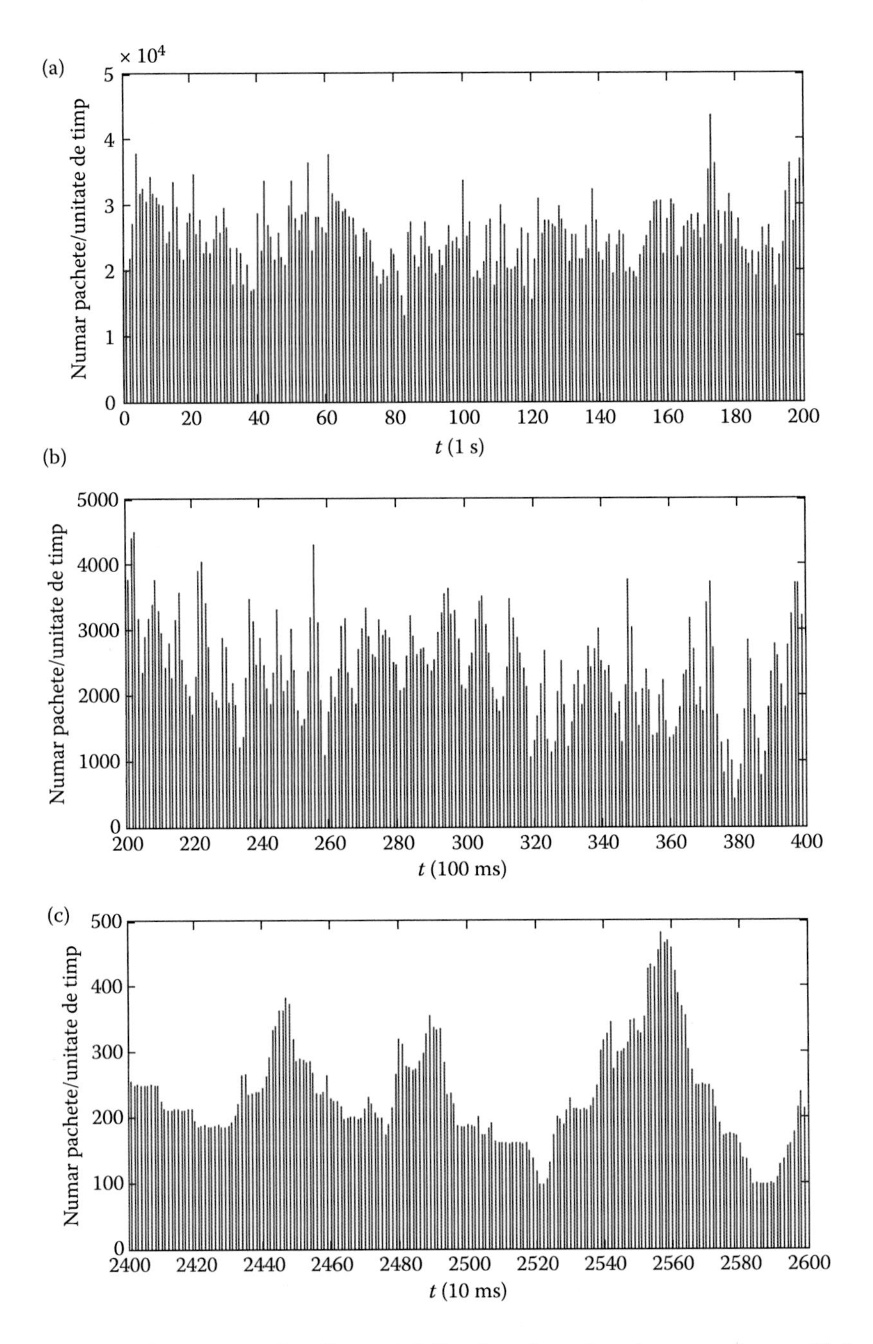

Figure 5.23 Illustration of traffic variability (number of packets per time unit) for three time scales with different quantization: 1 s, 100 ms, and 10 ms. Sets of 20 time units were enlarged consecutively (b) is red area from (a) and (c) is red area from (b).

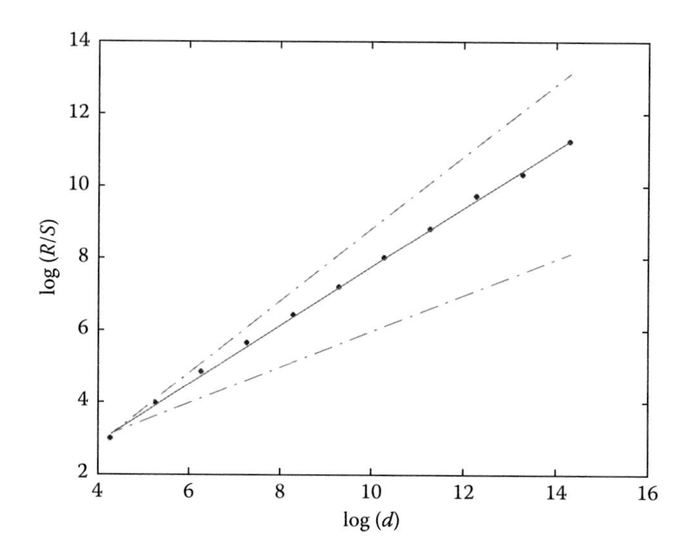

Figure 5.24 Domain rescaling diagram for aggregated traffic with a 10 ms interval. The estimation for the Hurst parameter is 0.8115.

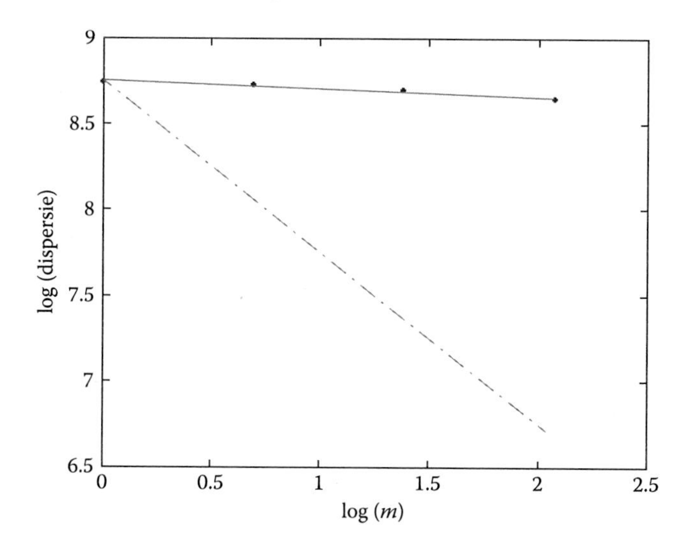

Figure 5.25 Dispersion–time diagram for aggregated traffic for 10 ms interval, $\hat{H} = 0.9761$.

value for the Hurst parameter is 0.8. In practice, the three graphical estimators output slightly higher values. Specifically, the domain rescaling provided the closest value (0.8115), dispersion–time diagram produced 0.9761, and the periodogram provided 1.1325. It is known that these last two indicators tend to overestimate the self-similarity coefficient (Karagiannis

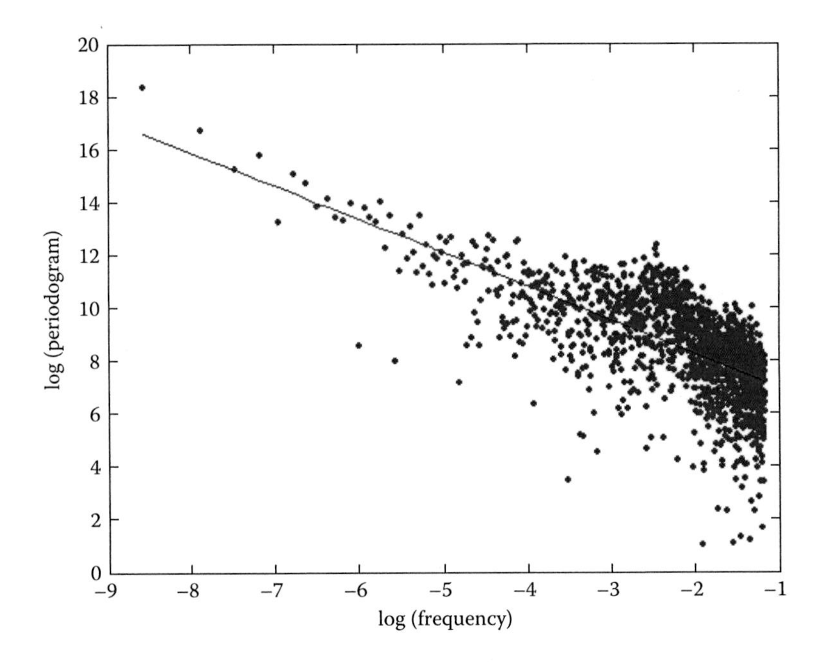

Figure 5.26 Periodogram for aggregated traffic for 10 ms interval, $\hat{H} = 1.1325$.

et al. 2003), but the clear conclusion is that the time series exhibits statistical self-similarity, and thus confirms the superposition model of ON–OFF sources, having renewal times given by a Pareto distribution leads to statistical self-similarity.

5.5 *Traffic simulation on combined topologies of networks and traffic sources*

In this section, the behavior of network traffic is studied, with particular focus on logical separating the node and source topologies. To this end, the simulation script generating algorithm has been modified to take into account the two topologies. Figure 5.27 presents a network in which the nodes are connected with black edges while the traffic sources and destinations are connected with red edges. Practically, two different topologies are obtained for two ISO-OSI layers: network layer (black) and the transport layer (red).

Using this approach, 27 traffic simulations were done. With the help of the algorithms described in the appendixes, all the above types of complex networks have been simulated in all possible arrangements. Thus, having three types of complex networks (random, small-world, and scale-free), the total number of arrangements is 9. On these topologies, three types of traffic generator have been used, all provided by the

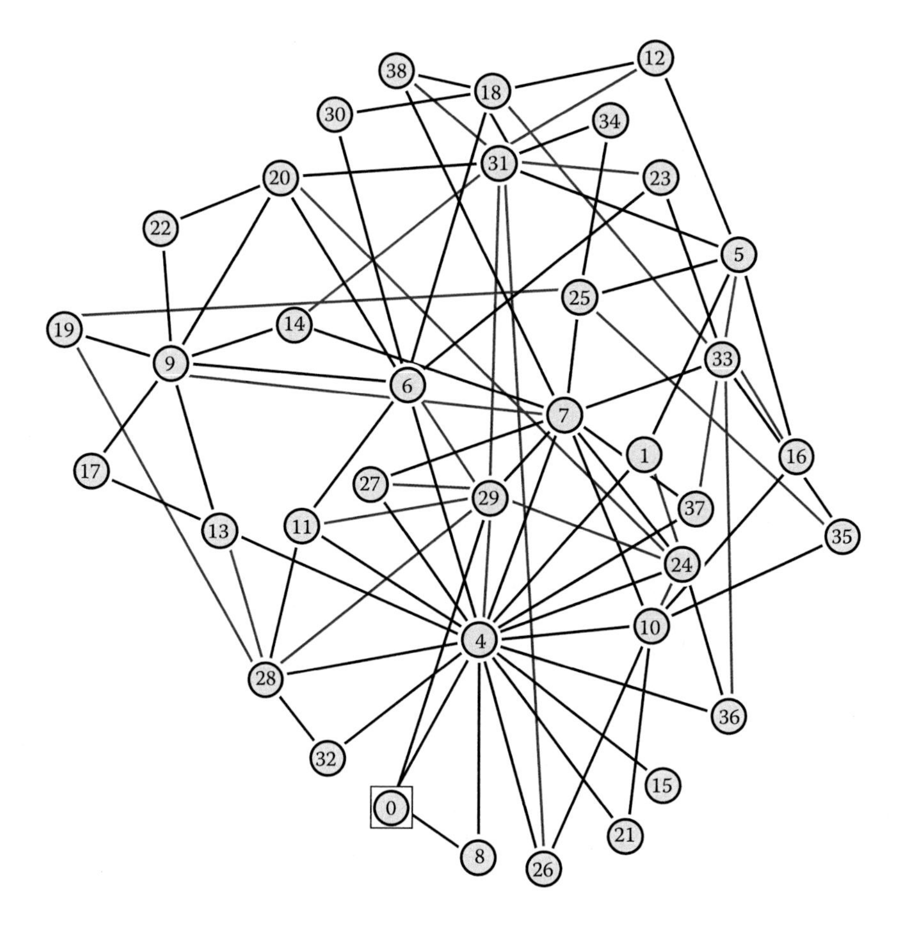

Figure 5.27 Physical (black) and logical (red) network topologies.

NS simulator: Pareto ON/OFF, Exponential ON/OFF, and CBR (Constant Bit Rate). Thus, the number of simulations equals 27.

In order to study the self-similar behavior of the traffic in each case, the Hurst parameter is estimated with the help of three graphical methods: domain rescaling, dispersion–time analysis, and the periodogram. The Hurst parameter estimation was done on aggregated packet number time series for three time intervals: 100, 10, and 1 ms. Time series have been extracted from the output files of the simulations through a counting process.

5.5.1 Details on the used topologies and traffic sources

For the purpose of obtaining consistent data, a simulation period of 200 s was chosen. To avoid needing an increased processing power, the number of nodes was kept to 100.

For random network topologies, the following parameters were used: $N = 100$ nodes, $N = 100$ traffic sources, and $p = 0.01$ connection probability.

For small-world topologies, the following parameters were used: $N = 100$ nodes, $N = 100$ traffic sources, $p = 0.01$ reconnection probability, and $k = 1$ starting degree.

For scale-free topologies, the following parameters were used: $N = 100$ nodes, $N = 100$ traffic sources, $m_0 = 4$ initial nodes, and $m = 1$ edges added at each step.

The topology arrangements are given in Tables 5.3 through 5.5 for each of the three traffic generators. In all of them, RND is the random, SW is the small-world, and SF is the scale-free topology.

The traffic generators have been configured so that to ensure a sufficient quantity of transferred packets during simulation. TCP agents were used as traffic agents on the transport layer. For the ON/OFF Pareto generator, the control parameters used are: packet size of 200, 500 ms burst time, 1 s idle time, 1000 kb transfer rate, and 1.4 shape factor. For the Exponential ON/OFF generator, the control parameters used are packet size of 210, 500 ms burst time, 1 s idle time, and a transfer rate of 1000 kb. For the CBR generator, the control parameters used are: packet size of 210 and 1000 kb transfer rate. For this type of generator, the random noise generating option was used during packet transfer.

Table 5.3 Simulations with Pareto traffic generators

Code	P1	P2	P3	P4	P5	P6	P7	P8	P9
Network layer	SF	SF	SF	SW	SW	SW	RND	RND	RND
Transport layer	RND	SW	SF	RND	SW	SF	RND	SW	SF
Sources	Pareto	Pareto	Pareto	Pareto	Pareto	Pareto	Pareto	Pareto	Pareto

Table 5.4 Simulations with Exponential traffic generators

Code	E1	E2	E3	E4	E5	E6	E7	E8	E9
Network layer	SF	SF	SF	SW	SW	SW	RND	RND	RND
Transport layer	RND	SW	SF	RND	SW	SF	RND	SW	SF
Sources	Exp	Exp	Exp	Exp	Exp	Exp	Exp	Exp	Exp

Table 5.5 Simulations with CBR traffic generators

Code	C1	C2	C3	C4	C5	C6	C7	C8	C9
Network layer	SF	SF	SF	SW	SW	SW	RND	RND	RND
Transport layer	RND	SW	SF	RND	SW	SF	RND	SW	SF
Sources	CBR	CBR	CBR	CBR	CBR	CBR	CBR	CBR	CBR

The size of the output files varied from a few hundreds of MB to a few tens of GB. These large dimensions indicate an intense traffic, and thus, it ensures a consistent estimation of the Hurst parameters.

5.5.2 *Hurst parameter estimation results*

For the purpose of highlighting the self-similar behavior of network traffic, three graphical estimators were used: domain rescaling (DR), dispersion time (DT), and periodogram (PG). The results are centralized in Table 5.6,

Table 5.6 Results of the Hurst parameter estimations for the 27 traffic simulations

Analysis codes	Granularity 100 ms			Granularity 10 ms			Granularity 1 ms		
	DR	DT	PG	DR	DT	PG	DR	DT	PG
P1	0.8519	0.8393	1.1718	0.8341	0.9803	1.1417	0.8902	0.9947	1.4989
P2	0.8179	0.8925	1.3894	0.8385	0.9445	1.2146	0.8627	0.9419	1.0137
P3	0.8242	0.8113	1.2780	0.8261	0.9782	1.2658	0.9317	0.9663	1.3005
P4	0.8747	0.8601	1.1803	0.8529	0.9810	1.1087	0.9009	0.9957	1.4885
P5	0.7885	0.7794	1.3023	0.8040	0.9782	1.2582	0.9219	0.9667	1.3142
P6	0.8230	0.8682	1.1449	0.8301	0.9828	1.1727	0.8910	0.9961	1.4952
P7	0.8650	0.8293	1.1887	0.8478	0.9815	1.2191	0.9441	0.9713	1.4583
P8	0.7550	0.8319	0.9265	0.7855	0.9776	1.1342	0.8973	0.9909	1.4412
P9	0.8263	0.8589	1.0263	0.8487	0.9807	1.1212	0.9948	0.9756	1.4121
E1	0.6240	0.8789	0.9673	0.7187	0.9824	1.2398	0.8349	0.9950	1.4978
E2	0.7084	0.9256	1.2234	0.7982	0.9693	1.2637	0.8663	0.9667	1.0909
E3	0.7011	0.8574	1.2520	0.7751	0.9804	1.3183	0.9124	0.9659	1.2730
E4	0.5747	0.8737	0.9652	0.6939	0.9831	1.2217	0.8205	0.9958	1.4813
E5	0.6394	0.8765	1.1450	0.7391	0.9817	1.3523	0.8944	0.9707	1.2828
E6	0.6741	0.8988	1.0855	0.7541	0.9878	1.2927	0.8565	0.9968	1.4889
E7	0.6213	0.8668	1.0268	0.7224	0.9831	1.2914	0.8827	0.9687	1.4406
E8	0.6059	0.8700	0.7050	0.7173	0.9816	1.2270	0.8760	0.9920	1.4332
E9	0.6421	0.8613	0.6038	0.7613	0.9801	1.1567	0.9575	0.9809	1.4366
C1	0.7977	0.9771	1.2468	0.8208	0.9813	1.3720	0.8257	0.9747	1.0498
C2	0.8642	0.9795	1.3425	0.9058	0.9918	1.4277	0.8906	0.9859	1.0089
C3	0.6296	0.8617	1.4070	0.6946	0.8212	1.4666	0.6606	0.6041	1.1840
C4	0.7704	0.9828	1.3216	0.8281	0.9968	1.7536	0.8379	0.9980	1.2688
C5	0.6614	0.8694	1.4073	0.7015	0.8089	1.4767	0.6645	0.6114	1.1883
C6	1.1238	0.9977	1.4408	1.0730	0.9997	1.6081	1.0092	0.9998	1.1693
C7	0.4661	0.4945	1.4403	0.4785	0.4775	1.4605	0.4812	0.4461	1.0904
C8	0.7040	0.9106	0.7897	0.7702	0.9935	1.6157	0.7742	0.9912	1.2430
C9	0.6286	0.7407	0.8042	0.7272	0.9734	1.4384	0.7402	0.9315	1.0907

for all the 27 performed distinct simulations. Using several estimators, it is normal for the output results to be close but not identical. Nevertheless, this provides an overview of the self-similar traffic scrutinized. Whether a self-similar characteristic exists is clear for all estimators.

The difference between self-similarity levels of two simulations proved to be tougher to prove. This is due to having different result from the three estimators for the same time series and also to the influence of the aggregation factors. Quantifying the number of incoming packets on fixed intervals of 100 ms, 10 ms, and then 1 ms affects even the values estimated through the same method (e.g., RS), on the same traffic simulation. Because of the significant number of samples, the time series constitutes a solid analysis base for the graphical estimators. Nevertheless, in the case of the periodogram, the Hurst parameter is systematically overestimated, reaching values higher than 1. It seems that for the same time series with the same granularity, the value estimated with the periodogram method is higher with 0.4–0.5 than the one produced with the other two estimators.

For a friendlier illustration of the estimated values, the estimations for the three analysis codes have been graphically presented: P (Pareto), E (Exponential), and C (Constant Bit Rate). Thus, Figure 5.28 depicts a comparison between the estimated values for the Hurst parameter for the nine traffic simulations using Pareto sources. Thus, Figure 5.29 depicts a comparison between the estimated values for the Hurst parameter for the nine traffic simulations using Exponential sources and Figure 5.30 depicts

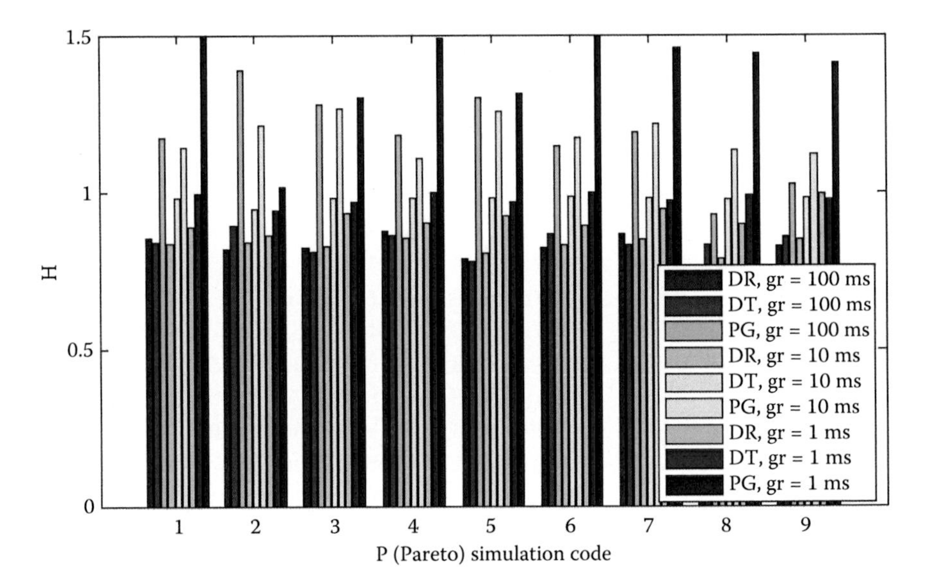

Figure 5.28 Comparison between the estimated values of the Hurst parameter for the nine traffic simulations using Pareto sources.

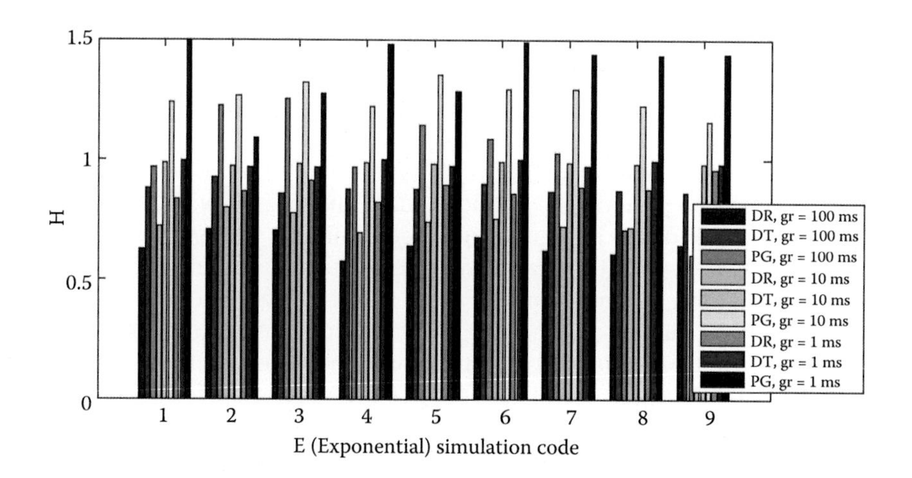

Figure 5.29 Comparison between the estimated values of the Hurst parameter for the nine traffic simulations using Exponential sources.

a comparison between the estimated values for the Hurst parameter for the nine traffic simulations using CBR sources.

In the case of the Pareto ON/OFF traffic generators, self-similarity was especially chosen, by attributing the value 1.4 to the distribution shape factor. As it was explained in Chapter 3, this corresponds to an analytical Hurst parameter of $H = (3 - \alpha)/2$, which is 0.8. The results confirmed a strong self-similar phenomenon for this simulation set with Pareto-distributed traffic.

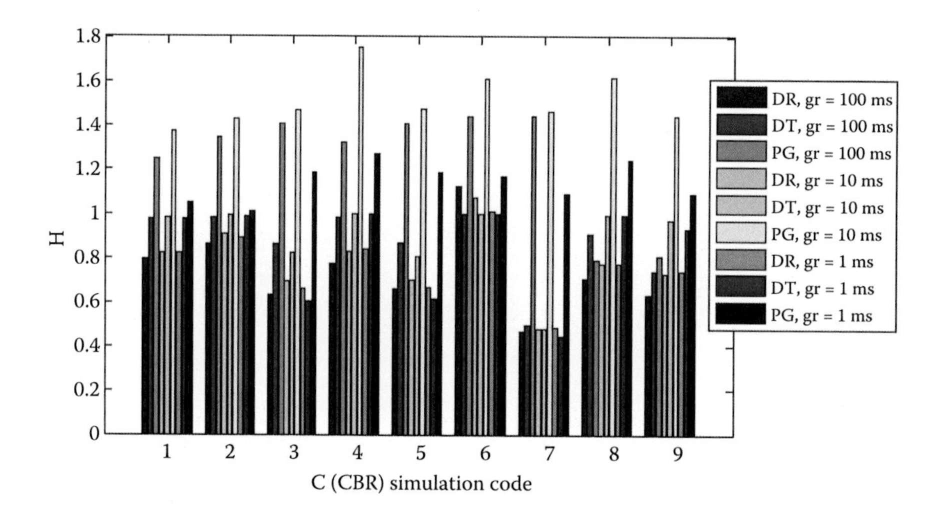

Figure 5.30 Comparison between the estimated values of the Hurst parameter for the nine traffic simulations using CBR sources.

From Figure 5.28, one can deduce that for all the studied granularities, the minimum value for the approximation is 0.8, so that the theory is verified. Yet, if we consider the granularity of 1 ms (orange, red, and brown bars), it can be seen that the minimal estimated values are sensibly larger than 0.8 for some topology arrangements. This fact seems to indicate some influence of the topology upon the traffic in the simulated network, even when the traffic is specifically designed to be self-similar.

Figure 5.31 depicts the scattering of the Hurst parameter estimations for the simulation set coded as P (Pareto traffic). It can be easily seen that the estimations cluster around the 0.8 value for all topological arrangements. One first possible conclusion is that when the nature of the traffic is explicitly self-similar, certain topologies can only amplify this phenomenon. The facts will be studied side by side in the following section.

In the case of Exponential ON/OFF traffic simulations, the estimated values proved to be surprisingly higher than the expected ones. Because the packets were sent from the queue following an exponential law, the traffic should have been more flat and produce lower values for the Hurst parameter. On the contrary, the results demonstrated the existence of statistical self-similarity but not on the same level as the Pareto traffic (Figure 5.29).

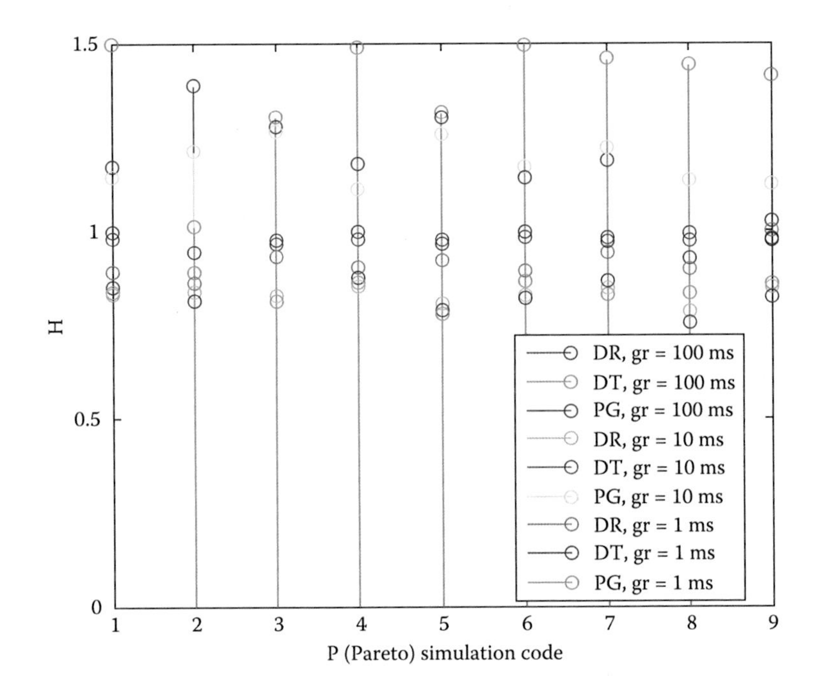

Figure 5.31 Hurst parameter estimations scattering for each Pareto simulation.

In the case of CBR traffic simulations (Figure 5.30), the topological influence proved to be even more acute than in the prior cases. More precisely, a minimum can be observed for the C7 set, this set being the combination of two random topologies. Due to the constant packet flow for CBR traffic, the self-similarity is significantly lower than in prior analyses. Practically, by keeping constant the topological generating parameters for each simulated network, differences in self-similarity levels could be highlighted for each simulated set.

By inspecting the three comparative diagrams of Hurst parameter estimations, both direct and indirect conclusions can be drawn. One first direct conclusion is that almost all traffic simulations reflect self-similar behavior in various degrees. Where Hurst estimations of over 0.75 were obtained, the long-time dependency of the respective time series was highlighted. This confirms the fact that the superposition of several processes represented in this case by traffic sources leads to long-time dependency and static self-similarity in the aggregated process (traffic). Another direct conclusion is that graphical estimators, in spite of the large size of the analyzed time series, provide different values for the Hurst parameter. It is known that the exact estimation is a difficult task, but it was expected for the different methods to provide similar values.

The secondary conclusions are related not as much with the Hurst exponent, but with the way of combining topologies that have distinct features. In other words, the focus is on an element that is yet new to scientific research, superimposing networks with topologies that are distinct on different layers. This approach opens a new way for studying networks from the fractal analysis standpoint. Although the current work only uses layers 3 and 4 (network and transport) from the OSI paradigm, the results provide a new perspective on the influence of the network structure upon its dynamic and behavior.

In other words, a new analysis method for networks in informatics is introduced, one that considers a multitude of perspectives instead of focusing only on the topology or on traffic component. Thus, the static and dynamic characteristics are brought together, conferring a global perspective, as a opposed to a local one, to the analysis.

5.5.3 *Influence of topology upon the traffic*

This section analyses several aspects related to the relation between topology and traffic. The original idea of determining whether a scale-free topology can induce self-similar traffic behavior is thus subject to experiments through simulations. By having the possibility of estimating the self-similarity level (given by the Hurst parameter), comparisons can be made between different topologies and different traffic scenarios. The conclusions of this study are reflected in a Romanian patent (Dobrescu et al. 2013) titled

"Procedure of pointing out self-similarity of Internet traffic," where the evidence for the influence of network topology on traffic behavior is provided.

For starters, a 1 ms granularity is considered to be the most relevant for a high definition estimation of the traffic self-similarity level. For the estimation to be as exact as possible, large data series are needed, and the highest quantization is used for simulations. Therefore, a traffic simulation of 200 s will comprise data series of 200,000 samples, which is enough for a good quality estimation.

In Figure 5.32, it can be seen that the estimated Hurst value is the largest for the Pareto P1 traffic type, for all three estimators. The smallest value is provided by the CBR C1 traffic. Note that the dispersion–time method produces the closest values, which could mean that the distribution of the departure time for the packets from a traffic source has a negligible effect on the self-similarity level of the aggregated traffic. Note also that the periodogram overestimates H in the cases of Pareto and exponential traffic.

A minimum and a maximum cannot be determined for the combination of scale-free nodes with small-world source topology (Figure 5.33). All estimators provide grouped close values, even in the case of the periodogram. This fact can justifies the question whether the convergence of the estimator results and the topology of the traffic sources are dependant.

From Figure 5.34, it can be deducted that the maximum value for H is produced for the Pareto traffic, while the minimum for the CBR traffic.

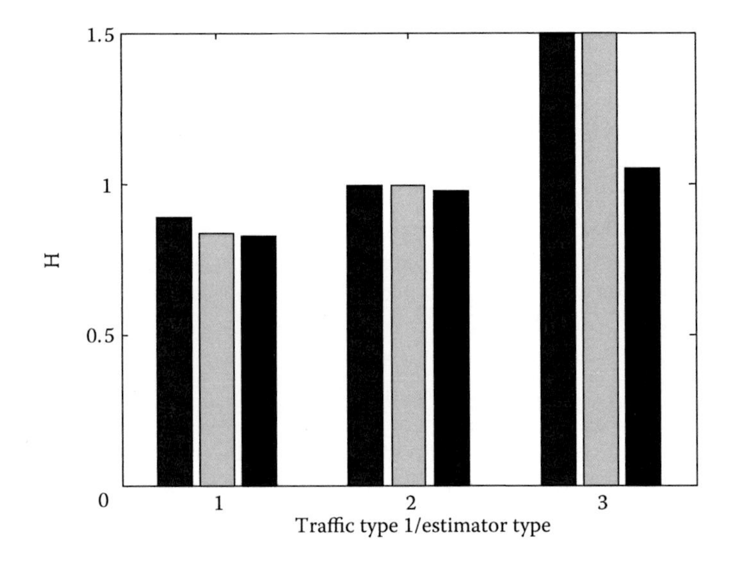

Figure 5.32 The Hurst parameter variation for a scale-free network topology combined with a random source topology; P1—blue, E1—green, C1—red; 1—domain rescaling, 2—dispersion–time, and 3—periodogram.

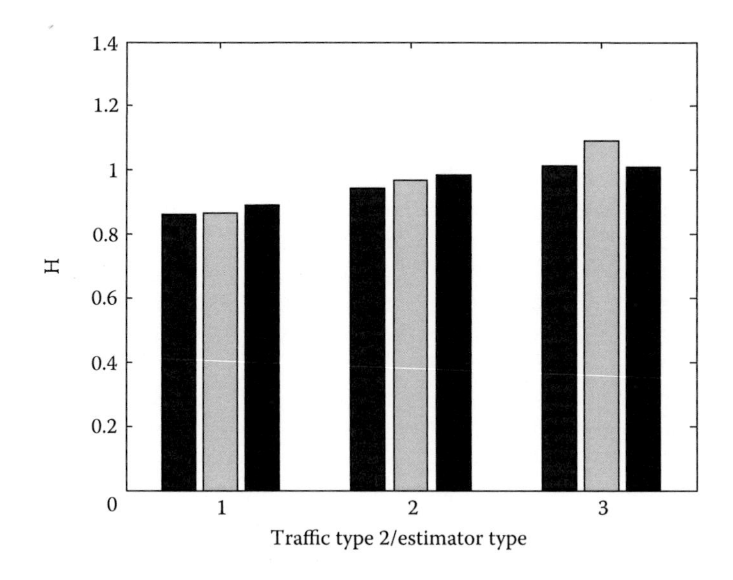

Figure 5.33 The Hurst parameter variation for a scale-free network topology combined with a small-world source topology; P2—blue, E2—green, C2—red; 1—domain rescaling, 2—dispersion–time, and 3—periodogram.

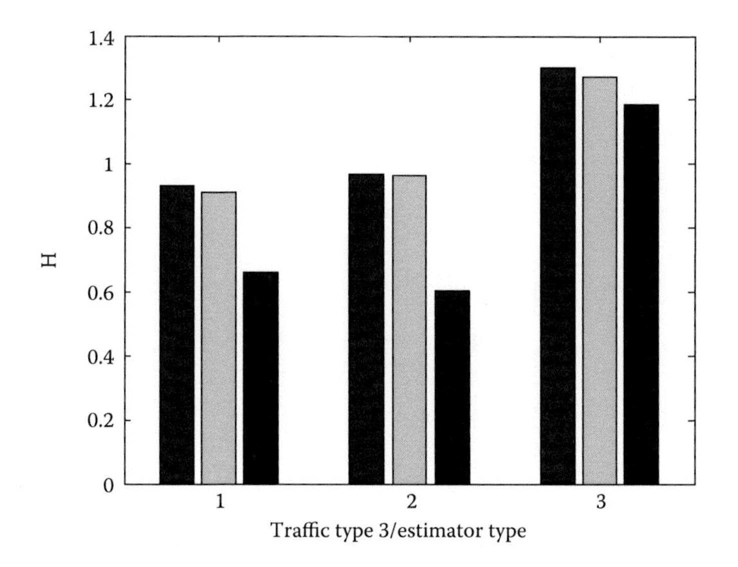

Figure 5.34 The Hurst parameter variation for a scale-free network topology combined with a scale-free source topology; P3—blue, E3—green, C3—red; 1—domain rescaling, 2—dispersion–time, and 3—periodogram.

Note that there is a significant difference (0.2) between estimations for the Pareto and the CBR traffic for domain rescaling and dispersion time.

In Figure 5.35, it can be seen that the dispersion–time method provides the same estimation independent of the chosen traffic. Also note the overestimation for the CBR traffic produced by the domain rescaling. The minimum and maximum values for the Hurst parameter cannot be determined for this topological combination.

Also in the case of Figure 5.36, no minimum and maximum values can be determined for the Hurst parameter. Note a large difference (0.2) between the Pareto and CBR traffic estimations for domain rescaling and dispersion time.

In Figure 5.37, one can remark that the Hurst parameter variation does not indicate clear minimum and maximum values and that the dispersion–time method produces identical estimations for all types of traffic sources.

For the case of the topological combination depicted in Figure 5.38, it can be seen that the maximum Hurst estimation is given by Pareto traffic while the minimum is provided by the CBR traffic. As in some previous cases, there is a significant difference (0.2–0.3) between the Pareto and CBR traffic for the domain rescaling and dispersion–time methods.

In Figure 5.39, it can be seen that there is a maximum for the Pareto traffic and a minimum for the CBR traffic. Also note that the dispersion–time method produces similar values for all three traffic types.

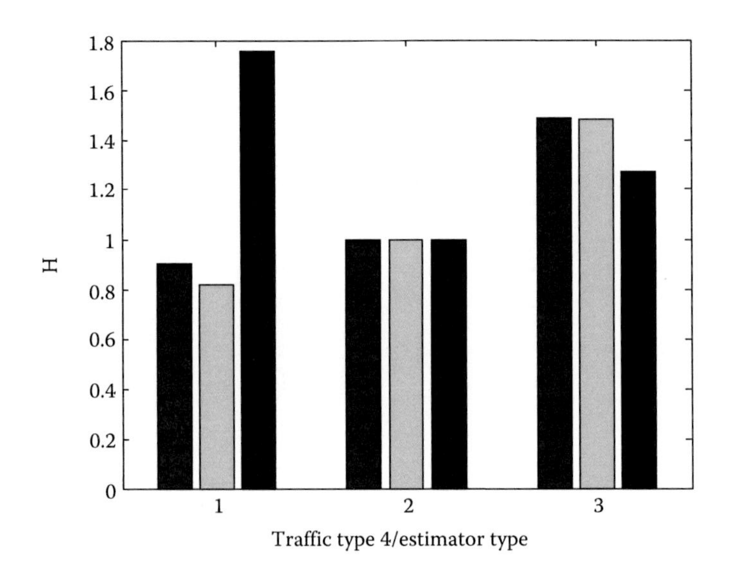

Figure 5.35 The Hurst parameter variation for a small-world network topology combined with a random source topology; P4—blue, E4—green, C4—red; 1—domain rescaling, 2—dispersion–time, and 3—periodogram.

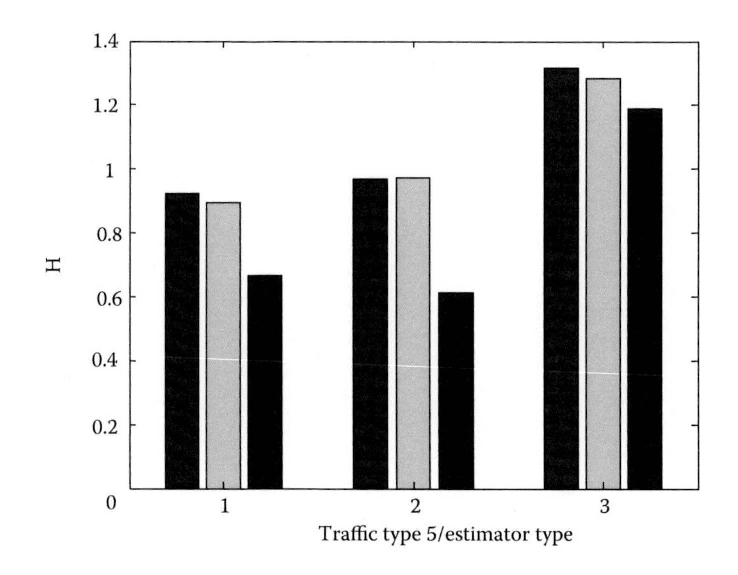

Figure 5.36 The Hurst parameter variation for a small-world network topology combined with a small-world source topology; P5—blue, E5—green, C5—red; 1—domain rescaling, 2—dispersion–time, and 3—periodogram.

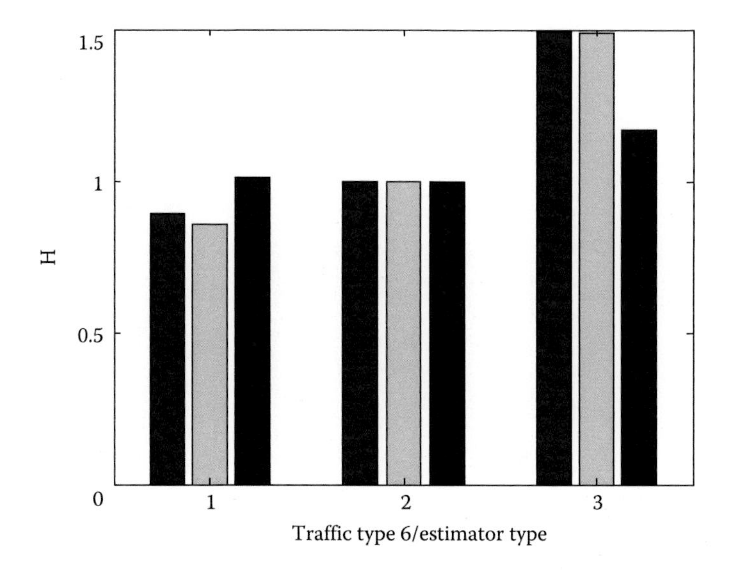

Figure 5.37 The Hurst parameter variation for a small-world network topology combined with a scale-free source topology; P6—blue, E6—green, C6—red; 1—domain rescaling, 2—dispersion–time, and 3—periodogram.

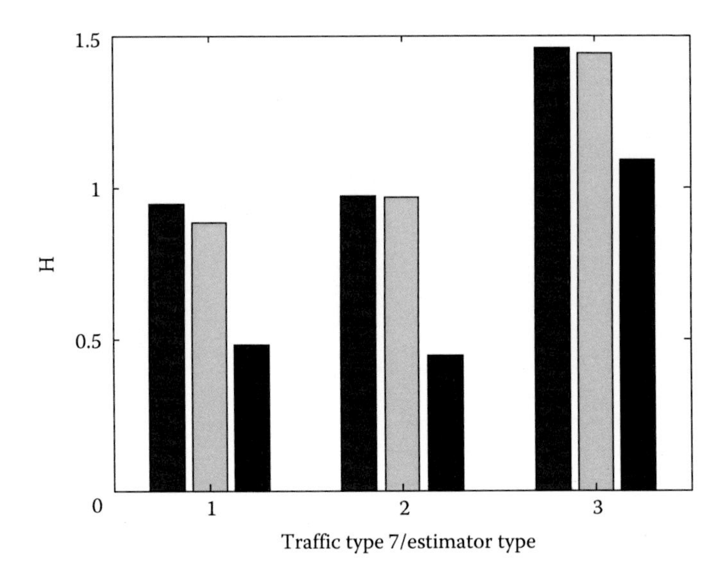

Figure 5.38 The Hurst parameter variation for a random network topology combined with a random source topology; P7—blue, E7—green, C7—red; 1—domain rescaling, 2—dispersion–time, and 3—periodogram.

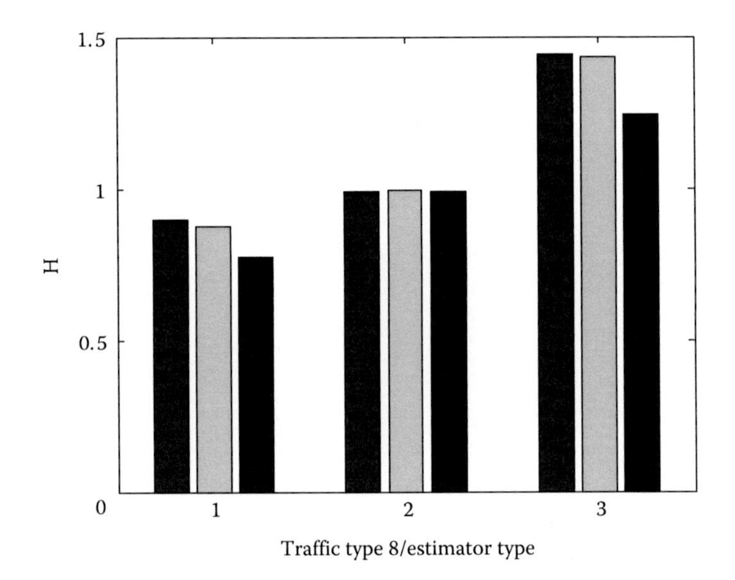

Figure 5.39 The Hurst parameter variation for a random network topology combined with a small-world source topology; P8—blue, E8—green, C8—red; 1—domain rescaling, 2—dispersion–time, and 3—periodogram.

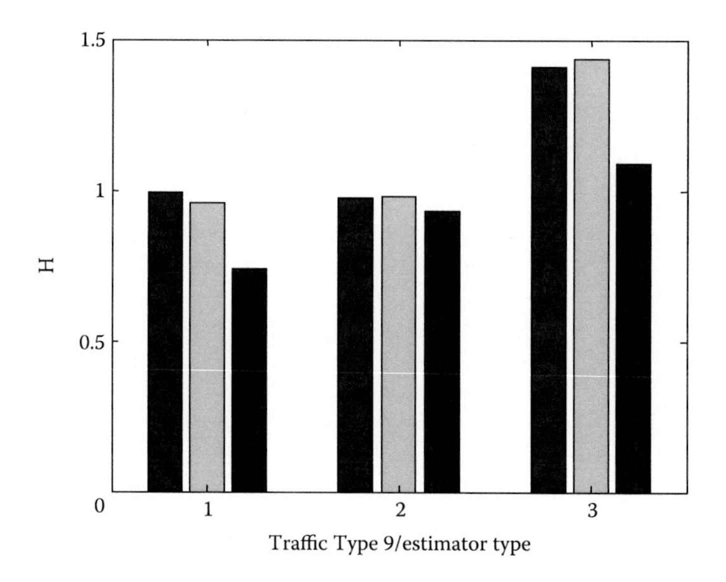

Figure 5.40 The Hurst parameter variation for a random network topology combined with a scale-free source topology; P9—blue, E9—green, C9—red; 1—domain rescaling, 2—dispersion–time, and 3—periodogram.

In the case of Figure 5.40, it can be seen that only a minimum Hurst variation can be indicated for the CBR traffic. A maximum cannot be determined.

For comparative analysis of the self-similarity level for the nine topological combinations, one single method had to be chosen from all the three available methods. Considering the limitations of each method, dispersion time was selected as the most representative one. For a self-similar traffic simulation, the domain rescaling proved to give more scattered results than the dispersion–time diagram compared to the analytically determined Hurst parameter. The periodogram was systematically overestimating H. In the meantime, in previous figures, it was shown that the dispersion–time method provides almost identical values for some cases.

Figure 5.41 depicts a Hurst estimation comparison for the nine analyzed topological combinations. The simulations were performed with ON/OFF Pareto traffic simulator. Thus, theoretically, superimposing individual sources with ON/OFF time parameters distributed after a long tail law, such as Pareto, should lead to statistical self-similarity in the aggregated traffic, which indeed can be noticed. All the values in Figure 5.41 are well over 0.9, which indicates a strong self-similar behavior.

Nevertheless, there is a variation between the estimations depending on the topological combination used. Maximum values characterize

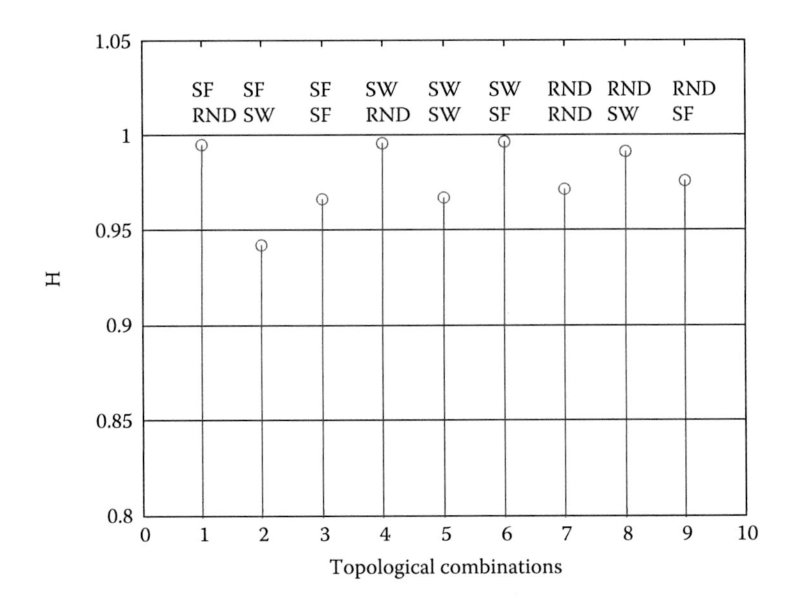

Figure 5.41 Comparative display of Hurst values for the nine topological combinations. (Estimated using the dispersion–time method for Pareto traffic generators.)

the following combinations (nodes-sources): scale-free—random, small-world—random, small-world—scale-free and random—small-world. The minimum is generated by the scale-free—small-world combination. Note that the topological impact upon the self-similarity level is higher for combinations of small-world nodes with one of the other two topologies. Combinations with two identical topologies produce smaller estimates. Therefore, self-similarity is stronger in an architecture with heterogeneous combinations and high node degrees (scale-free or small-world). A possible explanation is that for these cases the number of source-consumer links is higher.

Figure 5.42 shows a comparison between Hurst estimates for the nine traffic simulations performed with Exponential ON/OFF sources. In this case, the burst and idle time are considered to have an exponential distribution. This type of generator, although this is not the case, can be configured to act as a Poisson process. By having also in these simulations a superposition of individual sources, Hurst estimations rose to high values for almost every topological combination. Note that the maximums are similar to Figure 5.41. The combinations containing small-world and scale-free topologies prove that in this case the highest level of self-similarity is encountered in architectures with heterogeneous topological structures. Combinations between identical types produce slightly lower *H* values.

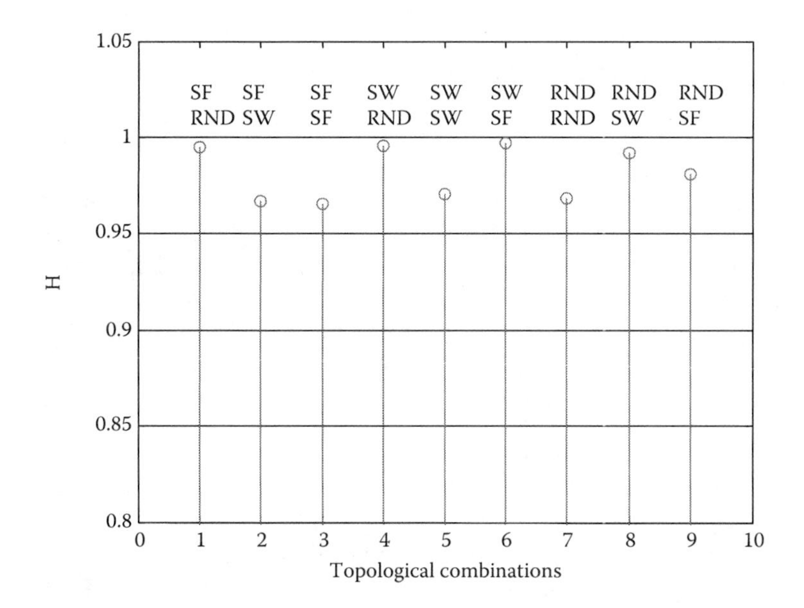

Figure 5.42 Comparative display of Hurst values for the nine topological combinations. (Estimated by using the dispersion–time method for Exponential traffic generators.)

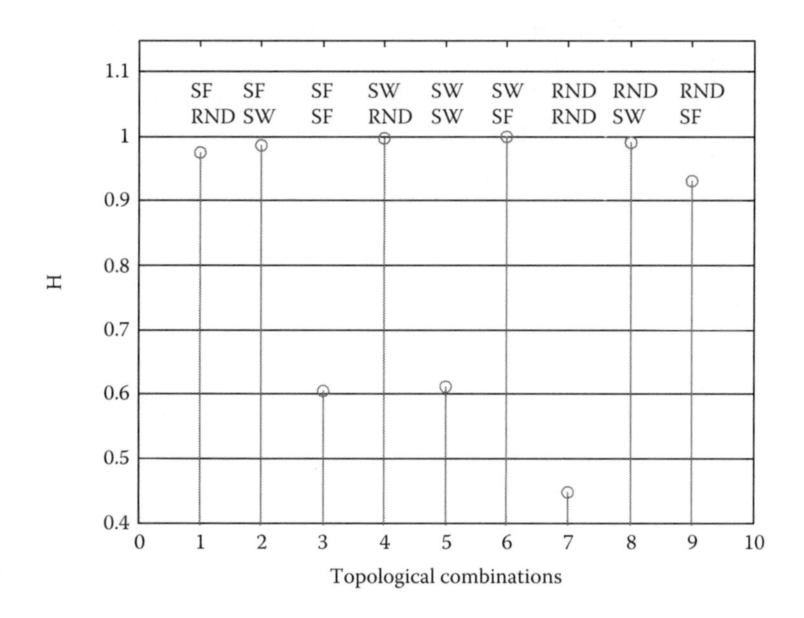

Figure 5.43 Comparative display of Hurst values for the nine topological combinations. (Estimated by using the dispersion–time method for CBR traffic generators.)

Figure 5.43 depicts the Hurst estimation diagram for the nine topological combinations having CBR traffic source. They are practically packets generators with a constant bit rate and can represent real video content streaming. By inspecting this diagram more closely, some important aspects become obvious. First, one can observe the lack of self-similarity for the cases that employ the same topology for both the nodes and the sources. If in the previous cases the phenomenon is present (Figures 5.41 and 5.42), in this case, the estimations have very small values between 0.4 and 0.6, which resembles a Poisson process rather than dependent one on the long term. The case of combining two random topologies leads to the lowest estimation of 0.44. Thus, one can note that the random topology does not influence whatsoever the self-similar behavior of the network traffic.

Maximum self-similarity levels are produced by the small-world—random, small-world—scale-free, followed by the scale-free—small-world and random—small-world combinations. Therefore, from Figures 5.41 through 5.43, one can deduce that a small-world topology for the nodes leads in every case to an increase of the self-similarity level. In the case of the CBR generators, every combination with a small-world topology (except combination number 5), maximizes the self-similarity level. Thus, the small-world topology has a greater impact upon self-similarity than a scale-free one. There are known similarities between the two topologies and almost all features are common, therefore their similar influences upon the long-term corresponding traffic are justifiable.

As a conclusion, topology plays an important role in shaping the network traffic. The fractal nature of the topology leads invariably to the increase of the traffic self-similarity. The huge difference in terms of Hurst estimation from a random topology to a small-world or scale free one clearly sustains this idea.

chapter six

Case studies

6.1 Hurst exponent analysis on real traffic

The following section presents the case study of an analysis of the Hurst exponent. It was done both to demonstrate the results of research and to highlight the value of the graphical estimators for time series with large horizons. The three previous presented graphical estimators are used: domain rescaling, time–dispersion diagram, and the periodogram. In order to implement these methods for determining the Hurst parameter, the MATLAB software package was used, which offers the user a powerful command interpreter as well as a series of predefined functions optimized for numerical computations.

The traffic to be analyzed was captured using the tcpdump utility (Brenner 1997). It spanned over approximately 5 h from a busy time interval of the day. The data capture was done between 11:38 and 16:47 on October 12, 2005 on the premises of the AccessNet, Bucharest (Rothenberg 2006). The network architecture comprises around 20 network stations connected to a router. The captured traffic was split into five parts of an hour each. Every sequence contains 36,000 measurements, and every measurement represents the number of packages sent through the network at every 100 ms. Note that, for time series of such a large horizon, the graphical estimators (domain rescaling and dispersion–time diagram) become quite powerful, thus rendering the self-similarity degree quite precise.

6.1.1 Traffic capture

The tcpdump was used for capturing packet-level traffic. The utility attaches to the network socket of the Linux kernel and copies each package that arrives at the network interface. More precisely, it allows for the network packages to be logged. This utility was used to obtain for each particular frame

- Time of arrival to kernel
- Source and destination addresses and port numbers
- Frame size (only user data are reported; in order to obtain real frame sizes, the header of each protocol should be added)
- Type of traffic (tcp, udp, icmp, etc.)

Table 6.1 The Hurst parameter estimated with the three estimator types

Set name	Time interval	No. of packets	Estimated Hurst parameter		
			R/S	D-T	Periodogram
Trace1	11:40–12:40	775.077	0.98	0.97	1.00
Trace2	12:40–13:40	547.373	0.98	0.92	0.90
Trace3	13:40–14:40	197.795	0.91	0.77	0.74
Trace4	14:40–15:40	230.468	0.82	0.82	0.84
Trace5	15:40–16:40	319.905	0.97	0.93	1.01

The time of arrival is practically the time when the kernel sees the packages and not the time of arrival at the network interface.

Note that for the first and last sets, the estimated Hurst parameter is close to 1, thus highlighting the strong self-similarity of the network traffic. For the other three sets, the estimated parameter is smaller, but different from 0.5. Judging by the time interval and the number of packets, one can conclude that the self-similarity increases when network usage is higher. In the lunch break (12–14 h) it is expected for the human activity to decrease, thus decreasing the traffic self-similarity (Table 6.1).

6.1.2 Graphical estimators representation

Hurst parameter estimation was performed only for time series composed of the number of packets that arrive in the 100 ms time interval. This chapter does not take into account the case of time series formed of the number of bytes transferred in the time unit. Anyway, estimation methods are identical in both cases.

Graphical estimator was used for graphical representation, that is, domain rescaling, dispersion–time diagram, and periodogram. Only six diagrams are provided (Figures 6.1 through 6.6), three for *Trace1* and another three for *Trace4*. Each diagram is represented on a fully logarithmic scale (both axes) and possesses a sequence of reference slopes. Thus, slopes corresponding to 0.5 and 1 values for the Hurst parameter have been drawn on the rescaled domain diagram. On the dispersion–time diagram, a line with –1 slope has been drawn, corresponding to a Hurst parameter of 0.5. The periodogram has no additional lines.

6.2 Inferring statistical characteristics as an indirect measure of the quality of service

6.2.1 Defining an inference model

In order to create a network model of a distributed environment such as the Internet, it is necessary to integrate the data collected from several

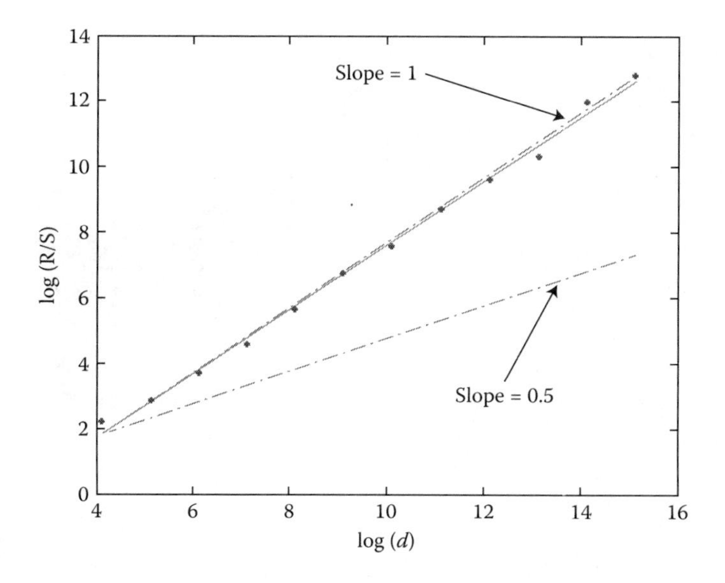

Figure 6.1 Rescaled domain diagram (R/S) for the set *Trace1.*

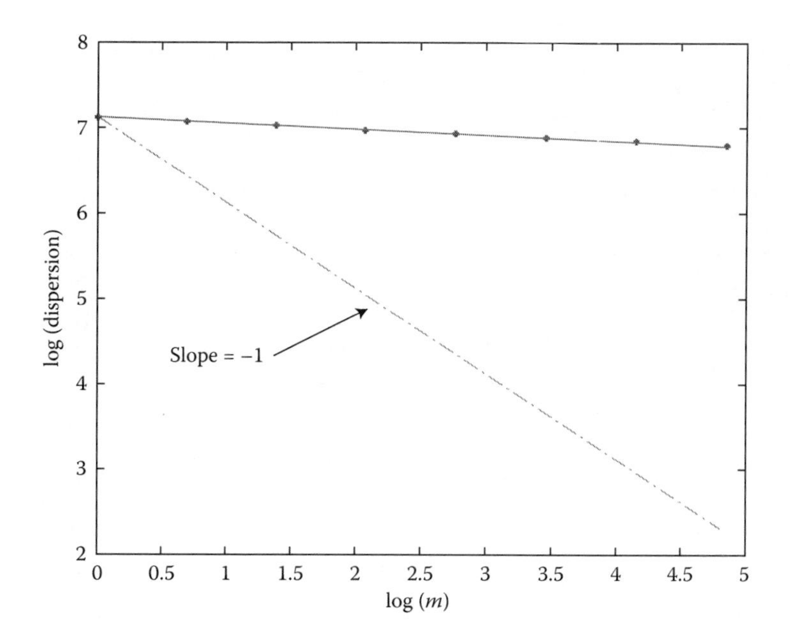

Figure 6.2 Dispersion–time diagram for the set *Trace1.*

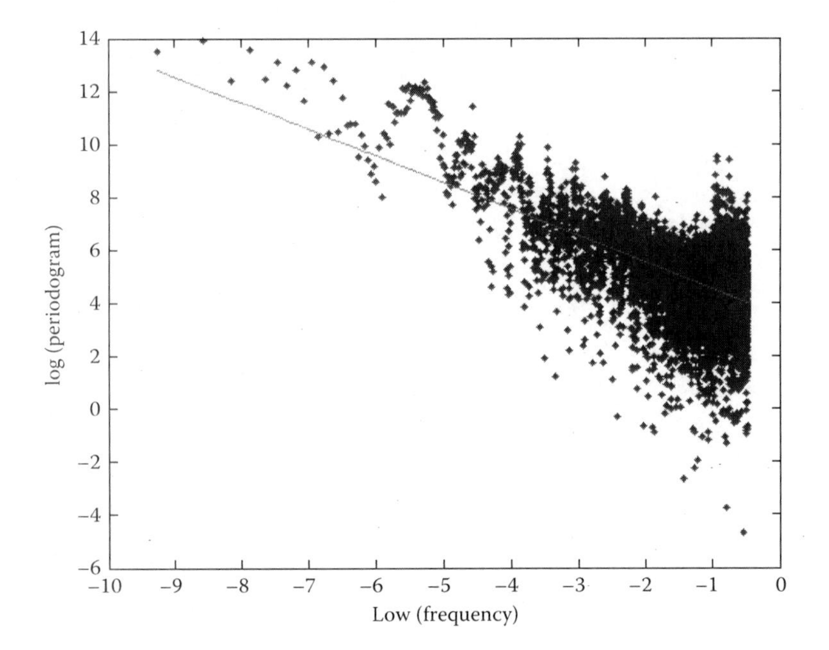

Figure 6.3 Periodogram for the set *Trace1*.

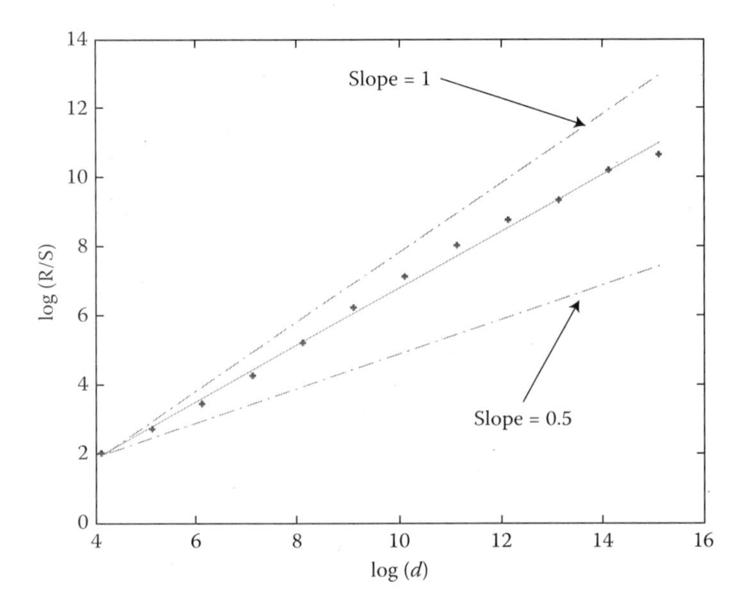

Figure 6.4 Rescaled domain diagram (R/S) for the set *Trace4*.

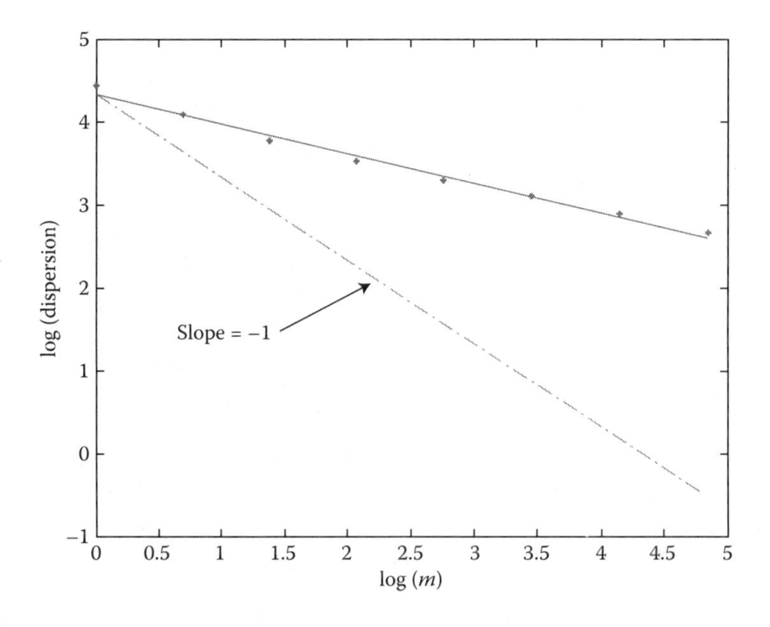

Figure 6.5 Dispersion–time diagram for the set *Trace4*.

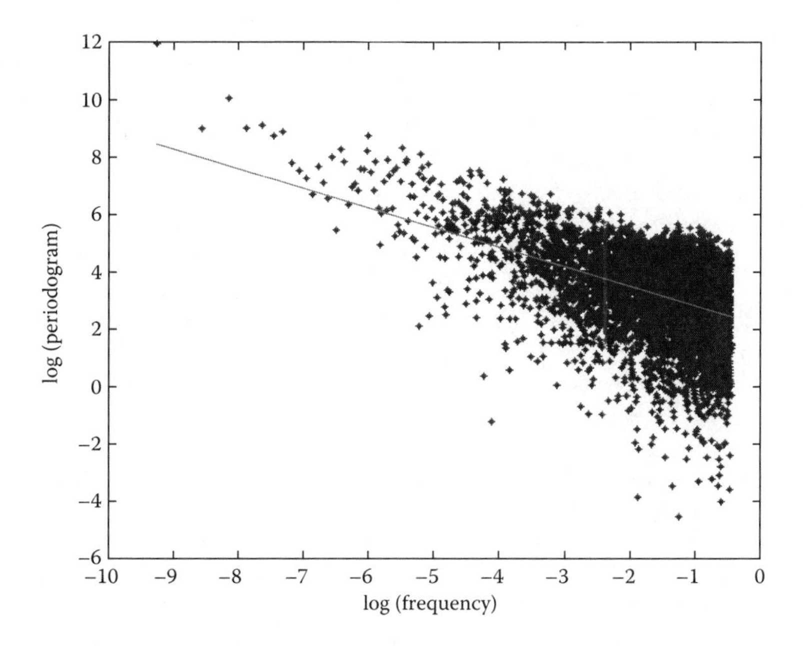

Figure 6.6 Periodogram for the set *Trace4*.

nodes. Thus, a complete image of the network is obtained for the traffic visualization level. For network management, it is essential to know the dynamic traffic characteristics, for example, for error detection (anomalies, congestion), for monitoring QoS or for selecting servers. In spite of all these, because of the very large size and of the access rights, it is costly and sometimes impossible to measure such characteristics directly. In order to tackle this problem, several methods and instruments can be used for inferring the unseen performance characteristics of the network. In other words, it is necessary to identify and develop indirect traffic measuring techniques, with estimations based on direct measurements in accessible points, in order to define the global behavior in the Internet. Because of the similarities between the inference characteristics of a network and those of a medical tomography, the concept of network tomography was conceived. The method implies the usage of a statistic inference scheme for the unseen variables based on the simultaneous measurement of other, more easy to observe, characteristics from another (sub)network. Thus, the network tomography can be seen as an inverse statistic problem that includes two types of inference: (1) inference of the internal network characteristics based on an end-to-end way (Bestavros et al. 2000) and (2) inference of the aggregated internal flows on an end-to-end way (Cao et al. 2000).

Without reducing the general level of the problem, the information network examples to be provided will refer to the Internet. The fractal nature of the Internet traffic has been determined through several measurements and statistic studies. This work presents and analyses a stochastic mono-fractal process defined as a limit of the integrated superposition of diffusion process with a differential and linear generator, which indeed is a characteristic of the fractal nature of the network traffic. Technically, the Internet is centered on the IP that supports multiple applications, whose characteristics are mixed with the resources of the network, that is, the network management packets are sent on the same channels as the data packets (there is no separate signaling network). Additionally, there are also flow control packets of the transport protocol (TCP). Thus, traffic analysis is a complex process that must satisfy the following minimal requirement:

- User traffic request evaluation, especially when having different service classes
- Network resource sizing: router processing capacity, datalink transmission flow, and interface buffer sizing
- Checking the QoS offered by the network: packet loss rate, end-to-end packet transfer delay for real-time applications, and defining the useful transport flow
- Testing the fitness level of performance models designed through analytical calculations or through simulation.

Network traffic measurements contain two categories: passive and active. For the majority of the following examples, the samples are obtained from passive probes, specifically adapted to provide traffic descriptive parameters.

The most common procedure for inferring traffic characteristics is the one using point to point links, based on the data collected from several routers via SNMP (Sikdar and Vastola 2001). Nevertheless, measuring static characteristics of aggregated flows still represents a problem open for research. The network traffic presents correlations between behaviors at different moments and different places, for various motives. It is already established that in the Internet one can observe daily self-similar patterns. Previous studies also demonstrated self-similarities that depend on the place of implementation (Wolman et al. 1999), nature of services (Lan and Heidemann 2001), and type of transiting data (Floyd and Paxson 2001).

The present approach is based on a dual modeling structure proposed in Lan and Heidemann (2002). In its simplest form, it is assumed that the network traffic (T) can be modeled as a function of three parameter sets: number of users (N), user behavior (U), and application-specific parameters (A). In other words, $T = f(N, U, A)$. User behavior parameters can be packet arrival distribution probability or the number of requested pages, while application-specific parameters can be object size distributions or the number of objects in a Web page. Experimental results show that behavior parameter distributions (U) display a correlation trend in the same network. This statement suggests that the user behavior parameter distributions at time t_2 can be modeled based on the ones at time t_1 on the same network. In addition, it is already demonstrated that the application-specific parameter distributions (A) of two networks with similar user populations are susceptible of being correlated. More precisely, it is observed that the application-specific parameter distributions of two similar networks tend to be correlated when the traffic is aggregated on a higher level. On a lower traffic level, the application-specific parameter distributions are still correlated but only in their cores. Because a higher level of aggregate traffic can be generated either by a greater number of users or by a longer measurement horizon, we can conclude that similar user populations tend to have similar behaviors in similar applications. On the contrary, for a lower aggregation level, it is more probable for some "dominant" flows to cause large variations in the distributions. Moreover, the traffic in similar networks has the same distribution model every day regarding the number of active users. In other words, the number of active users is also correlated for two similar networks in different moments of the day. The last two statements suggest that, by using a spatial correlation between two similar networks, the application-specific parameter distributions and the number of active users can be inferred based on the measurements performed on the other network.

Starting from the previous statements, the authors propose the following approaches for inferring the traffic in network n_1 based on the measurements of n_2, considering the two networks n_1 and n_2 as having similar user populations:

- The first step is to collect the traffic from both networks n_1 and n_2 over a certain period. Based on these initial measurements, the three traffic parameter sets can be established (N, U, A) for n_1 and n_2. Afterward, the derived statistics can be compared in order to determine if there exists a spatial correlation between n_1 and n_2.
- Once the similarity between n_1 and n_2 has been confirmed and the spatial correlations between them have been calculated, we can establish a model for each of the three traffic parameter sets (N, U, A) for n_1 at a given future time based only on measurements done on n_2. In particular, the number of active users (N) and the application-specific parameter distributions (A) of network n_1 can be inferred based on the measurements performed on network n_2 in the same time interval, while the application-specific parameter distributions (U) from n_1 can be inferred using the statistics derived from the initial measurements (done at step 1). Note that the first step must be performed only once.

Figure 6.7 illustrates this concept: the colored squares indicate the collected data, while in the striped squares have the inferred data. Based on the initial measurement from time t_1 that confirms the similarity between n_1 and n_2, further measurement in n_2 can be used to predict the traffic in n_1 at times t_2 and t_3.

6.2.2 Highlighting network similarity

Intuitively, two networks are "similar" if they have user populations with similar characteristics. For example, the traffic generated in two laboratories of the Faculty of Automatic Control and Computer Science may be similar because the students display common behavior when using the applications. In other words, it can be considered that two networks

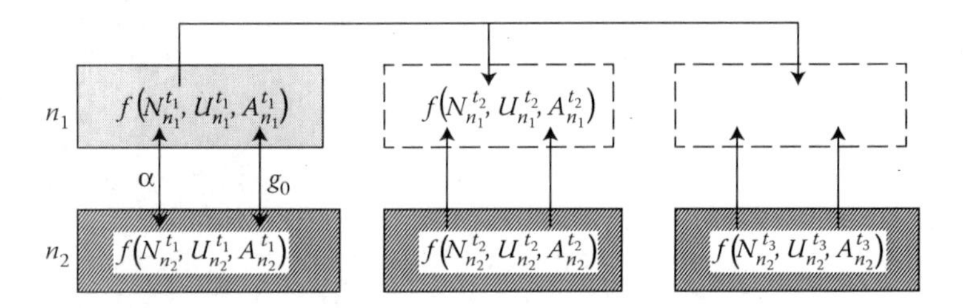

Figure 6.7 Traffic inference using network similarity.

are similar when there are correlations at the application level in the traffic. However, a formal procedure is needed in order to accomplish traffic parameters inference. First of all, the two networks will be checked for traffic similarities. Then, the distributions for user behavior and application-specific parameter will be calculated. Lastly, the statistical data of the traffic from both networks will be compared both qualitatively and quantitatively, in order to establish if the two are similar.

In order to compare the traffic quantitatively, the graphic of the cumulative distribution function (CDF) is analyzed for the distributions derived from the two networks. For a quantitative comparison, the two distributions are normalized. Three types of tests have been used in order to determine if the two distributions are significantly different regarding the average, variance, and shape. Practically, the Student *t*-test was used for comparing average, the *F* test for evaluating variance, and the Kolmogorov–Smirnov test for testing the shape. It is considered that the two distributions are strictly similar if all the three tests output a 95% trust level. Note that all the mentioned tests are exact, and thus hard to apply. In order to relax the testing procedure, approximate similarity is given by the function *s* that results as a linear combination of the average, variance, and shape differences: $s = w_1 \cdot m + w_2 \cdot v + w_3 \cdot D$,
where

$$m = \frac{|\mu(N_1) - \mu(N_2)|}{MAX(\mu(N_1), \mu(N_2)}$$

and

$$v = \frac{|\sigma(N_1) - \sigma(N_2)|}{MAX(\sigma(N_1), \sigma(N_2)}$$

In the previous formulas, μ denotes the average, v denotes the data variance, and D is the value of the Kolmogorov–Smirnov test that represents the largest absolute value between the two data sets. N_1 and N_2 are data samples taken from the two networks, while w_1, w_2, and w_3 are weights with positive values that allow the user to prioritize certain metrics. The possible values for m, V, and D are between 0 and 1. Intuitively, the two networks are more similar if the value for the function s is closer to 0.

It is clear that the application-specific parameter distributions of two networks are more susceptible of being correlated when the traffic is more intensely aggregated. In spite of this, one can observe that for a lower traffic aggregation level, there are some dominant flows that can cause large variations in the distributions tail. Such variations suggest the necessity of separately modeling the kernel and the tail, especially for traffic distributions with long distribution tails. While these dominant flows account for only a small number of the Internet flows, they consume a significant percent of bandwidth. In order to formally describe the aggregation level (G), it is defined as the product of the quantity of traffic generated by the sources (S) and the length of the measuring period (T), $G = S \times T$. Moreover, S can be described as a function of the number of users and the traffic volume generated by each user. Such a definition implies that the effect of dominant flows in the traffic tail can be reduced, if the user population is larger or if the data is collected over a longer time interval. Network similarity suggests the possibility of inferring traffic data in places where direct data acquisition is otherwise impossible. In other words, user behavior parameter distributions tend to be correlated in the same way as the application-specific parameter distributions defined denoted $T = f\,(N, U, A)$. By using such a correlation, the behavior parameter distribution can be modeled at t_2 based on measurement done at t_1 in the same network ($U_{n_1}^{t_1} \approx U_{n_2}^{t_2}$). The inference procedure obtained from Figure 6.7 can be refined as follows:

- The first step is to collect data traffic from both networks n_1 and n_2 for a certain period of time (from t_0). The three parameter sets (number of active users (N), user behavior parameter distributions (U), and application-specific parameter distributions (A)) can be calculated from traffic traces.
- The similarity between n_1 and n_2 suggests that it is highly probable for the number of users from network n_1 to be proportional to the number of users from n_2 at any moment ($N_{n1} = \alpha \times N_{n2}$ with α being a constant). By using traces collected at time t_0, the scale factor α can be calculated. Moreover, the similarity between the networks n_1 and n_2 implies the existence of some functions g_1, g_2, etc., named correlation functions, so that $\alpha_{1_{n1}} = g_1\left(\alpha_{1_{n2}}\right)$ and $\alpha_2 = g_2\left(\alpha_{2_{n2}}\right)$. After determining from the initial measurements of the spatial correlations between n_1 and n_2, the traffic from network n_1 can be predicted using only measurements from network n_2.
- Lastly, if the user behavior parameters are time correlated, they can be calculated for every moment in time in network n_1 based on the statistical data calculated initially at moment t_0. Now the relationship can be written as: $T_{n1}^{t} = f\left(N_{n1}^{t}, U_{n1}^{t}, A_{n1}^{t}\right) \approx f\left(\alpha \times N_{n2}^{t}, U_{n1}^{t0}, g\left(A_{n2}^{t}\right)\right)$.

6.2.3 *Case study: Interdomain characteristic interference*

The methodology described earlier was illustrated in a case study that wants to determine by inference the QoS inside the domain based on the measurements done at the end of one communication way. The network tomography technique is based on the assumption that end-to-end QoS time series data, denoted as Q_{e-e} are observable characteristics on every end in the domain. Additional to the Q_{e-e} there can be spatial QoS data series denoted Q_{spi}, $i = 1, \ldots, n$ that are measured on the same route in the interdomain and can be linked to Q_{e-e} for certain time intervals that are of interest for the estimation. The problem of determining the intermediate unobservable characteristics from the interdomain (hidden network) is thus formulated as follows: given time series Q_{e-e} and space series Q_{spi}—the hidden space data series $Q_{int} = Q_{e-e} - \Sigma Q_{spi}$, $i = 1, \ldots, n$ can be calculated.

The scenario for a network tomography is as follows (Figure 6.8):

1. A trust interval ($\mu_1 - \mu_2$) is chosen, which represents the difference between the two averages and contains all the values that will not be rejected in neither one of the two test hypothesis: H_0: $\mu_1 - \mu_2 = 0$ and H_1: $\mu_1 - \mu_2 \neq 0$.
2. The measured values that contributed to estimating the average are adjusted, in case they are quasi-independent (correlated, the self-correlation function, $ACF(i) \ll 1$ for any $i > 0$).
3. For normal or highly correlated distributions, the measured values are kept.

The experiments have been done on an emulation environment requiring only one PC running a Linux operating system configured as a router to create a wide variety of networks and also critical end-to-end performance conditions for different voice over IP (VoIP) traffic classes. Because there was no additional traffic, maximum and minimum values of the average are almost identical. Time series estimated for the interdomain are given in Figure 6.9.

In case the load of the network is high, spatial QoS values can be ignored, while estimating the interdomain delay with the help of the model that reflects such a delay based on an end-to-end pattern. The estimated end-to-end delay is presented in Figure 6.10.

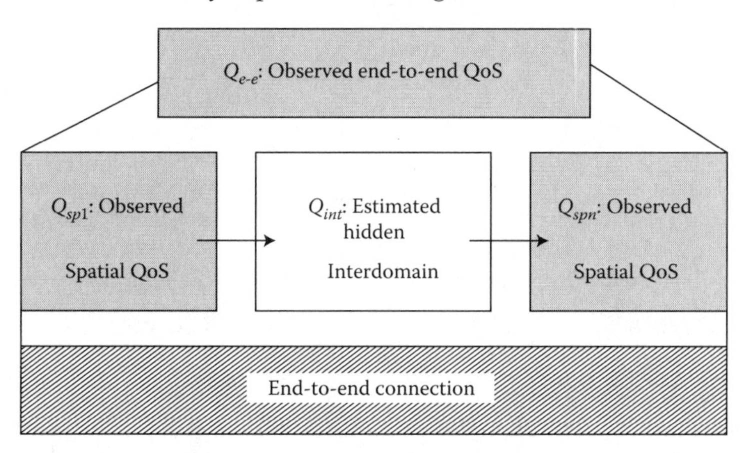

Figure 6.8 Block diagram of the network tomography.

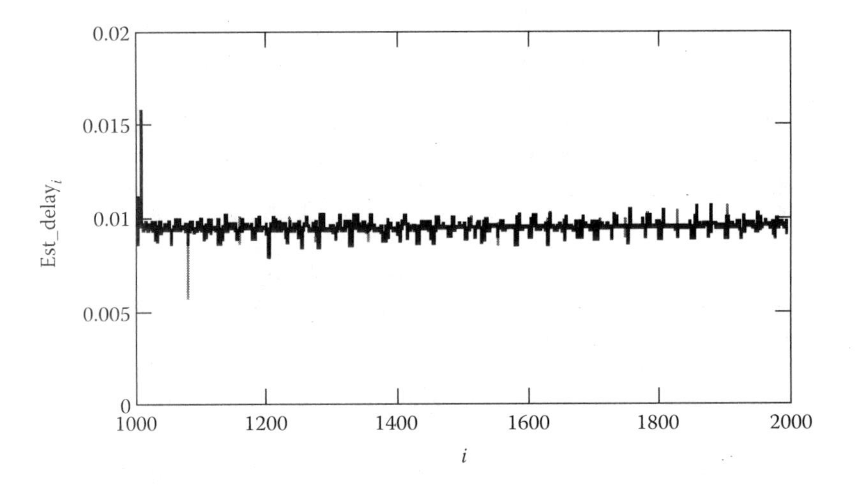

Figure 6.9 Interdomain delay estimation.

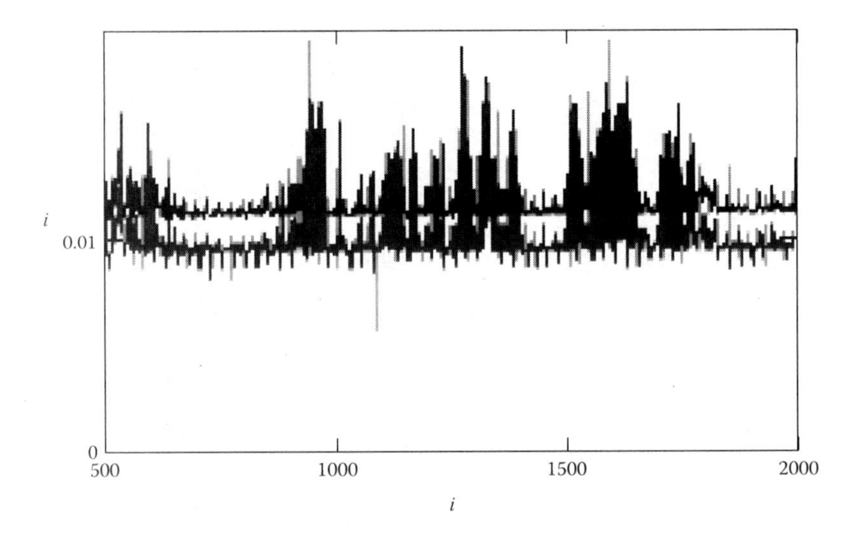

Figure 6.10 "Hidden" delay estimation in high traffic environments.

6.3 Modeling nonlinear phenomena in complex networks and detecting traffic anomalies

6.3.1 Introduction

The potential of fractal analysis for evaluating traffic characteristics, especially detecting traffic anomalies, is demonstrated in this chapter by simulation in a virtual environment. To this end, some original methods are proposed: an Internet traffic model based on graphs of scale-free networks, one method for determining self-similarity in generated traffic, a method that compares the behavior of two subnetworks and a powerful experimental testing method (using a parallel computing platform) running the *pdns* (parallel/distributed network simulator) on different subnetworks with a scale-free topology.

While the number of nodes, protocols, used ports, and new applications (e.g., Internet multimedia applications) keep increasing, it is more evident for the network administrators that there are not enough instruments for measuring and monitoring traffic and topology in high-speed networks. Even the simulation of such networks is becoming a challenge. Presently, there is no instrument to precisely simulate the Internet traffic.

There are two type of network monitoring: active and passive. During active monitoring, the routers of a domain are polled periodically in order to collect statistics about the general state of the system. This intense data volume is stored so that to extract useful information for monitoring. In the case of passive monitoring, the network is analyzed only on its contour edges. The parameters are determined by applying mathematical formulas on the set of collected data. In this case, the major problems are related to storing and processing the traffic that passes through the nodes on the contour.

Numerous network management applications have traffic processing methods for packets or bytes, differentiate according to the header fields in classes that depend on the application requirements, such as service development, security applications, system engineering, consumer billing, etc.

While for active measuring it is necessary to inject probes in the network traffic and then to extrapolate the network performance based on the performance of the injected traffic, the passive systems are based on observing and logging the network traffic. One characteristic of passive measuring systems is that they generate a large volume of measured data, obtained through inference techniques that exploit traffic self-similarity.

The network tomography (notion coined in (Vardi 1996)) tackles inference problems. It comprises of statistic inference schemes of unobservable characteristics from a network by simultaneously measuring other observable ones. The network tomography includes two typical forms: (1) inference of internal network characteristics based on end-to-end measurements (Coates and Nowak 2001; Pfeiffenberger et al. 2003) and (2) inference of the traffic behavior on the end-to-end path based on aggregated internal flows such as traffic intensity estimation at sender–receiver level based on analyzing link-level traffic measurements (Cao et al. 2000; Coates et al. 2002).

Fractal analysis seems to be suitable for both methods: direct and indirect measurement of network characteristics and anomalies. Previous work (Dobrescu et al. 2004a) is focalized on using fractal techniques in video traffic analysis. In this framework, two analytical methods have been used: self-similarity (Dobrescu et al. 2004b) and multifractal analysis (Dobrescu and Rothenberg 2005a).

The goal of this chapter is to demonstrate, through virtual environment simulations, the potential of fractal analysis techniques for measuring, analyzing, and synthesizing network information in order to determine the normal network behavior, to be used later for detecting traffic anomalies. To this end, the authors propose a model of Internet traffic based on graphs of scale-free networks where one can apply an inference based on self-similarity and fractal characteristics. This model allows for splitting a large network in several networks with quasi-similar behavior and then for applying fractal analysis for detecting traffic anomalies in the network.

The current state of rapid development of advanced technology has led to the emergence and spread of ubiquitous networks and packet data, which gradually began to force the system switching network and at the same time increase the relevance of the traffic self-similarity. The fractal analysis can solve several problems, such as the calculation of the burst for a given flow, the development of algorithms and mechanisms providing quality of service in terms of self-similar traffic and the determination of parameters and indicators of quality of information distribution.

6.3.2 *Self-similarity characteristic of the informational traffic in networks*

6.3.2.1 *SS: Self-similar processes*

Mandelbrot introduces an analogy between self-similar and fractal processes (Mandelbrot and Van Ness 1968). For incremental processes, $X_{s,t} = X_t - X_s$, the self-similarity is defined by the equation:

$$X_{t0,t0+rt} = r^H \cdot X_{t0,\,t0+t}, \quad \forall t_0,\ t, \forall r > 0 \tag{6.1}$$

Mandelbrot builds its own self-similar process (*fBm*—fractional Brownian movement) starting from two properties of the Brownian movement (*Bm*): it has independent increments and it is self-similar with a Hurst parameter $H = 0.5$.

Denoting *Bm* as $B(t)$ and *fBm* as $B_H(t)$, below is given a simplified version of the Mandelbrot *fBm*:

$$B_H(0) = 0, \quad H \in [0,\,1] \text{și}$$

$$B_H(t) = \frac{1}{\Gamma(H+0.5)}\left\{\int_{-\infty}^{0}\left[(t-s)^{H-1/2} - (-s)^{H-1/2}\right]dB(s) + \int_{0}^{t}(t-s)^{H-1/2}dB(s)\right\} \tag{6.2}$$

A self-similar process is one with long-range dependence (LRD) if $\alpha \in (0,1)$ and $C > 0$ exist so that:

$$\lim_{k \to \infty} \frac{\rho(k)}{Ck^{-\alpha}} = 1 \tag{6.3}$$

where $\rho(k)$ is the kth order self-correlation.

When presented in logarithmic coordinates, Equation 6.3 is called the correlogram of the process and has an asymptote at $-\alpha$. Note that self-similar processes exist that are not LRD, while LRD processes also exist that are not self-similar. Nevertheless, *fBm* with $H > 0{,}5$ is both self-similar and LRD.

In a paper from 1993, Leland et al. report discovering the self-similar nature of local-area network (LAN) traffic in general, and of Ethernet traffic, to be more precise. Note that all methods used in (Leland et al. 1993) (and in numerous previous works) detect and estimate LRD rather than self-similarity. The only proof given by self-similarity is the visual inspection of the data series at different time scales. Since then, self-similarity (LRD) has been reported for different kinds of traffic: LAN, WAN, variable byte rate video, SS7, HTTP, etc.

The lack of access in high-speed, highly aggregated links, as well as the lack of devices capable of measuring traffic parameters have hindered until now the study of Internet informational traffic. In principle, in this case, the traffic can differ quantitatively from the types enumerated earlier, because of factors like: too high aggregation level, end-to-end traffic policing and shaping, and a RTT (round trip time) too large for TCP sessions. Currently, some researchers stated that the Internet traffic aggregation causes a convergence toward a Poisson limit. For reasons previously presented and based on several remarks on a rather short timescale, the authors do not agree on the following: on a short timescale, the effects of network transport protocols are believed to dominate the traffic correlations; and on a long timescale, nonstationary effects such as daily load graphs become significant.

6.3.2.1.1 Self-similarity in network operation In almost all fields of network communications, it is important that the traffic characteristics are understood. In a packet exchange environment, there are no reserved resources for a connection, which would allow for the packets to reach a node with a higher rate than the processing one.

Although, in general, there can be many possible operations, they are split in just two classes: serialized (sending bits outside the link) and internal processing (all the processing is done inside the node: classifying, routing tables parsing, etc). The serialization delays are determined by the speed of the output line. Internal processing delays are random, but data structures and advanced algorithms are employed in order to obtain deterministic performances. As a consequence, the truly random part of the delay is the time a packet spends waiting. The internal processing of a node is relatively well known (through analysis or direct measurements). In contrast, incoming traffic flows are neither fully known nor controllable. Thus, it is of great importance to accurately model these flows and to predict the impact they may have upon the network performance.

It is already demonstrated that a high variability associated with LRD or SS processes can strongly deteriorate network performance, generating queue delays higher than the ones predicted with traditional models. Figure 6.11 illustrates this aspect. The average length of a queue obtained through queue simulation is very different from the one obtained through a simulation model (M/D/1).

In the case of the simple node model, the "D" (determinist) and "1" (no parallel service) hypotheses are feasible, so we would like to identify "M" (Markovian, Poisson exponential process) as one discrepancy cause. In the following section, the aim is to demonstrate that there is a "fBm/s/1" queue distribution model.

6.3.2.1.2 *Fractal analysis of Internet traffic* When the measurements have fractal properties then this characteristic is expressed through some observations. Self-similarity is a statistic property. Suppose there are (very long) data series, with finite average and variance (covariance of the stationary statistic processes). Self-similarity implies a fractal behavior: it retains its aspect over several scales.

The main characteristics, deduced from the self-similarity, are long-range dependency (LRD), slowly decreasing dispersion. The dispersion–time graph shows a slow decrease. Moreover, the self-correlation function for aggregate processes cannot be identified against the original processes.

Consider that there is available a set of data expressed as a time series. In order to find long-range dependencies, data obtained from the simulation model must be evaluated on different timescales. The same number

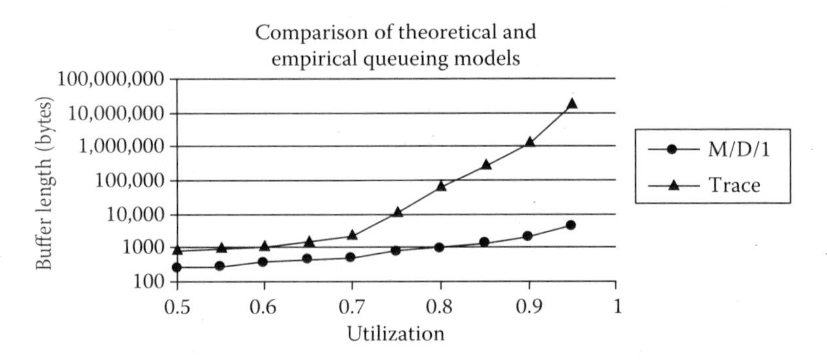

Figure 6.11 Representation limit for simulating a queue with a classical model.

of cells sample versus time must be cut in lags. Cuts should be performed several times, varying, in the meantime, the number and the cut length. The length of n lags falls in the scope future observations as well as the number of observations in a lag. For each n, a number of lags is randomly selected. This lag number must be identical for all values of n. For the selected lags, two parameters are calculated:

- $R(n) = \max(0, W_1, W_2,\ldots, W_n) - \min(0, W_1, W_2,\ldots, W_n)$ cu $W_i = (X_1 + X_2 + \cdots + X_i) - i\bar{x}(n)$ that is the lag sample range
- $S(n)$ – dispersion of $\{X_1 + X_2 + \cdots + Xn\}$ of a lag

For the short-range dependence sets, the expected value for $E[R(n)/S(n)]$ is approximately equal to $c_0 n^{1/2}$. On the contrary, for the sets with long-range dependency, $E[R(n)/S(n)]$ is approximately $c_0 n^H$ with $0{,}5 < H < 1$. H is the Hurst parameter and c_0 is a constant of relative minor importance.

Hurst designed a normalized, dimensionless measure for characterizing variability, named the rescaled domain (R/S). For an observation data set $X = \{X_n, n \in Z^+\}$ with a sample average $\overline{X}(n)$, sample dispersion $S^2(n)$, and domain $R(n)$, the adjusted rescaled domain or the R/S statistics are given by

$$\frac{R(n)}{S(n)} = \frac{\max(0,\Delta_1,\Delta_2\ldots\Delta_n) - \min(0,\Delta_1,\Delta_2\ldots\Delta_n)}{S(n)}$$

where

$$\Delta_k = \sum_{i=1}^{k} X_i - k\overline{X} \quad \text{for } k = 1,2,\ldots,n$$

For many natural phenomena, $E[R(n)/S(n)] \sim cn^H$ when $n \to \infty$, c being a positive constant, independent of n. By introducing logarithm on both sides, the result is $\log\{E[R(n)/S(n)]\} \sim H \log(n) + \log(c)$ for $n \to \infty$. Thus, the value of H can be estimated by graphically representing $\log\{E[R(n)/S(n)]\}$ as a function of $\log(n)$ (POX diagram), and approximating the resulted points with a straight line of slope H.

The R/S method is not precise; it does only estimate the self-similarity level of a time series.

Figure 6.12 contains an example of a POX diagram. H is the Hurst parameter. If there is no fractal behavior, then H is close to 0.5, while a value higher than 0.5 suggests self-similarity and fractal character.

6.3.3 *Using similarity in network management*

6.3.3.1 *Related work on anomalies detection methods*

One of the main goals of fractal traffic analysis is to show the potential to apply signal processing techniques to the problem of network anomaly

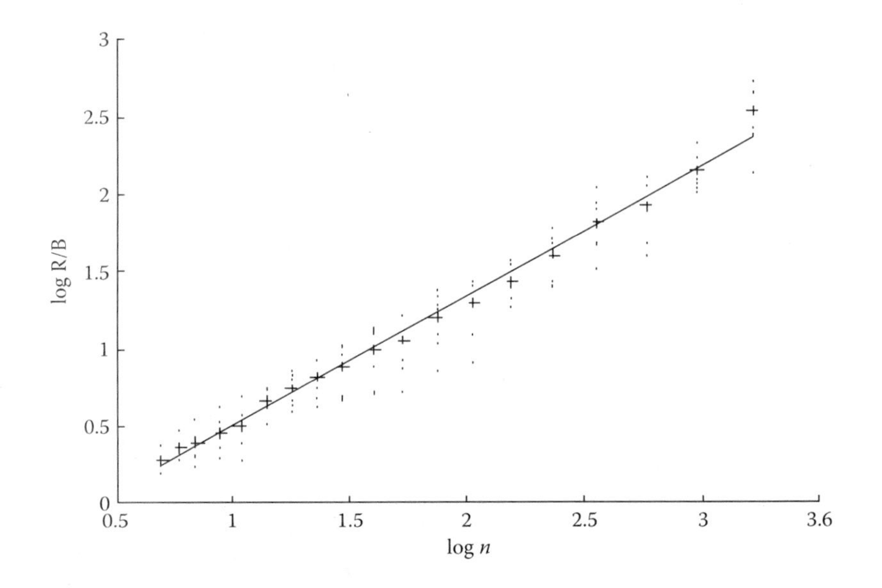

Figure 6.12 POX diagram example.

detection. Application of such techniques will provide better insight for improving existing detection tools as well as provide benchmarks to the detection schemes employed by these tools. Rigorous statistical data analysis makes it possible to quantify network behavior and, therefore, more accurately describe network anomalies. In this context, network behavior analysis (NBA) methods are necessary. Information about a single flow does not usually provide sufficient information to determinate its maliciousness. These methods have to evaluate the flow in the context of the flow set. There are several different approaches that can be classified into groups depending on the field they utilize for the detection.

6.3.3.1.1 Statistical-based techniques These techniques are widely used for network detection. Flow data represent the behavior of the network by statistical data, like a traffic rate, packet, and flow count for different protocols and ports. Statistical methods typically operate on the basis of a model and keep one traffic profile representing the normal or learned network traffic and a second one representing the current traffic. The anomaly detection is based on comparison of the current profile to the learned one, significant deviation of these two profiles means network anomaly and a possible attack (Garcia-Teodoro et al. 2009).

6.3.3.1.2 Knowledge-based techniques This category includes so-called expert system approach. Expert systems classify the network data according to a set of rules, which are deduced based on the training data. Specification-based anomaly methods use a model, which is manually constructed by a human expert, in terms of a set of rules (the specifications) that determine the normal system behavior.

6.3.3.1.3 Machine learning-based techniques Several tools can be included in this category, the most representative being the following.

Bayesian networks, which are probabilistic graphical models that represent a set of random variables and their conditional dependencies in a form of directed acyclic graph. The Bayesian network learns the casual relations between attributes and class labels from the training data set before it can classify unknown data.

Markov models include two different approaches: Markov chains and hidden Markov models. A Markov chain is a set of interconnected through certain transition probabilities, which determine the topology and the capabilities of the model. The probability of transition is determined during training phase and reflects the normal behavior of the system. Anomalies are detected by comparing the probability of observed sequences with a fixed threshold. The second hidden Markov model assumes that the system of interest is a Markov process with hidden transitions and states. Only the so-called productions are observable.

Decision tree, which can efficiently classify anomalous data. The root of a decision tree is the first attribute with test conditions that split input data toward one of the internal nodes depending on the characteristics of the data. The decision tree has to be first trained with known data before it can be used to classify unknown data.

Fuzzy logic techniques, which are derived from fuzzy set theory. Fuzzy methods can be used as an extension to other approaches like in case of the intrusion detection using fuzzy hidden Markov models. The main disadvantage is high resource requirements.

Clustering and outer detection, which are methods used to separate a set of data into groups which members are similar or close to each other according to given distance metric.

The abovementioned methods for anomaly detection in computer networks have been studied by many researchers.

One of the first surveys about anomaly detection was presented in Coates et al. (2002), which proposed a taxonomy in order to classify anomaly detection systems. On the other hand, it focused on classifying commercial products related to behavioral analysis and proposed a system that collects data from SNMP objects and organizes them as a time series.

An autoregressive process is used to model those time series. Then, the deviations are detected by a hypothesis test based in the method generalized likelihood ratio (GLR). The behavior deviations that were detected in each SNMP object are correlated later according to the objects' characteristics. The first component of the algorithm applies wavelets to detect abrupt changes in the levels of ingress and egress packet counts. The second one searches for correlations in the structures of ingress and egress packets, based on the premise of traffic symmetry in normal scenarios. The last component uses a Bayes network in order to combine the first two components, generating alarms.

In a notable paper (Li et al. 2008), the authors analyzed the traffic flow data using the technique known as principal component analysis (PCA). It separates the measurements into two disjoint subspaces: the normal subspace and the anomalous subspace, allowing the anomaly detection for characterizing the normal network-wide traffic using a spatial hidden Markov model (SHMM), combined with topology information. The CUSUM algorithm (Cumulative Sum) was applied to detect the anomalies. This work proposes the application of simple parameterized algorithms and heuristics in order to detect anomalies in network devices, building a lightweight solution. Besides, the system can configure its own parameters, meeting network administrator's requirements and decreasing the need for human intervention in management.

Brutlag (2000) used a forecasting model to capture the history of the network traffic variations and to predict the future traffic rate in the form of a confidence band. When the variance of the network traffic continues to fall outside of the confidence band, an alarm is raised. Barford et al. (2002) employed a wavelet-based signal analysis of flow traffic to characterize single-link byte anomalies. Network anomalies are detected by applying a threshold to a deviation score computed from the analysis. Thottan and Ji take management information base (MIB) data collected from routers as time series data and use an auto-regressive process to model the process. Network anomalies are detected by inspecting abrupt changes in the statistics of the data (Thottan and Ji 2001). Wang et al. (2002) take the difference in the number of SYNs (beginning messages) and FINs (end messages) collected within one sampling period as time series data and use a nonparametric cumulative sum method to detect SYN flooding by detecting the change point of the time series. While these methods can detect anomalies, which cause unpredicted changes in the network traffic, they may be deceived by attacks that increase their traffic slowly.

Let us look at other studies related to the use of the machine learning techniques to detect outliers in data sets from a variety of fields. Ma and Perkins present an algorithm using support vector regression to perform online anomaly detection on time-series data (Ma and Perkins 2003). Ihler et al. present an adaptive anomaly detection algorithm that is based on a Markov-modulated Poisson process model, and use Markov chain Monte Carlo methods in a Bayesian approach to learn the model parameters (Ihler et al. 2006).

6.3.3.2 Anomaly detection using statistical analysis of SNMP–MIB

Detecting anomalies is the main reason for monitoring the network. Anomalies represent deviations from the normal behavior. For a basic definition, anomalies can be characterized by correlating changes in measured data during abnormal events (Thottan and Ji 2003). The term "transient changes" refer to abrupt modifications of measured data. Anomalies can be classified into two categories. The first category is related to performance problems and network failure. Typical examples of performance anomalies are file server defects and congestions. In certain cases, software problems can manifest as network anomalies, such as protocol implementation errors that determine increases or decreases of the traffic load characteristic. The second category of anomalies is related to security problems (Wang et al. 2002).

Probing instruments facilitate an instantaneous measurement of the network behavior. Network test samples are obtained with specialized instruments such as *ping* or *traceroute* that can used to obtain specific parameters such as packet loss or end-to-end delays.

These methods do not necessitate cooperation between network services providers, but imply a symmetrical route between source and destination. On the Internet, this assumption cannot be guaranteed and thus data obtained through the probing mechanism can be limited in the case of anomaly detection. Other methods have been used for determining the traffic behavior: data routing protocols (Aukia et al. 2000), packet filtering (Cleveland et al. 2000), and data from network management protocols (Stallings 1994). This last category includes simple network management protocol (SNMP) that assures the communication mechanism between a server and hundreds of SNMP agents in the network equipment. The SNMP server holds certain variables, management information base (MIB) ones, in a database. Because of the accuracy provided by the SNMP, it is considered an ideal data source for detecting network anomalies. The most used anomaly detection methods are the rule-based approach (Franceschi et al. 1996), finite state machine models (Lazar et al. 1992), pattern matching (Clegg 2004), and statistical analysis (Duffield 2004). Among these methods, statistical analysis is the only one capable of continually following the network behavior and does not need significant recalibration. For this reason, the proposed anomaly detection method employs statistical analysis based on correlation and similarity in SNMP MIB variable data series.

Network management protocols provide information about network traffic statistics. These protocols support variables that correspond to traffic counts at the device level. This information from the network devices can be passively monitored. The information obtained may not directly provide a traffic performance metric but could be used to characterize network behavior and, therefore can be used for network anomaly detection. In the following will be presented the details of the method for anomaly detection proposed by Thottan and Ji (2003).

Using this type of information requires the cooperation of the service provider's network management software. However, these protocols provide a wealth of information that is available at very fine granularity. The following subsections will describe this data source in greater detail.

6.3.3.2.1 Simple Network Management Protocol (SNMP) SNMP works in a client–server paradigm. The protocol provides a mechanism to communicate between the manager and the agent. A single SNMP manager can monitor hundreds of SNMP agents that are located on the network devices. SNMP is implemented at the application layer and runs over the UDP. The SNMP manager has the ability to collect management data that is provided by the SNMP agent but does not have the ability to process this data. The SNMP server maintains a database of management variables called the management information base (MIB) variables. These variables contain information pertaining to the different functions performed by the network devices. Although this is a valuable resource for network management, we are only beginning to understand how this information can be used in problems such as failure and anomaly detection.

Every network device has a set of MIB variables that are specific to its functionality. MIB variables are defined based on the type of device as well as on the protocol level at which it operates. For example, bridges that are data link-layer devices contain variables that measure link-level traffic information. Routers that are network-layer devices contain variables that provide network-layer information. The advantage of using SNMP is that it is a widely deployed protocol and has been standardized for all different network devices. Due to the fine-grained data available from SNMP, it is an ideal data source for network anomaly detection.

6.3.3.2.2 SNMP–MIB Variables The MIB variables fall into the following groups: system, interfaces (*if*), address translation (*at*), Internet protocol (*ip*), Internet control message protocol (*icmp*), transmission control protocol (*tcp*), user datagram protocol (*udp*), exterior gateway protocol (*egp*), and simple network management protocol (*snmp*). Each group of variables describes the functionality of a specific protocol of the network device.

Depending on the type of node monitored, an appropriate group of variables can be considered. If the node being monitored is a router, then the *ip* group of variables are investigated. The *ip* variables describe the traffic characteristics at the network layer. MIB variables are implemented as counters. Time-series data for each MIB variable is obtained by differencing the MIB variables at two subsequent time instances called the polling interval.

There is no single MIB variable that is capable of capturing all network anomalies or all manifestations of the same network anomaly. Therefore, the choice of MIB variables depends on the perspective from which the anomalies are detected. For example, in the case of a router, the *ip* group of MIB is chosen, whereas for a bridge, the *if* group is used.

As the network evolves, each of the methods for statistical analysis described above requires significant recalibration or retraining. However, using online learning and statistical approaches, it is possible to continuously track the behavior of the network. Statistical analysis has been used to detect both anomalies corresponding to network failures, as well as network intrusions. Interestingly, both of these cases make use of the standard sequential change point detection approach.

Abrupt changes in time-series data can be modeled using an autoregressive (AR) process. The assumption here is that abrupt changes are correlated in time, yet are short-range dependent. In our approach, we use an AR process of order to model the data in a 5-min window. Intuitively, in the event of an anomaly, these abrupt changes should propagate through the network, and they can be traced as correlated events among the different MIB variables. This correlation property helps distinguish the abrupt changes intrinsic to anomalous situations from the random changes of the variables that are related to the network's normal function. Therefore, network anomalies can be defined by their effect on network traffic as follows: Network anomalies are characterized by traffic-related MIB variables undergoing abrupt changes in a correlated fashion. Abrupt changes in the generated traffic are detected by comparing the variance of the residues obtained from two sequential data windows: learning and testing (Figure 6.13). Residues are obtained by imposing an AR model on the time series in each window.

It can be seen that in the learning windows there is a change, while in the test one the correlogram remains similar.

6.3.3.3 *Subnetwork similarity testing*

It can be said that two networks are similar if there are correlations at the application level of the traffic. In order to demonstrate similarity, the method proposed by Lan and Heidemann (2005) is used, which employs a global weight function s to determine if two distributions are significantly

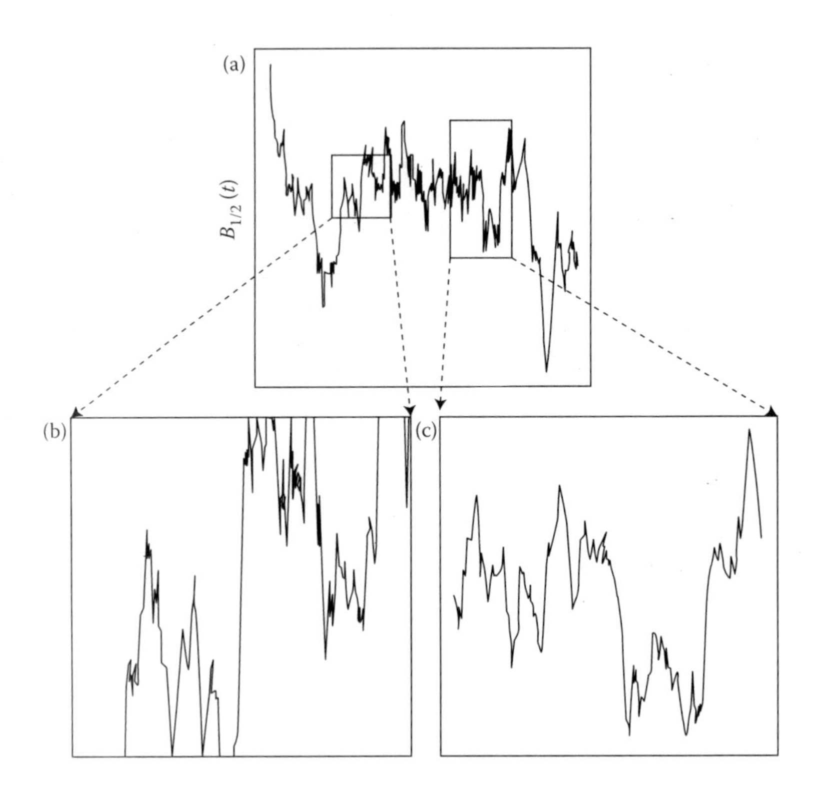

Figure 6.13 Detecting abrupt changes.

similar regarding average, dispersion, and shape: $s = w_1 \cdot A + w_2 \cdot B + w_3 \cdot C$, where

$$A = \frac{|\mu(N_1) - \mu(N_2)|}{\text{Max}(\mu(N_1), \mu(N_2))} \quad B = \frac{|\sigma(N_1) - \sigma(N_2)|}{\text{Max}(\sigma(N_1), \sigma(N_2))}$$

where μ is the average, σ is the dispersion, and C is the Kolmogorov–Smirnov distance (the greatest absolute difference between the cumulative distributions of the two data sets).

N_1 and N_2 are samples that come from two different networks, and w_1, w_2, and w_3 are positive, user-defined weights that allow he or she to prioritize some particular metrics (average, dispersion, or shape). Intuitively, two networks are similar if s is close to zero. This method was tested considering the Web traffic simulated on an SFN Internet model (Ulrich and Rothenberg 2005). After generating a large network model, a federalization is performed and the similarities are checked in pairs of subnetworks.

6.3.4 Test platform and processing procedure for traffic analysis

6.3.4.1 Traffic generation

Traffic generation is an essential part of the simulation. Randomly, one to three simultaneous traffic connections are initiated from the client networks, and, for simplicity, ftp sessions are used to randomly select destination servers. It is decided that the links connecting routers must have speed superior to that of the link connecting clients to routers, for example, server-router 1 Gbps, client-router 10 Mbps, router-router 10, 100 Mbps or 1 Gbps depending on the router type. The code generated respecting these two conditions is added to the network description file in order to be processed by the simulator.

A modular solution is used, which permits reusing various components from different parts of the simulator. The same network type generated by the initial script can be used for both single-CPU and distributed simulation, thus offering a comparison between the two simulation setups.

NS2 is a network simulator based on discrete events, discussed already in Sections 4.3.2.3 and 5.1. The current version of NS (version 2) is written in C++ and OTcl (Chung 2011). Multi CPU simulations used the *pdns* extension of NS2, which has a syntax close to NS2. The main difference is in the number of extensions necessary for parallelizing so that the various instances of *pdns* to be able to communicate one to another and create the global image of the network to be simulated. All simulation scenarios run for 40 s.

6.3.4.2 Cluster description

In order to obtain a parallel/distributed environment, it was necessary to build a cluster running Linux. The cluster can run applications using a virtualization environment. Parallel virtual machine (PVM) was used for developing and testing applications. It is a framework of software packages that facilitate the creation of a machine running on multiple CPUs by using available network connections and specific libraries (Grama et al. 2003). The cluster comprises of a main machines and a total of six cluster nodes. The main machine provides all necessary services for the cluster nodes, that is, IP addresses, boot, NFS for data storage, tools for updating, managing, and controlling the images used by the cluster nodes, as well as the access to the cluster. The main computer provides an image of the operating system to be loaded on each of the cluster nodes, because they do not have their own storage. Because the images reside in the memory of each cluster, some special steps are necessary to reduce the dimension of the image and to provide as much free memory as possible for running simulation processes. The application partition is read-only, while the data partition is read-write and accessible to users on all machines, in a similar manner, providing transparent access to user data. In order to access the cluster, the user must connect to a virtual server on the main machine that can behave like a cluster node when it is not needed as an extra computer.

6.3.4.3 *Network federalization*

In order to use the *pdns* simulator, the network must be subdivided in quasi-independent subnetworks (Wilkinson and Pearson 2005). Each *pdns* instance addresses a specific subnetwork, thus the dependencies between them are minimal; there must be as few links as possible between nodes from different subnetworks.

The approach of simulating on a federalized network is used. A federalization algorithm was devised and implemented in order to separate the starting network in several smaller subnetworks. The algorithm generates n components. The *pdns* script takes as input the generated network and the federalization. Depending on the network connectivity, they are assigned different roles—router or end-user—and corresponding traffic scenarios are associated. Also, a different approach of partitioning a NS script in several *pdns* is employed, using *autopart* (Riley et al. 2004), a tool for simulating the partitioning based on the METIS partitioning package (METIS). *Autopart* takes a NS2 script and creates a number of *pdns* scripts that are ready to run in parallel on several machines.

6.3.5 *Discussion on experimental results of case studies*

6.3.5.1 *Implementing and testing an SFN model for Internet*

A scale-free network (SFN) is a model where the inference based on self-similarity and fractal characteristics can be applied. Scale-free networks are complex networks, in which a few nodes are heavily connected while the majority of the nodes have a very small number of connections. Scale-free networks have a degree distribution of $P(k) = k^{(-\lambda)}$ where λ can vary from 2 to 3 for the majority of real networks.

Numerous networks display a scale-free behavior including the Internet.

An algorithm was designed and implemented for generating those subsets of scale-free networks that are close to real ones, such as the Internet. This application is able to handle a very large collection of nodes, to control the generation of network cycles and the number of isolated nodes. It was written in Python, thus making it portable. The algorithm starts with a manually created network of a few nodes. By using increase and preferential attachment algorithms new ones are added. An original component is introduced calculating ahead a number of nodes for each level. Preferential attachment is followed by applying the restriction of having the optimal number of nodes per degree.

Table 6.2 Simulations time for a network model of 100,000 nodes and high traffic

	Number of nodes used in cluster					
Runs	1	2	3	4	5	6
1	Fail	319	338	135	173	165
2	Fail	343	357	140	176	171
3	Fail	347	351	134	177	166
4	Fail	316	347	139	177	165
5	Fail	308	320	138	178	163
Avg.	–	326.6	342.6	137.2	176.2	166

For tests, it was decided to run 40 s traffic simulations for a scale-free networks model with 10,000 nodes. On such a scale, one-node processing is not possible, because the cluster node runs out of memory. In order to obtain valid results, the simulation was run on a different, more stronger machine, with more memory and virtual memory. Table 6.2 shows the time used for simulations(s) for a network model of 100,000 nodes and high traffic.

This scenario was simulated for five times in similar load conditions. During this simulation, it was noted that by adding more nodes (in our case, more than 4), the simulation process is slower. Another remark is that the 2-CPU simulation is swifter than the 3-CPU one, although the optimal number of nodes is not 2.

6.3.5.2 *Detecting network anomalies*

The statistical technique presented was applied for detecting anomalies for SNMP MIB. A summary of the results obtained through statistical techniques is presented in Table 6.3.

The types of observed anomalies are the following: server disfunctionalities (errors, failures), protocol implementation errors, network access problems. For example, in the case of a server error, the defect was

Table 6.3 Summary of results obtained through statistical techniques

Type of anomaly	Mean time of prediction (min)	Mean time of detection (min)	Mean time between false alarms (min)
Server failures	26	30	455
Access problems	26	23	260
Protocol error	15	–	–
Uncontrollable process	1	–	235

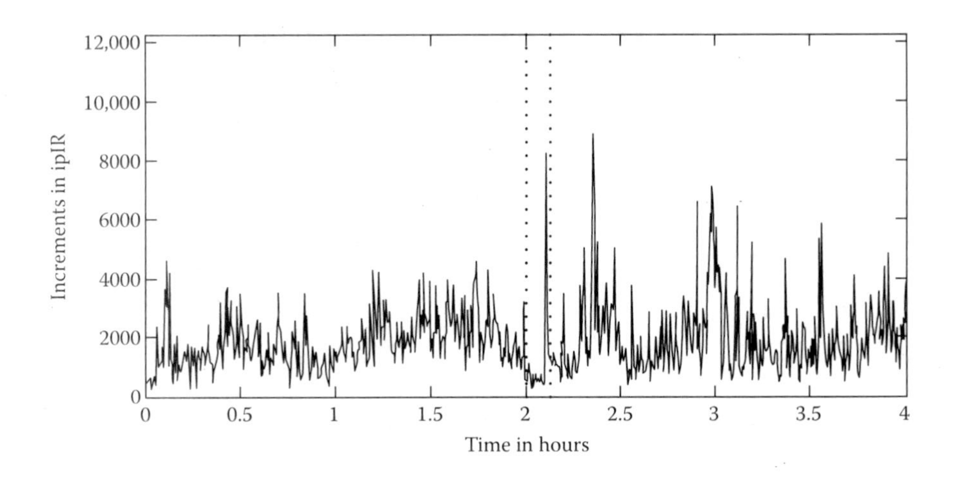

Figure 6.14 MIB anomaly of type ipIR.

predicted almost 7 min before incidence. In order to detect the anomalies from the point of view of a router, only the IP level is used. Figure 6.14 shows an anomaly of the MIB variable called *ipIR* (representing the total number of datagrams received for all router interfaces). The defect period is indicated by the dotted vertical line. Note the existence of a drop in the average traffic level reflected in the ipIR variable right after the failure.

Predicting an anomaly implies two situations: (a) In the case of server failures or network access problems, prediction is possible by analyzing the MIB data. (b) In the case of unpredictable errors such as protocol implementation error, early detection is possible as compared to the existing mechanisms such as syslog and defect reports. Any anomaly declaration that does not correspond to a label is declared to be false. The quantities used in the study are shown in Figure 6.15 (after Thottan and Ji 2003). The persistence criterion implies that the alarms need to remain for consecutive lags for an anomaly to be declared.

In the experiments, some example of defects were detected, coming from two networks: an enterprise and a campus network. Both networks have been actively monitored and were designed for the needs of the user. The types of anomaly that were observed were: server errors, protocol implementation errors, and access problems. The majority of these are caused by an abnormal usage activity, with the exception of the protocol implementation errors. Thus, all these cases affect the MIB data characteristic.

6.3.5.3 *Overall conclusions*

Although high-speed traffic on Internet links has an asymptotic self-similarity, its correlation structure at small timescales allows for modeling it

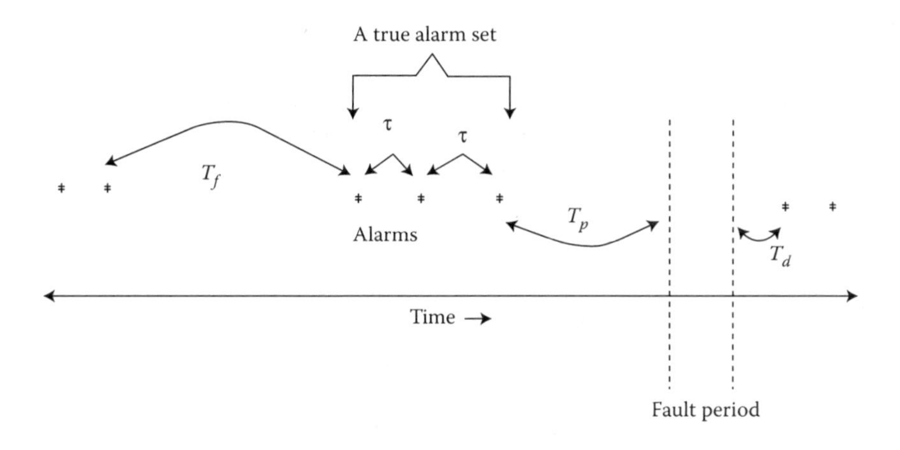

Figure 6.15 Quantities used in performance analysis.

as a self-similar processes (as fBm). Based on simulations done on SFN Internet model, one can conclude that the Internet traffic has self-similar properties even when it is heavily aggregated. Experiments revealed the following results:

1. Self-similarity is adaptable to the network traffic and it is not based on a heavy tail transfer distribution.
2. The region of scaling the traffic self-similarity is divided into two timescale regions: short-range dependencies (SRD), determined by the bandwidth, the size of the TCP window and the packet size, and a long-range dependence (LRD), determined by the statistics with a long tail characteristic.
3. In LRD, an increase in the bandwidth does not increase the performance.
4. There is a significant advantage in using fractal analysis methods for solving anomaly detection problems.

An accurate estimation of the Hurst parameter for MIB variables offer a valuable anomaly indicator obtained from the *ipIR* variable. So, by increasing the prediction capability, it is possible to reduce the time prediction time and to increase the network availability. Regarding the parallel processing methods, it is very important for the network model to be correctly federalized, because there is an optimum point between separating and balancing the federalization—the more separate the networks are, the more unbalanced they become.

The method of testing self-similarity between networks suggests the possibility of inferring traffic data in places where measurements are inaccessible. Considering the results of the discussed method, it has been found to have competitive performance for network traffic analysis, similar with those claimed by Breunig et al. (2000) when using LOF algorithm for computation of similarity between a pair of flows that contain a combination of categorical and numerical features. In comparison, our technique enables meaningful calculation without being computationally intensive, because it directly integrates data sets of previously detected incidents.

One can conclude that the anomaly detection algorithm is also effective in detecting behavior anomalies on a network, which typically translates into malicious activities such as DoS traffic, worms, policy violations, and inside abuse.

6.3.6 Recent trends in traffic self-similarity assessment for Cloud data modeling

Existing modern Cloud data storage, like Cloud data warehousing, do not use its potential capabilities functionality due mainly to the complexity in behavior of network traffic. The quality of service can be increased by building a converged multiservice network. Such a network ought to provide an unlimited range of services to provide flexibility for the management and creation of new services.

The current state of rapid development of high technology has led to the emergence and spread of ubiquitous networks, packet data, which gradually began to force the system switching network and at the same time increase the relevance of the traffic self-similarity. The fractal analysis can solve several problems, such as the calculation of the burst for a given flow, the development of algorithms and mechanisms providing quality of service in terms of self-similar traffic, and the determination of parameters and indicators of quality of information distribution.

6.3.6.1 Anomalies detection in Cloud

Deploying high-assurance services in the Cloud increases security concerns. To address these concerns, a range of security measures must be put in place. An important measure is the ability to detect when a Cloud infrastructure, and the services it hosts, is under attack via the network, for example, from a distributed denial of service (DDoS) attack. A number of approaches to network attack detection exist, based on the detection of anomalies in relation to normal network behavior. An essential characteristic of Cloud computing is the use of virtualization technology (Mell and Grance 2011).

Virtualization supports the ability to migrate services between different underlying physical infrastructure. In some cases, a service must be migrated between geographically and topologically distinct data centers (process known as wide-area migration). When a service is migrated within a data centre, the process is defined as local-area migration. In both cases, there are multiple approaches to ensure the network traffic that is destined for a migrated service is forwarded to the correct location.

Importantly for anomaly detection, service migration may result in observable effects in network traffic—this depends on where network traffic that is to be analyzed is collected in the data center topology, and the type of migration that is carried out. For example, if a local-area migration is executed, the change in traffic could be observable at top-of-rack (ToR) and aggregate switches, but not at the gateway to the data center. Since it is common practice to analyze traffic at the edge of a data center, local-area migration is opaque at this location. However, for wide-area migration, traffic destined to the migrated service will be forwarded to a different data centre, and will stop being received at the origin after the migration process has finished. This will result in potentially observable effects in traffic collected at the edge of a data center, and hamper anomaly detection techniques, so it is important to explore the effect that wide-area service migration has on anomaly detection techniques when network traffic that is used for analysis is collected at the data center edge.

An anomaly detection tool (ADL) that analyses the traffic data for observing similar patterns, along with the injected anomalies and migration is described in Adamova et al. (2014). For the experiments, the authors used network flow data that was collected from the Swiss National Research Network—SWITCH—a medium-sized backbone network that provides Internet connectivity to several universities, research labs, and governmental institutions. The ADL has three inputs: an anomaly vector, which specifies at what times anomalies were injected, and two time-series files—a baseline file and an evaluation file. A time-series file contains entries that summarize each 5 min period of the traffic and is created by extracting the traffic features from a set of preselected services and anomalies. The traffic features are volume-based (the flow, packet, and byte count) and distribution-based (the Shannon entropy of source and destination IP and port distributions). These features are commonly used in flow-based anomaly detection. The baseline time series are created for training data; these are free from anomalies and migration. Conversely, evaluation time series contain anomalies and may contain migration, depending on the experiments we wish to conduct.

Figure 6.16 illustrates the tool chain used for testing the performance of the anomaly detection tool.

It can be seen that using this tool chain one can create evaluation scenarios that include a range of attack behavior, and service migration activity of different magnitudes. These scenarios can then be provided as input to different flow-based anomaly detection techniques. Among

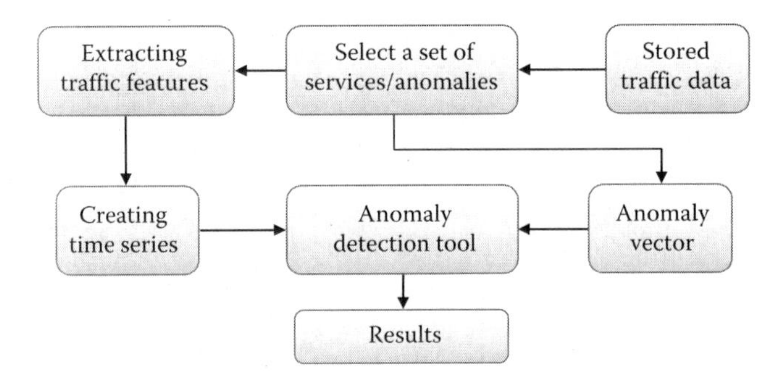

Figure 6.16 The tool chain used in ADL testing.

them, two are based on traffic self-similarity: spectral analysis, based on PCA, and clustering using expectation-maximization (EM). When using PCA, anomalies are detected based on the difference of the baseline and evaluation profiles. Detection of anomalous behavior based on clustering is realized by assigning data points that have similar features to cluster structures. Items that do not belong to a cluster, or are part of a cluster that is labeled as containing attacks, are considered anomalous.

6.3.6.2 *Combining filtering and statistical methods for anomaly detection*

Traffic anomalies such as attacks, flash crowds, large file transfers, and outages occur fairly frequently in the Internet today. Despite the recent growth in monitoring technology and in intrusion detection systems, correctly detecting anomalies in a timely fashion remains a challenging task.

A new approach for anomaly detection in large-scale networks is based on the analysis of traffic patterns using the traffic matrix (Soule et al. 2005). In the first step, a Kalman filter is used to filter out the "normal" traffic. This is done by comparing future predictions of the traffic matrix state to an inference of the actual traffic matrix that is made using more recent measurement data than those used for prediction. In the second step, the residual filtered process is then examined for anomalies. Since any flow will traverse multiple links along its path, it is intuitive that a flow carrying an anomaly will appear in multiple links, thus increasing the evidence to detect it.

A traffic matrix is a representation of the network-wide traffic demands. Each traffic matrix entry describes the average volume of traffic, in a given time interval, that originates at a given source node and is headed toward a particular destination node. Since a traffic matrix is a representation of traffic volume, the types of anomalies, which can be detected via analysis of the traffic matrix, are volume anomalies (Lakhina et al. 2004). Examples of events that create volume anomalies are denial-of-service attacks (DOS), flash crowds, and alpha events (e.g., nonmalicious large file transfers), as well as outages (e.g., coming from equipment failures).

Obtaining traffic matrices was originally viewed as a challenging task since it is believed that directly measuring them is extremely costly as it requires the deployment of monitoring infrastructure everywhere, the collection of fine granularity data at the flow level, and then the processing of large amounts of data. However, in the last few years many inference-based techniques have been developed that can estimate traffic matrices reasonably well given only per-link data such as SNMP data (that is widely available). Let note that usually these techniques focus on estimation and not prediction (Zhang et al. 2003).

A better approach for traffic matrix estimation seems to be that based on predictions of future values of the traffic matrix. A traffic matrix is a dynamic entity that continually evolves over time, thus estimates of a traffic matrix are usually provided for each time interval (most previous techniques focus on 5-min intervals), and the same interval for the prediction step of the traffic matrix will be considered in the following. The procedure consists in two stages. In the first stage, 5 minutes after the prediction is made, one obtain new link-level SNMP measurements, and then what the actual traffic matrix should be is estimated. In the second stage, one examines the difference between the last prediction (made without the most recent link-level measurements) and the estimation (made using the most recent measurements). If the estimates and predictor are good, then this difference should be close to zero. When the difference is necessary to analyze this residual further to determine whether or not an anomaly alert should be generated. In this connection, we recommend to use a generalized likelihood ratio test to identify the moment an anomaly starts, by identifying a change in mean rate of the residual signal. This method is a particular application of known statistical techniques to the anomaly detection domain and will be detailed in the following.

The traffic matrix (TM) that we discuss represents an interdomain view of traffic, and it expresses the total volume of traffic exchanged between pairs of backbone nodes, both edge and internal. Despite the massive growth of the Internet, its global structure is still heavily hierarchical, with a core made of a reduced number of large autonomous systems (ASs), among them some huge public Clouds (Casas et al. 2010). Our contribution regards first TM modeling and estimation, for which a new statistical model and a new estimation method to analyze the origin–destination (OD) flows of a large-scale TM from easily available link traffic measurements, and then regards the detection and localization of volume anomalies in the TM.

The traffic matrix estimation (TME) problem can be stated as follows: assuming a network with m OD flows and r links, $X_t = [x_t(1), \ldots, x_t(m)]^T$ represents the TM organized as a vector, where $x_t(k)$, $k = 1 \ldots m$ stands for the volume of traffic of each OD flow at time t. The routing matrix $R \in \Re^{rxm}$ indicates which links are traversed by each OD-flow. The element $R_{ij} = 1$ if OD flow j takes link i and 0 otherwise. Finally, the vector $Y_t = [y_t(1), \ldots, y_t(r)]^T$, where $y_t(i)$, $i = 1 \ldots r$ represents the total aggregated volume of traffic from those OD-flows that traverse link at time t represents the SNMP measurements vector:

$$Y_t = R \cdot X_t \tag{6.4}$$

The TME problem consists in the estimation the value of X_t from R and Y_t.

Different methods have been proposed in the last few years to solve the TME problem. They can be classified into two groups. In the first group are included methods that rely exclusively upon SNMP measurements and routing information to estimate a TM, whereas in the second group are included mixed methods which additionally consider the availability of partial flow measurements used for calibration purposes and exploit temporal and spatial correlations among the multiple OD-flows of the TM to improve estimation results.

A mixed TME method named recursive traffic matrix estimation (RTME) will be discussed, which not only uses Y_t to estimate X_t, but also takes advantage of the TM temporal correlation, using a set of past SNMP measurements $\{Y_{t-1}, Y_{t-2}, \ldots, Y_1\}$ to compute an estimate $\hat{X}_{t|t} = E(X_t \mid Y_t, \ldots Y_1)$.

Let first consider the model that is assumed in Soule et al. (2005). In this model, authors consider the OD-flows of the TM as the hidden states of a dynamic system. A linear state space model is adopted to capture the temporal evolution of the TM, and the relation between the TM and the SNMP measurements given by Equation 6.4 is used as the observation process:

$$\begin{cases} X_{t+1} = AX_t + W_{t+1} \\ Y_t = RX_t + V_t \end{cases} \tag{6.5}$$

In Equation 6.5, matrix X_t characterizes the evolution of the TM. Matrix A is the transition matrix that captures the dynamic behavior of the system, and W_t is an uncorrelated zero-mean Gaussian white noise that accounts both for modeling errors and randomness in the traffic flows. Through the routing matrix the observed links traffic Y_t is related to the unobserved state X_t. The measurement noise V_t is also an uncorrelated zero-mean Gaussian white noise process. Given this model, it is possible to recursively derive the least mean-squares linear estimate of Xt given $\{Y_{t-1}, Y_{t-2},..., Y_1\}$, $\hat{X}_{t|t} = E(X_t \mid Y_t,...Y_1)$ by using the standard Kalman filter (KF) method. The KF is an efficient recursive filter that estimates the state X_t of a linear dynamic system from a series of noisy measurements $\{Y_{t-1}, Y_{t-2}, \ldots, Y_1\}$. It consists of two distinct phases, iteratively applied, and the update phase.

The prediction phase uses the state estimate from the previous time-step $\hat{X}_{t|t}$ to produce an estimate of the state at the current time-step $t + 1$, usually known as the "predicted" state $\hat{X}_{t+1|t} = E(X_{t+1} \mid Y_t,...Y_1)$, according to the equation:

$$\begin{cases} \hat{X}_{t+1|t} = A\hat{X}_{t|t} \\ P_{t+1|t} = AP_{t|t}A^T + Q_w \end{cases} \tag{6.6}$$

where $P_{t|t}$ and $P_{t+1|t}$ are the covariance matrices of the estimation error $e_{t|t} = X_t - \hat{X}_{t|t}$, and the prediction error $e_{t+1|t} = X_{t+1} - \hat{X}_{t+1|t}$, respectively.

In the update phase, the measurements vector at current time-step Y_{t+1} is used to refine the prediction $\hat{X}_{t+1|t}$, computing a more accurate state estimate for current time-step $t + 1$,

$$\begin{cases} \hat{X}_{t+1|t+1} = \hat{X}_{t+1|t} + K_{t+1}(Y_{t+1} - R\hat{X}_{t+1|t}) \\ P_{t+1|t+1} = (I - K_{t+1}R)P_{t+1|t} \end{cases} \tag{6.7}$$

where $K_{t+1} = P_{t+1}|_t R^T(RP_{t+1}|_t R^T + Q_v)^{-1}$ is the optimal Kalman gain, which minimizes the mean-square error: $E(||e_{t+1}|_{t+1}||^2)$.

Combining Equations 6.6 and 6.7, the recursive Kalman filter-based estimation (RKFE) method recursively computes $\hat{X}_t^{RKFE} = \hat{X}_{t|t}$ from

$$\hat{X}_t^{RKFE} = (I - K_tR)A\hat{X}_{t-1}^{RKFE} + K_tY_t \tag{6.8}$$

In his doctoral thesis (Hernandez 2010), Casas Hernandez reported some drawbacks of this model, especially in the case where A is a diagonal matrix, because the first equation in Equation 6.5 is valid only for centered data and false if data are not centered.

In order to have a correct state space model for the case of a diagonal state transition matrix A, he proposed a new model, considering the variations of X_t around its average value m_X, that is, $X_t^c = X_t - m_X$ the system (6.5) becomes

$$\begin{cases} X_{t+1}^c = AX_t^c + W_{t+1} \\ Y_t = RX_t^c + V_t + Rm_X \end{cases} \tag{6.9}$$

This state space model is correct for centered TM variations, for both static and dynamic mean, but a problem still remains, because the Kalman filter does not converge to the real value of the traffic matrix if noncentered data Y_t is used in the filter. This problem can be easily solved considering a centered observation process Y_t^c, by adding a new deterministic state to the state model. Let us define a new state variable $U_t = [X_t^c m_X]^T$. In this case, Equation 6.9 becomes

$$\begin{cases} X_{t+1}^c = \begin{bmatrix} A & 0 \\ 0 & I \end{bmatrix} U_t + \begin{bmatrix} W_{t+1} \\ 0 \end{bmatrix} = CU_t + \Psi_{t+1} \\ Y_t = \begin{bmatrix} R & R \end{bmatrix} U_t + V_t = BU_t + V_t \end{cases} \tag{6.10}$$

where 0 is the null matrix of correct dimensions. This new model has twice the number of states, augmenting the computation time and complexity of the Kalman filter. However, it presents several advantages

- It is not necessary to center the observations Y_t.
- Matrix A can be chosen as a diagonal matrix, which corresponds to the case of modeling the centered OD-flows as autoregressive AR(1) processes. This is clearly much easier and more stable than calibrating a nondiagonal matrix such that $(I - A)m_X = 0$.
- The Kalman filter estimates the mean value of the OD-flows m_X, assumed constant in Equation 6.10.
- This model allows to impose a dynamic behavior to m_X, improving the estimation properties of the filter.

Using model, Equation 6.10 with the Kalman filtering technique produces quite good estimation results. However, this model presents a major drawback: it assumes that the mean value of the OD-flows m_X is constant in time. It can be improved by adopting a dynamic model for m_X, in order to allow small variations of the OD-flows mean value:

$$m_X(t+1) = m_X(t) + \zeta_{t+1} \tag{6.11}$$

where $m_X(t)$ represents the dynamic mean value of X_t and ζ_t is a zero-mean white Gaussian noise process with covariance matrix Q_ζ. In this context, Equation 6.10 becomes

$$\begin{cases} U_{t+1} = \begin{bmatrix} A & 0 \\ 0 & I \end{bmatrix} U_t + \begin{bmatrix} W_{t+1} \\ \zeta_{t+1} \end{bmatrix} = CU_t + \Theta_{t+1} \\ Y_t = [R \quad R] U_t + V_t = BU_t + V_t \end{cases} \tag{6.12}$$

Hernandez (2010) presented experimental results, which confirm that this model provides more accurate and more stable results.

One can conclude that considering a variable mean value $m_X(t)$ produces better results, both as regards accuracy and stability. In all evaluations, the stable evolution of the error shows that the underlying model remains valid during several days when considering such a transition matrix. Therefore, if the underlying model remains stable, it is not necessary to conduct periodical recalibrations, dramatically reducing measurement overheads.

Let us now explain how TM, as a tool for statistical analysis, can be used for optimal anomaly detection and localization and consequently for correctly managing the large and unexpected congestion problems caused by volume anomalies. Volume anomalies represent large and sudden variations in OD-flow traffic. These variations arise from unexpected events such as flash crowds, network equipment failures, network attacks, and external routing modifications among others. Besides being a major problem in itself, a volume anomaly may have a significant impact on the overall performance of the affected network. The focus is only on two main aspects of traffic monitoring: (1) the rapid and accurate detection of volume anomalies and (2) the localization of the origins of the detected anomalies.

The first issue corresponds to the anomaly detection field, with a procedure that consists of identifying patterns that deviate from normal traffic behavior. In order to detect abnormal behaviors, accurate and stable traffic models should be used to describe what constitutes an anomaly-free traffic behavior. This is indeed a critical step in the detection of anomalies, because a rough or unstable traffic model may completely spoil the correct detection performance and cause many false alarms. In this chapter, we focus on network-wide volume anomaly detection, analyzing network traffic at the TM level. In particular, we use SNMP measurements to detect volume anomalies in the OD-flows that comprise a TM. The TM is a volume representation of OD-flow traffic, and thus the types of anomalies that we can expect to detect from its analysis are volume anomalies. As each OD-flow typically spans multiple network links, a volume anomaly in one single OD-flow is simultaneously visible on several links. This multiple evidence can be exploited to localize the anomalous OD-flows.

The second addressed issue is the localization of the origins of a detected anomaly. The localization of an anomaly consists in inferring the exact location of the problem from a set of observed anomaly indications. This represents another critical task in network monitoring, given that a correct localization may represent the difference between a successful or a failed countermeasure. Traffic anomalies are exogenous unexpected events (flash crowds, external routing modifications, and external network attacks) that significantly modify the volume of one or multiple OD flows within the monitored network. For this reason, the localization of the anomaly consists in finding the OD-flows that suffer such a variation, referred to from now on as the anomalous OD-flows. The method that we discuss locates the anomalous OD-flows from SNMP measurements, taking advantage of the multiple evidence that these anomalous OD-flows leave through the traversed links.

The method proposed before for anomaly detection uses the recursive Kalman filtering technique, but instead of computing an estimated TM, it uses the Kalman filter to detect volume anomalies as large estimation errors. Since the Kalman filter method can properly track the evolution of the TM in the absence of anomalies, estimation errors should be small most of the time. A volume anomaly is then flagged when the estimation error exceeds certain detection threshold. The method analyses the estimation error $\varepsilon(t) = X_t - \hat{X}_{t|t}$, where $\hat{X}_{t|t}$ represents the estimated TM using all the SNMP measurements until time t, namely $\{Y_{t-1}, Y_{t-2}, \ldots, Y_1\}$. The problem with this approach is that the real value of the TM, namely X_t, cannot be directly measured. Fortunately, the Kalman filter equations 6.6 and 6.7 allow to us compute the estimation error indirectly, using the inferred process ι_{t+1}. The inferred process represents the difference between the observed SNMP measurements vector Y_{t+1} and its predicted value, obtained from the predicted TM $\hat{X}_{t+1|t}$:

$$\iota_{t+1} = Y_{t+1} - R\hat{X}_{t+1|t} \tag{6.13}$$

Under the Kalman filtering hypotheses, the innovation process is a zero-mean Gaussian process, whose covariance matrix Γ_{t+1} can be easily derived from Equations 6.5 through 6.7:

$$\Gamma_{t+1} = RP_{t+1|t}R^T + Q_{v_{t+1}} \tag{6.14}$$

where $P_{t+1|t}$ is the covariance matrix of the prediction error $e_{t+1|t} = X_{t+1} - \hat{X}_{t+1|t}$ and $Q_{v_{t+1}}$ is the covariance matrix of the observation noise process V_{t+1}.

Finally, we discuss the computational complexity for applying the Kalman-based method for the detection of anomalies (KDA). The KDA complexity corresponds to that of the standard Kalman filter recursive equations. The method must store in memory an $m \times m$ state transition diagonal matrix that models the evolution of the anomaly free traffic matrix, the routing matrix R, and the noise covariance matrices associated with the observation and the evolution processes, the last being also a diagonal matrix. This accounts for a total of $2(r^2 + m)$ variables in memory. The recursive nature of the Kalman filter implies to keep in memory two additional matrices, the $m \times r$ Kalman gain matrix and the

$m \times m$ prediction error covariance matrix. The use of the KDA for online anomaly detection implies to update the Kalman gain, the estimation covariance error, and the residual error. This involves matrix multiplications and inversions, and thus the associated cost is $O(m^3)$. One can see that the KDA method is largely more expensive than most of the algorithms for anomaly detection, which comes directly from using the Kalman filter with large matrices, but the detection and localization algorithms are more efficient than current methods, allowing a robust growth of the network monitoring field. Its utilization will confirm that the results of decision theory applied to the field of network monitoring will provide methods that rely on coarse-grained, easily available SNMP data to detect and locate network-wide volume anomalies in OD-flows traffic. This is a main advantage in order to develop light monitoring systems without the necessity of direct flow measurement technology, particularly in the advent of the forecast massive traffic to analyze in the near future.

6.4 *Optimization of quality of services by monitoring Cloud traffic*

6.4.1 *Monitoring the dynamics of network traffic in Cloud*

Traffic monitoring is a fundamental function block for Cloud services, so well contoured that can define a specific concept, Monitoring as a Services (MaaS). One of the key challenges for wide deployment of MaaS is to provide better balance among a set of critical quality and performance parameters, such as accuracy, cost, scalability, and customizability.

On a high level, a monitoring system (in any large-scale network) consists of n monitoring nodes and one central node, that is, data collector. Each monitoring node has a set of observable attributes $A_i = \{a_i \mid i \in [1, m]\}$. We consider an attribute as a continuously changing variable, which gives a new value as output in every unit time. For simplicity, we assume all attributes are of the same size a. Each node I has a capacity b_i (considered as a resource constraint of node i) for receiving and transmitting monitoring data. Each message transmitted in the system is associated with a per-message overhead C, so the cost of transmitting a message with x values is $C + ax$. This cost model considers both per-message overhead and the cost of payload (Meng 2012). Figure 6.17 shows the structure of a monitoring planner, which organizes the monitoring nodes (according a given list of node–attribute pairs) into a cluster of monitoring trees where each node collects values for a set of attributes. The planner considers the aforementioned per-message overhead as well as the cost of attributes transmission to avoid overloading certain monitoring nodes

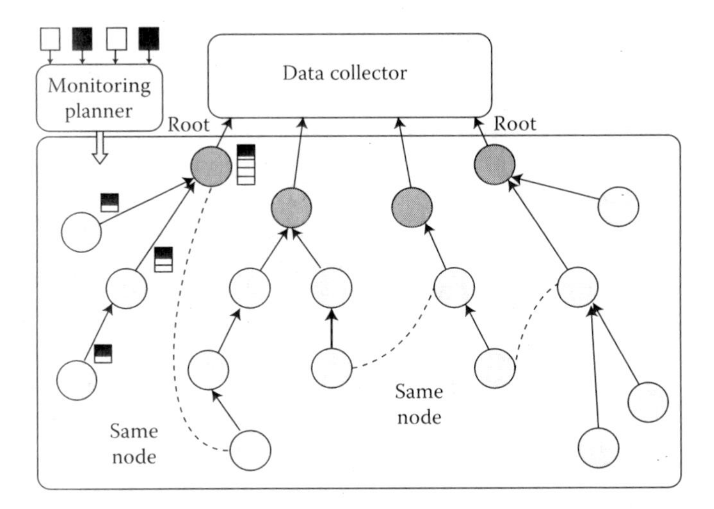

Figure 6.17 Structure of monitoring planning. (After Meng, S. *Monitoring as a Service in the Cloud*, PhD thesis, Georgia Institute of Technology, 2012.)

in the generated monitoring topology. In addition, it also optimizes the monitoring topology to achieve maximum monitoring data delivery efficiency. Within a monitoring tree T, each node i periodically sends an update message to its parent. As application state monitoring requires collecting values of certain attributes from a set of nodes, such update messages include both values locally observed by node i and values sent by i's children, for attributes monitored by T. Thus, the size of a message is proportional to the number of monitoring nodes in the subtree rooted at node i. This process continues upward in the tree until the message reaches the central data collector node.

From the users' perspective, monitoring results should be as accurate as possible, suggesting that the underlying monitoring network should maximize the number of node–attribute pairs received at the central node. In addition, such a monitoring network should not cause the excessive use of resource at any node. Accordingly, we define the monitoring planning problem (MP) as follows:

For a set of node–attribute pairs ω_q for monitoring $\Omega = \{\omega_1, \ldots, \omega_p\}$ where $\omega_q = (I, j), i \in N, j \in A, q \in [1, p]$, and resource constraint b_i for each associated node, find a parent $f(i\ j)$, $\forall i, j$, where $j \in A_i$ such that node i forwards attribute j to node $f(i,j)$ that maximizes the total number of node–attribute pairs received at the central node and the resource demand d_i of node i, satisfies $d_i \leq b_i$, $\forall i \in N$.

The monitoring planning problem raises two fundamental issues: (1) how to determine the number of monitoring trees and the set of attributes on each and (2) how to determine the topology for nodes in each monitoring tree under node level resource constraints. Therefore, constructing monitoring trees subject to resource constraints at nodes is a nontrivial problem and the choice of topology can significantly impact node resource usage.

A particular solution for the monitoring planning problem, namely a provisioning planning method for prediction-based multitier Cloud application is presented in the following.

Many Infrastructure-as-a-Service (IaaS) users are interested in deploying a multitier web application in the Cloud to meet a certain performance goal with minimum virtual instance renting cost. There are some difficulties in achieving this goal, first because the information with regarding the performance capacities and rental rates of different virtual server instances offered by the market is sometimes insufficient, second because the relation between cluster configuration (e.g., the number of member nodes, the distribution of workloads among member nodes) and performance (load balance, scalability, and fault tolerance) is application-dependent and must be carefully chosen for each particular application.

To meet a given performance goal, users often overprovision a multitier Web application by renting high-end virtual server instances and employing large clusters. Overprovisioning introduces high instance renting cost, which may make Cloud deployment a less desirable option compared with traditional deployment options. To avoid this drawback, one can propose a prediction-based provisioning planning method, which can find the most cost-effective provisioning plan for a given performance goal by searching the space of candidate plans with performance prediction. This procedure utilizes historical performance monitoring data (including self-similar patterns of time series) and data collected from a small set of automatic experiments to build a composite performance prediction model that takes as input different application workloads, types of virtual server instances, and cluster configuration, and gives the predicted performance as the output. This method was proposed by Meng, in his doctoral thesis (Meng 2012), and only the structure and the main advantages are presented here.

The proposed method avoids exhaustively performing experiments on all candidate deployments to build a performance prediction model by using a two-step performance prediction procedure. Instead of directly predicting the performance of an arbitrary deployment (target), it first predicts the performance on a known deployment (base) and then predicts the performance differences between the target deployment and the base deployment. It combines the predicted base performance and the predicted performance changes to obtain the performance on the target deployment. To achieve efficiency, the procedure predicts the performance change based on the deployment difference between the base deployment and the target deployment within each tier. By considering each tier independently, the method avoids the need for exploring all deployments that represent combinations of deployment changes cross-tiers.

The overall flow of the method has the following steps:

1. An user submits a request to deploy a multitier Cloud application in an Infrastructure-as-a-Service Cloud environment. The request also describes the expected range of workloads and expected performance.
2. The application is first deployed in an overprovisioned setting.
3. While the application running in the Cloud infrastructure, its workloads and performance are monitored and the corresponding monitoring data are stored.
4. The collected workloads and performance data are used to train a cross-tier performance model.
5. The application is replicated for a set of automatic experiments that deploy the application with different provisioning plans and measure the corresponding performance with different workloads. The goal of the automatic experiments is to learn the performance characteristics of different deployment options (e.g., virtual machine types and the number of virtual machines in a cluster).
6. The workloads and performance data collected in the automatic experiments are used to train a per-tier performance model.
7. The method explores all candidate provisioning plans and predicts the corresponding performance (for the user-specified workload range) using both the cross-tier and the per-tier performance model.
8. Among all candidate provisioning plans, the one that meets the user-specified performance goal and has the lowest virtual machine instance renting cost is selected as the suggested deployment for the user.

The implementation of the method implies three algorithms: (1) a prediction algorithm that takes workloads and deployment as input, and gives as output the predicted application performance; (2) an algorithm that captures the changes of perceived workloads across different deployments; and (3) a planning algorithm that explores all candidate provisioning plans and outputs the optimal one.

6.4.1.1 *Prediction algorithm*

The idea of the prediction techniques to first predict the response time for a given workload on an overprovisioned deployment (referred to as the base deployment), and then modify the predicted response time considering changes introduced by the difference between the overprovisioned deployment and the actual targeted deployment. Correspondingly, two performance models are employed to accomplish this task: a cross-tier performance model, which captures the relation between workload and response time for the base deployment, and a per-tier performance model that captures the relation between deployment changes (to the base deployment) and corresponding changes of the response time.

A cross-tier model has the following form: $\Theta_c(w) \rightarrow r$, where w is the workload and r is the average response time of requests. The cross-tier model takes workload as input and outputs the response time on the base deployment. As a nonparametric technique, it does not specify a certain linear between w and r, but produces a nonlinear relation between w and r that best fits the observed performance data.

A per-tier model has the form: $\Theta_p^t(w,v,c) \rightarrow r_d$, where t denotes the object tier, v is the virtual machine type, c is the cluster size, and r_d is the change of response time compared with the base deployment. The per-tier model is actually a set of models where each model is trained for a particular tier. Each per-tier model takes the workload, the type, and the number of virtual machine used at the object tier as input, and sends as output the changes of response time introduced by this tier over that of the base deployment.

To predict the response time for a target deployment and a given workload, first one uses the per-tier model to estimate the differences of response time introduced at each tier due to the deployment differences between the target deployment and the based deployment.

The work flow of the prediction process has the following steps:

Step 1. Select a candidate plan
Step 2. Predict the base performance using the cross-tier model
Step 3. Use the per-tier differential performance model to predict the performance change at the current tier
Step 4. Combining the predicted base performance and the predicted performance changes at all tiers
Step 5. Test if all tiers were explored. If no, return to step 3. If yes, go to step 6
Step 6. Test if all candidate plans were explored. If no, return to step 2. If yes, go to step 7
Step 7. Output the candidate plan that meets the performance goal with the lowest cost

The cross-tier model and the per-tier model are trained separately in two steps. The training of the cross-tier model requires only performance monitoring data on the base deployment. Specifically, the training data set should include the request rates spanning from light workloads to peak workloads and the corresponding average response time. Typically, the base deployment is overprovisioned to ensure the request response time meets the performance goal. The base deployment is also used as contrasts to generate training data for the per-tier model.

The per-tier models are trained in a tier-by-tier basis based on performance data collected on a series of automatic experiments. Specifically, first a duplicate of the based deployment is created and referred as the background deployment. For a per-tier model on tier t, we vary the configuration of tier t on the background deployment by changing the virtual machine type and the number of virtual machines, and leave the configuration of other tiers unchanged (same as the configuration in the base deployment). This leads to mn different background deployments where m is the total number of virtual machine types and n is the maximum number of virtual machines in tier t. For each resulting background deployment (with virtual machine type $v(t)$ and virtual machine number $c(t)$ in tier t), different levels of workloads are introduced (from light level to peak level just as those in the cross-tier model training dataset) to the deployment and the difference of response time r_d between the background deployment and the base deployment for each level of workload w is recorded. The resulting data points (w, $v(t)$, $c(t)$, r_d) are used to train the per-tier model Θ_p^t.

6.4.1.2 Algorithm for capturing cross-deployment workload changes

The above prediction method makes an implicit assumption that the actual workloads perceived at each tier do not change across different deployments. However, the perceived workload at a tier may not be the same as the workload introduced to the application due to prioritization, rate limiting mechanisms implemented at different tiers. For instance, an application may drop certain low-priority requests when a certain tier becomes a performance bottleneck, which in turn causes the change of workload at other tiers. Even for applications without prioritization mechanisms, a bottleneck tier may limit the overall system throughput and introduce changes to the workload on other tiers.

Performance prediction without considering such workload changes may lead to significant prediction accuracy loss. As another example, a database tier of a Web application configured with a single low-end virtual machine can be a performance bottleneck when the Web application is fed with a peak workload w_p. As a result, the actual workloads perceived at each tier w' is often less than w_p as a certain amount of requests are queued due to database overloading. Clearly, using the data (w_p, v, c, r_d) for training would introduce error to the per-tier model.

To address this issue, a throughput model Θ_h^t for a tier t is introduced with the following form: $\Theta_h^t(w,v,c) \rightarrow w'$, where w′ is the actual workload perceived by all tiers. When making performance predictions, we apply the throughput model to obtain the predicted workload at each tier, and use the lowest predicted workload as the input of the per-tier model. Specifically, with the throughput model, the per-tier model has the following form: $\Theta_p^t(\min \Theta_h^t(w,v(t),c(t),v,c) \rightarrow r_d$, where the input workload w is replaced with the actual workload predicted by the throughput model. Note that the data used for training the throughput model is (w, v, c, w') and w' can be easily measured by counting the number of responses within a time window.

6.4.1.3 Provisioning planning algorithm

In order to find the optimal provisioning plan for an application, it is enough only exploring all candidate provisioning plans and estimating the cost (such as virtual machine renting fee) and performance (obtained by the prediction method) of each candidate plan. The optimal plan is the one with the lowest cost and performance that satisfies the performance goal. In addition, the performance prediction model, once trained, can be repeated used for different planning tasks with different performance goals.

The entire procedure for prediction-based multitier Cloud applications has better performance than single-deployment performance modeling. Existing approaches based on single-deployment often apply queuing models and regression techniques, which use instrumentation at the middleware level to obtain critical model parameters such as per-request service time. This often limits the applicability of these approaches in the Cloud environment where middleware may not be a valid option for Cloud users. These approaches build models for a specific hardware/software deployment (fixed machines and cluster configurations) and focus on the impact of workload changes on performance.

The resulting models often produce poor prediction results on a different hardware/software deployment. On the contrary, the above discussed multitier deployment method not only considers workload changes, but also deployment changes. This cross-deployment feature is important for Cloud application provisioning due to the large number of available deployment options (e.g., different virtual machines types and different cluster configurations). Another distinct feature of this approach is that it utilizes a per-tier model to capture the performance difference introduced by deployment changes at each tier. This allows us to predict performance changes for any combination of deployment changes at different tiers without collecting performance data from the corresponding deployment, which saves tremendous amount of experiment time and cost.

The discussed method is only one step toward a truly scalable and customizable MaaS solution. Many other issues need to be investigated for MaaS to be a successful service for Cloud state management. Some of these important open problems in providing MaaS are mentioned as further research topics: monitoring heterogeneity (important for tracking anomalies and locating root causes, especially in distributed environments); flow-based monitoring (useful in distributed multitier applications, for tracking the execution of a request across nodes); smart Cloud management (by utilizing performance monitoring data to automate performance-driven Cloud application provisioning); exploring all automation opportunities (Cloud applications evolve over its lifetime in the form of reconfiguration, feature enrichment, automatic bug localization and fixing, new functionalities implementation, disturbance-free patching, scheduling, etc.); security and privacy (virtual machines running on the same host are vulnerable to performance attacks, so MaaS should also provide new monitoring functionalities to address such security and privacy issues in Cloud environments).

6.4.2 *Coping with traffic uncertainty for load balancing in Cloud*

Load balancer is a key element in resource provisioning for highly available Cloud solutions and its performance depends on the load offered by the traffic. These resources are expected to provide high performance, high availability, and secure and scalable solutions to support hosted applications. Many content-intensive applications show elastic load behavior that might scale up beyond the potential processing power of a single server. At the same time, service providers need a dynamic capacity of server resources to deploy as many servers as needed seamlessly and transparently. In Cloud, server load balancing structures multiple servers as a single virtual server.

Cloud balancing approach extends the architectural deployment model of traditional load balanced servers to be used in conjunction with global load balanced servers off the Cloud (Fayoumi 2011). This approach increases the application routing options, which are associated with decisions based on both technical and business goals of deploying load-balanced global application. Because many empirical studies have shown that heavy-tailed Pareto distribution, typical for self-similar traffic, provides a good fit for a number of important characteristics of communication networks and Web servers, it is justified to consider that the sizes of user requests and the service times of packets handled in Cloud satisfy heavy-tailed distributions.

The performance of the Internet itself depends, in large part, on the operation of the underlying routing protocols. Routing optimization becomes increasingly challenging due to the dynamic and uncertain nature of current traffic. The currently utilized routing protocols in IP networks compute routing configurations based on the network topology and some rough knowledge of traffic demands (e.g., worst-case traffic, average traffic, and long-term forecasts), without considering the current traffic load on routers and links or possible traffic misbehaviors and therefore without paying special attention to the traffic uncertainty issue. Large-scale networks are usually overprovisioned, and routing modifications due to traffic variations do not occur that often. Research studies in the field of routing optimization agree that today's approach may no longer be suitable to manage current traffic patterns and more reliable routing mechanisms are necessary to perform and maintain networks functions in the case of component failures, in particular in the case of unexpected traffic events, represented by volume anomalies.

Given a single known traffic matrix (TM), routing optimization consists in computing a set of origin–destination paths and a traffic distribution among these paths in order to optimize some performance criterion, usually expressed by means of a cost function. Traditional routing optimization approaches have addressed the problem relying on either a small group of representative TMs or estimated TMs to compute optimal and reliable routing configurations. These techniques usually maintain a history of observed TMs in the past, and optimize routing for the representative traffic extracted from the observed TMs during a certain history window. In this sense, we shall refer to traditional algorithms as prediction-based Routing. The most well-known and traditionally applied prediction-based routing approach is to optimize routing for a worst-case traffic scenario, using for example the busy-hour TM seen in the history of traffic. This may not be a cost-effective solution, but operators have traditionally relied on the overprovisioning of their networks to support the associated performance loss. Better solutions are (1) to optimize routing for an estimated TM; this approach can provide more efficient solutions, but it highly depends on the goodness of the estimation technique and (2) to optimize routing for multiple TMs simultaneously, using for example a finite number of TMs from a previous day, from the same day of a previous week, etc. In both approaches, it must use different methods to find routing configurations with good, average, and worst-case performance over these TMs, knowing well that the worst case is only among the samples used in the optimization, and not among all possible TMs. All these traditional approaches tend to work reasonably well, but they certainly require a lot of management effort to ensure robustness against unexpected traffic variations.

Two new different approaches have emerged to cope with both the traffic increasing dynamism and uncertainty and the need for cost-effective solutions: robust routing (RR) (Mardani and Giannakis 2013) and dynamic load balancing (DLB) (Katiyar and Grover 2013). RR approach copes with traffic uncertainty in an offline preemptive fashion, computing a fixed routing configuration that is optimized for a large set of possible traffic demands. On the contrary, dynamic load balancing delivers traffic among multiple fixed paths in an online reactive fashion, adapting to traffic variations.

In robust routing, traffic uncertainty is taken into account directly within the routing optimization, computing a single routing configuration for all traffic demands within some uncertainty set where traffic is assumed to vary. This uncertainty set can be defined in different ways, depending on the available information: busy-hour traffic, largest values of links load previously seen, a set of previously observed traffic demands, etc. The criteria to search for this unique routing configuration is generally to minimize the maximum link utilization over all demands of the corresponding uncertainty set. Accurate estimation of origin-to-destination (OD) traffic flows provides valuable input for network management tasks. However, lack of flow-level observations as well as intentional and unintentional anomalies pose major challenges toward achieving this goal. Leveraging the low intrinsic-dimensionality of OD flows and the sparse nature of anomalies, a robust balancing algorithm allows to estimate the nominal and anomalous traffic components, using a small subset of (possibly anomalous) flow counts in addition to link counts and can exactly recover sufficiently low-dimensional nominal traffic and sparse enough anomalies when the traffic matrix is vague. The results offer valuable insights about the measurement types and network scenarios giving rise to accurate traffic estimation.

Dynamic load balancing copes with traffic uncertainty and variability by splitting traffic among multiple paths online. In this dynamic scheme, each origin–destination pair of nodes within the network is connected by several a priori configured paths, and the problem is simply how to distribute traffic among these paths in order to optimize a certain cost function. DLB is generally defined in terms of a link–cost function, where the portions of traffic are adjusted by each origin–destination pair of nodes in order to minimize the total network cost. For the sake of efficient resource utilization, load balancing system problem needs more attention in Cloud computing, due to its nature of on-demand computing. Cloud Computing considers shared pool of configurable computing resources, which requires proper resource distribution among the tasks; otherwise in some situations resources may be overutilized or underutilized. The dynamic load balancing approach, in which each agent plays very important role, which is a software entity and usually defined as an independent software program that runs on behalf of a network administrator, reduces the communication cost of servers, accelerates the rate of load balancing, which indirectly improves the throughput and response time of the Cloud.

A comparative analysis between robust load balancing algorithms and dynamic load balancing mechanisms leads to the following conclusions. The proponents who promote DLB mechanisms highlight among others the fact that it represents the most resource-efficient solution, adapting to current network load in an automated and decentralized fashion. On the other hand, those who advocate the use of RR claim that there is actually no need to implement supposedly complicated dynamic routing mechanisms, and that the incurred performance loss for using a single routing configuration is negligible when compared with the increase in complexity. An interesting characteristic of RR relies on the use of a single fixed routing configuration, avoiding possible instabilities due to routing modifications. In practice, network operators are reluctant to use dynamic mechanisms and prefer fixed routing configurations, as they claim they have a better pulse on what is going on in the network.

The first shortcoming that we identify in a RR paradigm is the associated cost efficiency. Using a single routing configuration for long time periods can be highly inefficient. The definition of the uncertainty set in RR defines a critical tradeoff between performance and robustness: larger sets allow to handle a broader group of traffic demands, but at the cost of routing inefficiency; conversely, tighter sets produce more efficient routing schemes, but subject to poor performance guarantees.

For the case of volume anomalies, a dynamic extension of RR known as reactive robust routing (RRR) can obviate this drawback. The RRR approach uses the sequential volume anomaly detection/localization method discussed in the previous section to rapidly detect and localize abrupt changes in traffic demands and decide routing modifications. A load balancing approach for RRR, in which traffic is balanced among fixed paths according to a centralized entity that controls the fractions of traffic sent on each path can solve the issue.

The second drawback that we identify in current RR is related to the objective function it intends to minimize. Optimization under uncertainty is generally more complex than classical optimization, which forces the use of simpler optimization criteria such as the maximum link utilization (MLU), that is, minimize the load of the most utilized link in the network. The MLU is by far the most popular traffic engineering objective function, but clearly it is not the most suitable network-wide optimization criterion; setting the focus too strictly on MLU often leads to worse distributions of traffic, adversely affecting the mean network load and thus the total network end-to-end delay, an important QoS indicator (Hernandez 2010). It is easy to see that the minimization of the MLU in a network topology with heterogeneous link capacities may lead to poor results as regards global network performance. To avoid this issue, it is possible to minimize the mean link utilization instead of the MLU. The mean link utilization provides a better image of network-wide performance, as it does not depend on the particular load or capacity of each single link in the network but on the average value. Despite this advantage, a direct minimization of the mean link utilization does not assure a bounded MLU, so for its utilization is necessary to limit MLU by a certain utilization threshold.

In DLB, traffic is split among fixed *a priori* established paths in order to optimize a certain cost function. For example, a link cost function based on measurements of the queuing delay can lead to better global performance from a QoS perspective. DLB presents a desirable property—that of keeping routing adapted to dynamic traffic. However, DLB algorithms present a tradeoff between adaptability and stability, which might be particularly difficult to address under significant and abrupt traffic changes.

As regards a comparative study between RR and DLB, the best method seems to be the dynamic approach, which consists of computing an optimal routing for each traffic demand i and evaluate its performance with the following traffic demand $i + 1$, on the basis of a given time series of traffic demands. This dynamic approach allows us to avoid two important shortcomings of DLB simulation:

1. Adaptation in DLB is iterative and never instantaneous.
2. In all DLB mechanisms, paths are set a priori and remain unchanged during operation, because each new routing optimization may change not only traffic portions but paths themselves.

Finally, a comprehensive analysis of different plausible solutions to the problem, including traditional prediction-based routing optimization, robust routing optimization, and dynamic load balancing led to the following conclusions:

1. Using a single routing configuration is not a cost-effective solution when traffic is relatively dynamic. Traditional prediction-based routing optimization may provide quite inefficient or even infeasible routing configurations when traffic is uncertain and difficult to forecast. Stable robust routing optimization offers performance guarantees against traffic uncertainty, but the associated tradeoff between robustness and routing efficiency may be particularly difficult to manage with a single routing configuration.
2. It is clear that some form of dynamism is necessary, either in the form of reactive robust routing and load balancing (RRLB) or dynamic load balancing (DLB). RRLB computes a nominal operation robust routing configuration, and provides an alternative robust routing configuration (using the same paths than in normal operation) for every possible single OD-flow anomalous situation. In order to detect these anomalous situations, link load measurements have to be gathered and processed by a centralized entity. On the

other hand, DLB gathers the same measurements but also requires updating load balancing in a relatively small timescale. The added complexity is then to distribute these measurements to all ingress routers and to update the load balancing online.

3. The additional complexity involved in DLB is not justified when traffic variability is not very significant. In the case of large volume anomalies, DLB algorithms generally provide better results than RRLB after convergence, but they present an undesirable transient behavior. On the other hand, RRLB algorithms do not suffer from this problem, basically because the load balancing fractions are computed offline in a robust basis, taking advantage of the goodness of the robust routing paradigm.
4. The use of DLB algorithms becomes very appealing when volume anomalies are difficult to locate. The transient behavior that they present under large traffic modifications can be effectively controlled, or at least alleviated, by simple mechanisms, which results in a somewhat bigger maximum link utilization, but a generally much better global performance.
5. A local performance criterion such as the maximum link utilization does not represent a suitable objective function as regards global network performance and QoS provisioning. The maximum link utilization is widely used in current network optimization problems, thus this should be considered in enhanced future implementations.
6. By using a simple combination of performance indicators such as the maximum and the mean link utilization, one can obtain a robust routing configuration that definitely outperforms current implementations from a global end-to-end perspective, while achieving very similar results as regards worst-case link utilization, In fact, it means that objective optimization functions can be kept simple, and yet better network-wide performance can be attained.
7. A major drawback of RR is its inherent dependence on the definition of the uncertainty set: larger sets allow to handle a broader group of traffic demands, but at the cost of routing inefficiency; conversely, tighter sets produce more efficient routing schemes, but subject to poor performance guarantees.

6.4.3 *Wide-area data analysis for detection changes in traffic patterns*

Traffic monitoring is crucial to network operation to detect changes in traffic patterns often indicating signs of flash crowds, misconfigurations, DDoS attacks, etc. Flow-based traffic monitoring has been most widely used for traffic monitoring. A "flow" is defined by a unique 5-tuple (source and destination IP addresses, source and destination ports, and protocol), and used to identify a conversation between nodes. A challenge for flow-based traffic monitoring is how to extract significant flows in constantly changing traffic and produce a concise summary.

A simple and common way to extract significant flows is by observing traffic volume and/or packet counts, and reporting the top ranking flows. However, a significant event often consists of multiple flows, leading to the concentration of flows originating from one node. To this end, individual flows with common attributes in 5-tuple can be classified into an aggregated flow. A limitation is that it can identify only predefined flows and cannot detect unexpected or unknown flows.

To overcome the limitation of predefined rules, automatic flow aggregation has been proposed (Kato et al. 2012). The basic idea is to perform flow clustering on the fly in order to adapt aggregated flows to traffic changes. Although the idea is simple, it is not easy to search the best matching aggregated flow in a five-dimensional space, especially because traffic information is continuously generated so that the clustering process needs to keep up with incoming traffic information in near real time.

To solve the problem of flow aggregation requires analyzing data that is continuously created across wide-area networks, and in particular in Cloud computing. Queries on such data often have real-time requirements that need the latency between data generation and query response to be bounded. If the monitoring system is based on a single data center, where data streams are processed inside a high-bandwidth network, probably it will not perform well in the wide area, where limited bandwidth availability makes it impractical to backhaul all potentially useful data to a central location. Instead, a system for wide-area analytics must prioritize which data to transfer in the face of time-varying bandwidth constraints. Such a system must incorporate structured storage that facilitates aggregation, combining related data together into succinct summaries. On the other hand, it must also incorporate degradation that allows trading data size against fidelity.

For performing wide-area data analysis, a large amount of data is stored at edge locations that have adequate compute and storage capacity, but there is limited or unpredictable bandwidth available to access the data. Today's analytics pipelines lack visibility into network conditions, and do not adapt dynamically if available bandwidth changes. As a result, the developer must specify in advance which data to store or collect based on pessimistic assumptions about available bandwidth, which ultimately leads to overprovisioning compared to average usage, and so capacity is not used efficiently.

The goal of a specialized wide-area data analysis system (WDAS) is to enable real-time analysis by reducing the volume of data being transferred. Storing and aggregating data where it is generated helps, but does not always reduce data volumes sufficiently. Thus, such system includes besides aggregation also degradations. Integrating aggregation and degradation into a streaming system require addressing three main challenges:

- Incorporating storage into the system while supporting real-time aggregation. Aggregation for queries with real-time requirements is particularly challenging in an environment where data sources have varying bandwidth capacities and may become disconnected.
- Dynamically adjusting data volumes to the available bandwidth using degradation mechanisms. Such adaptation must be performed on a timescale of seconds to keep latency low.
- Allowing users to formulate policies for collecting data that maximize data value and that can be implemented effectively by the system. The policy framework must be expressive enough to meet the data quality needs of diverse queries. In particular, it should support combining multiple aggregation and degradation mechanisms.

In meeting these challenges, the main function of WDAS is that it adjusts data quality to bound the latency of streaming queries in bandwidth-constrained environments. Previous and current systems, by ignoring bandwidth limitations, force users to make an unappealing choice: either be optimistic about available bandwidth and backhaul too much data, or they can be pessimistic about bandwidth and backhaul only limited data. By integrating durable storage into the dataflow and supporting dynamic adjustments to data quality, WDAS allows a user to focus on two different tradeoffs: deciding which query results are needed in real time and which inaccuracies are acceptable to maintain real-time performance.

6.4.3.1 WDAS design

For WDAS design, one must take into account several essential rules: (i) the system integrates data storage at the edge to allow users to collect data that may be useful for future analysis without necessarily transferring it to a central location; (ii) users can define ad-hoc queries as well as standing queries, which can be used to create centralized data structures to optimize the performance of common queries; and (iii) a standing query has a hard real-time requirement: if the query cannot handle the incoming data rate, then queues and queuing delays will grow, which means that the system must keep latency bounded.

Since WDAS aims to provide low-latency results on standing queries, the designer can use the basic computation model of traditional stream-processing systems. A worker process runs on each participating compute node. A query is implemented by a set of linked dataflow operators that operate on streams of tuples. Each operator is placed on a particular host and performs some transformations on the data. The system routes tuples between operators, whether on the same host or connected by the network.

6.4.3.2 Aggregation

Integrating structured storage into a streaming operator graph solves several problems. Ideally, WDAS should handle several different forms of aggregation in a unified way. Windowed aggregation combines data residing on the same node across time. Tree aggregation combines data residing on different nodes for the same period. This can be applied recursively: data can be grouped within a data center or point of presence, and then data from multiple locations is combined at a central point. Both these forms of aggregation can be handled by inserting data into structured storage and then extracting it by appropriate queries. A solution for structured storage is a cube interface similar with that of OLAP data warehouses (Rabkin et al. 2014). However, cubes in data warehouses typically involve heavy precomputation, incurring high ingest times to support fast data. In contrast, for WDAS one can uses cubes for edge storage and aggregation and for that only maintain a primary-key index. This reduces ingest overhead and allows cubes to be effectively used inside a streaming computation. Integrating storage with distributed streaming requires new interfaces and abstractions. Cubes can have multiple sources inserting data and multiple unrelated queries reading results. In a single-node or data center system, it is clear when all the data has arrived, but in a wide-area system, synchronization is costly, particularly if nodes are temporarily unavailable.

A good solution seems to be the practical methodology for recursive multidimensional flow aggregation proposed in Kato et al. (2012). The key insight is that, once aggregated flow records are created, they can be efficiently reaggregated with coarser temporal and spatial granularity. The advantage is that an operator can monitor only coarse-grained aggregated flows to identify anomalies and, if necessary, he can look into more fine-grained aggregated flows by changing aggregation granularity. The algorithm ensures a two-stage flow aggregation: the primary aggregation stage focuses on efficiency while the secondary aggregation stage focuses on flexibility.

The primary aggregation stage reads raw traffic records and produces rudimentary aggregated flow records on the fly. The secondary aggregation stage reads the output of the primary aggregation stage, and produces concise summaries for operators. In both aggregation stages, from the 5-tuple of attributes, source and destination IP addresses are used as main attributes, and protocol and ports as subattributes, following the operational practices. Although it is possible to treat each attribute equally, it is more expensive and the use case is fairly limited. To identify significant flows, both traffic volume and packet counts are used, in order to detect both volume-based and packet count-based anomalous flows.

6.4.3.3 Degradation

In order to maximize data quality, it is desirable to apply data reduction only when necessary to conserve bandwidth. Keeping this in mind, WDAS includes specialized tunable degradation operators. The system automatically adjusts these operators to match available bandwidth, thus minimizing the impact on data quality. Because the system dynamically changes degradation levels, it needs a way to indicate the fidelity of the data it is sending downstream. Therefore, the system has the ability to send metadata messages both upstream and downstream, which are ordered with respect to data messages. This mechanism is useful to signal congestion and also to implement multiround protocols within the operator graph.

Degradation operators can be either standard operators that operate on a tuple-by-tuple basis, or cube subscribers that produce tuples by querying cubes. A degradation operator is associated with a set of degradation levels, which defines its behavior on data streams. Even with partial aggregation at sources, some queries will require more bandwidth than is available. To keep latency low, WDAS allows queries to specify a graceful degradation plan that trades a little data quality for reduced bandwidth. Similarly, quantitative data can be degraded by increasing the coarseness of the aggregation or using sampling. Since degradations impose an accuracy penalty, they should only be used to the extent necessary. This is achieved by using explicit feedback control, through which WDAS can detect congestion before queues fill up, enabling it to respond in a timely manner. Moreover, this feedback reflects the degree of congestion, allowing to tune the degradation to the right level. This avoids expensive synchronization over the wide area, but it also means that sources may send data at different degradation levels, based on their bandwidth conditions at the time.

The best degradation for a given application depends not only on the statistics of the data, but also on the set of queries that may be applied to the data. As a result, the system cannot choose the best degradation without knowing this intended use. Some guidelines to use degradations in different circumstances are as follows:

- For data with synopses that can be degraded, degrading the synopsis will have predictable bandwidth savings and predictable increases in error terms.
- If the dimension values for the data have a natural hierarchy, then aggregation can effectively reduce the data volume for many distributions.
- For long-tailed distributions, a filter can remove the large, but irrelevant tail.

- A global value threshold gives exact answers for every item whose total value is above a threshold; a local threshold will have worse error bounds, but can do well in practice in many cases.
- Consistent samples help to analyze relationships between multiple aggregates.
- Users can specify an initial degradation with good error bounds, and a fall-back degradation to constrain bandwidth regardless.
- The operators have flexibility in how they estimate the bandwidth savings; levels can be (i) dynamically changed by the operator; (ii) statically defined by the implementation; and (iii) configurable at runtime.

Anyway, it is obvious that no single degradation technique is always best; a combination of techniques can perform better than any individual technique. By combining multiple techniques in a modular and reusable way using adaptive policies, it is possible to create a powerful framework for streaming wide-area analysis.

6.4.4 Monitoring Cloud services using NetFlow standard

The advantages of Cloud services, such as reduced costs, easy provisioning, and high scalability, led to heightened interest from organizations in migrating services (especially file storage and e-mail) to Cloud providers. However, such migration also has its drawbacks. Cloud services have been repeatedly involved major failures, including data center outages, loss of connectivity, and performance degradation. The goal of this section is to investigate to what extent the performance of Cloud services can be monitored using standard monitoring programs.

In the early 2000s, the IPFIX Working Group was formed in the IETF, to standardize protocols to export IP flows. While originally targeting some key applications (Zseby et al. 2009), such as accounting and traffic profiling, flow data have proven to be useful for several network management activities. A reference architecture for measuring flows is presented in Sadasivan et al. (2009). Three main components are part of this architecture

1. IPFIX devices that measure flows and export the corresponding records. IPFIX devices include at least one exporting process and, normally, observation points and metering processes. Metering processes receive packets from observation points and maintain flow statistics. Exporting processes encapsulate flow records and control information (e.g., templates) in IPFIX messages, and send the IPFIX messages to flow collectors.
2. Collectors that receive flow records and control information from IPFIX devices and take actions to store or further process the flows.
3. Applications that consume and analyze flows.

Because flow records are the primary source of information in applications analysis and since any flow-based application requires flow data of good quality to perform its tasks satisfactorily, it is essential to understand how flows are measured in practice.

Figure 6.18 illustrates the typical tasks of an IPFIX device (Sadasivan et al. 2009). First, packets are captured in an observation point and timestamped by the metering process. Then, the metering process can apply functions to sample or filter packets. Filters select packets deterministically based on a function applied to packet contents. Samplers, in contrast, combine such functions with techniques to randomly select packets. Sampling and filtering play a fundamental role in flow monitoring because of the continuous increase in network speeds.

By reducing the amount of packets to be examined by metering processes, sampling and filtering make flow monitoring feasible under higher network speeds. On the other hand, these techniques might imply loss of information, which restrict the usage of flow data substantially. Packets that pass the sampling and filtering stages update entries in the flow cache, according to predefined templates. Flow records are held in the cache until they are considered expired, following the reasons described next.

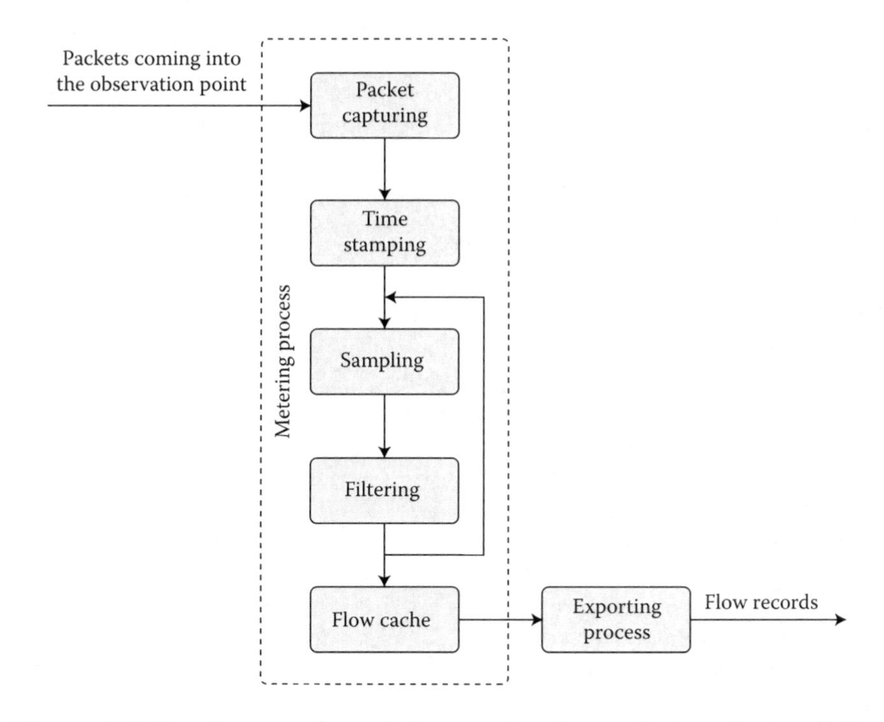

Figure 6.18 Functions performed by a network monitoring device. (After Sadasivan, G. et al., Architecture for IP flow information export, *RFC 5470*, 2009.)

Expired records are made available by the metering process to exporting processes, where they are combined into IPFIX messages and sent out to flow collectors.

IPFIX devices are also called flow exporters—for example, when only exporting processes are present, and their use in flow monitoring is defined as a flow export technologies. Cisco NetFlow is one of the widely deployed flow export technologies, compatible with the standardization effort IPFIX.

Network traffic monitoring standards such as sFlow (Phaal et al. 2001) and NetFlow/IPFIX (Claise 2004; Leinen 2004) have been conceived at the beginning of the last decade. Both protocols have been designed for being embedded into physical network devices such as routers and switches where the network traffic is flowing. In order to keep up with the increasing network speeds, sFlow natively implements packet sampling in order to reduce the load on the monitoring probe. While both flow and packet sampling is supported in NetFlow/IPFIX, network administrators try to avoid these mechanisms in order to have accurate traffic measurement. These principles do not entirely apply in Cloud computing. Physical devices cannot monitor virtualized environments because inter-VM traffic is not visible to the physical network interface. However, virtualization software developers have created virtual network switches with the ability to mirror network traffic from virtual environments into physical Ethernet ports where monitoring probes can be attached. Recently, virtual switches such as VMware vSphere Distributed Switch or Open vSwitch natively support NetFlow/sFlow for inter-VM communications (Deri et al. 2014), thus facilitating the monitoring of virtual environments. Network managers need traffic monitoring tools that are able to spot bottlenecks and security issues while providing accurate information for troubleshooting the cause. This means that while NetFlow/sFlow can prove a quantitative analysis in terms of traffic volume and TCP/UDP ports being used, they are unable to report the cause of the problems.

Because modern services use dynamic TCP/UDP ports, the network administrator needs to know what ports map to what application. The result is that traditional device-based traffic monitoring devices need to move toward software-based monitoring probes that increase network visibility at the user and application level. As this activity cannot be performed at network level (i.e., by observing traffic at a monitoring point that sees all traffic), software probes are installed on the physical/virtual servers where services are provided. This enables probes to observe the system internals and collect information (e.g., what user/process is responsible for a specific network connection) that would be otherwise difficult to analyze outside the system's context just by looking at packets.

Network administrators can then view virtual and Cloud environments in real time. The flow-based monitoring paradigm is by nature unable to produce real-time information. Flows statistics such as throughput can be computed in flow collectors only for the duration of the flow. This means that using the flow paradigm, network administrators cannot have a real-time traffic view due to the latency intrinsic to this monitoring architecture and also because flows can only report average values.

Flow measurements have been successfully employed in a variety of applications that require scalable ways to collect data in high-speed networks, such as to detect network intruders, to perform traffic engineering, and to discover connectivity problems in production networks. NetFlow, in particular, is a widely deployed technology for measuring flows, often available at enterprise and campus networks. Therefore, NetFlow seems to be an alternative to provide immediate and scalable means for customers to monitor their services in the Cloud, without the burdens of instrumenting end-user devices and, more importantly, without any interference from Cloud providers. However, although the use of NetFlow to monitor Cloud services is intuitively appealing, NetFlow measurements are known to be unrelated to high-level metrics usually employed to report the performance of applications, such as availability and response time.

In the rest of this section, the use of NetFlow to monitor Cloud services will be analyzed, on the basis of a method proposed by Drago (2013) that (i) normalizes NetFlow data exported under different settings and (ii) calculates a simple performance metric to indicate the availability of Cloud services.

Method description. Figure 6.19 illustrates a deployment scenario of the method, assuming that a set of routers is exporting flow records while handling the traffic from a group of end users, with aim of using these data to monitor Cloud services. In this example, users in an enterprise network interact with two Cloud services (i.e., Service A and B in the figure). The link between the enterprise network and the Internet is monitored

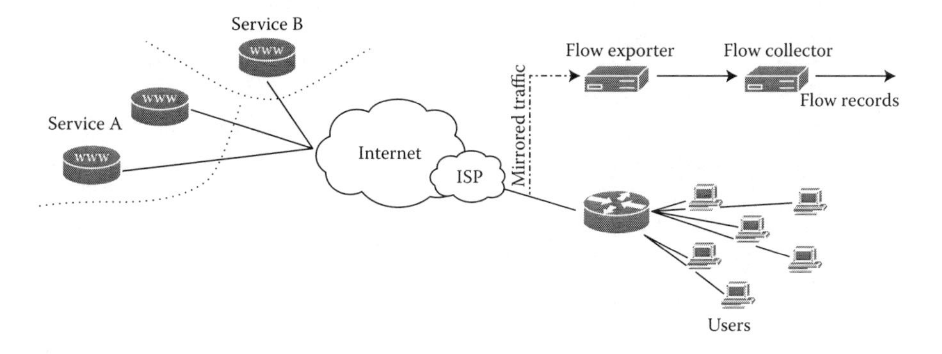

Figure 6.19 Deployment scenario of flow monitoring method. (After Drago, I. *Understanding and Monitoring Cloud Services*, PhD thesis, Centre for Telematics and Information Technology, 2013.)

by a flow exporter. The exporter processes the traffic mirrored from the network and forwards flow records to collectors, from where the records are accessible to analysis applications.

This example makes clear that a single enterprise does not have access to all traffic related to the services, but it can still observe a significant fraction, generated by its own users. Users trying to contact unavailable services will generate traffic with particular characteristics that can be exploited to reveal the status of the Cloud services. Considering such traffic as unhealthy, it is a need for a method to estimate a health index, defined as the fraction of healthy traffic to a service in a given time interval. Applying this procedure means to receive batches of NetFlow records from flow collectors and compute the health index for selected services by performing the following steps:

Step 1. The traffic related to the services is filtered using either predefined lists of server IP addresses or names of IP owners.

Step 2. The data are normalized to remove the effects of parameter settings of flow exporters (for example, the effects of packet sampling or flow expiration policies). This is achieved by estimating the total number of TCP connections n and the number of healthy TCP connections n_h for each service. A TCP connection is considered healthy if it could exchange transport layer payload—that is, if the connection has a complete TCP handshake. Next, before reaching the step 3, the procedure follows two routes, depending on the flow records nature: nonsampled or packet-sampled. These intermediary substeps will be described later in detail.

Step 3. The health index $h_i = n_h/n$, determined as the fraction of healthy TCP connections in the network, is returned for each service. The health index is correlated to the service availability as follows. When a service is fully available, client TCP connections are expected to be normally established, making the index to approach 1; when the service becomes partially or fully unavailable, failed client connection attempts decrease the index. An upper bound for determining the actual service status is necessary. The health index decreases only if a service is unavailable. A high index, however, does not guarantee that a service is operating, since the method does not take application layer errors into account. Now let see how the method handles nonsampled and packet-sampled flows.

6.4.4.1 *Nonsampled flow data*

When packet sampling is disabled, all packets are taken into account, and flow records provide a summary of the observed TCP flags. Since different TCP connections are not reported in a single flow record, it is possible to monitor the status of TCP connections by identifying the flow records that belong to each connection. Hence, the normalization of nonsampled flow data is executed in two steps: (1) records that report information about the same TCP connection are aggregated and (2) the information about TCP flags are evaluated to determine n and n_h.

Aggregating method means that flow records with identical keys—that is, sharing the same IP addresses and port numbers—are merged until the original TCP connection is considered complete. Since NetFlow is usually unidirectional, flow records from both traffic directions are also aggregated. The critical step for this heuristic is to determine when a TCP connection is complete, such that later records sharing the same key start a new connection. Normally, a TCP connection is complete if an idle timeout has elapsed or if records with TCP FIN flag set have been observed from both end-points and a short timeout has elapsed. Therefore, the new records sharing the same key are part of a new TCP connection, even if the time interval between the TCP connections is shorter than the previous timeout parameter.

Finally, the evaluation of the aggregated records is performed. The total number of TCP connections n as well as the number of healthy connections n_h can be trivially counted from the aggregated flow records. More precisely, all TCP connections are counted as healthy, except for failed attempts (when clients send a TCP SYN packet and do not obtain any reply) or for rejected TCP connections (when servers reset connections without completing the handshake).

6.4.4.2 Packet-sampled flow data

The flow data exported under sampling are normalized as follows. First, it estimates the total number of TCP connections $\hat{n}$ and the number of healthy TCP connections $\hat{n}_h$ directly from the raw packet-sampled flow records. Owing to sampling, those numbers are realizations of random variables. In order to compare the quantities in a meaningful way, confidence intervals are determined and taken into account: if the intervals overlap, or if $\hat{n}_h > \hat{n}$ there is not enough evidence of unhealthy TCP connections in the network. Therefore, we make $\hat{n}_h = \hat{n}$ and the index is reported to be 1; otherwise, the health index is calculated using $\hat{n}$ and $\hat{n}_h$.

For estimating the original number $\hat{f}$ of TCP connections from raw packet-sampled flows, we recommend the method described in Drago (2013). If packets are sampled independently with probability $p = 1/N$, then $\hat{f} = Nk_S$, where k_s is the number of observed records with SYN flag set, is proven to be an unbiased estimator for the number f of TCP flows, under the assumption of one single SYN packet per TCP connection. In a similar reasoning, mode is estimated the number of healthy TCP connections. Because healthy connections must have a complete TCP handshake, we assume that these connections have exactly one SYN packet from both originators and responders, while unhealthy TCP connections have exactly one SYN packet from originators, but none from responders.

Let k_{orig} and k_{resp} be the number of observed flow records with SYN flag set from originators and responders, respectively, in a time interval. Then, $\hat{n} = Nk_{orig}$ and $\hat{n}_h = Nk_{resp}$ are unbiased estimations of the total number of TCP connections n and of the number of healthy TCP connections n_h in that time interval.

The expected values of $\hat{n}$ and $\hat{n}_h$ are equal if a service is available. However, a system can only be considered unhealthy if the difference between $\hat{n}$ and $\hat{n}_h$ is statistically significant. We evaluate the difference between $\hat{n}$ and $\hat{n}_h$ by estimating the probability distributions of n and n_h and calculating the confidence intervals. If confidence intervals for a given level of significance show a high probability that $n_h < n$, then $\hat{n}$ and $\hat{n}_h$ are returned. Otherwise, we make $\hat{n}_h = \hat{n}$ and the service is considered healthy.

The method assumes that n and n_h are constant over r consecutive time intervals, which is a reasonable approximation for highly loaded services in short time intervals.

There are some conclusions to formulate in order to understand not only the advantages, but the limitations of using NetFlow to monitor Cloud services and so provide appropriate guidelines for using this tool.

1. Most procedures require to filter traffic based on IP addresses and IP owners. These methods perform well only if flow records are filtered properly, otherwise they are ineffective.
2. All methods must identify flows generated by an application. In particular, the limited information normally exported by NetFlow makes it hard to isolate the traffic of Cloud services in flow datasets, but this limitation can be overcomed by augmenting flow measurements with other information that can be passively observed in the network as well.
3. One cannot ignore the application layer semantics. Even if a method is informative to reveal the most severe availability problems, in which clients cannot establish communication with providers, it is very common that servers are still able to handle part of users' requests when suffering performance degradation. In such situations, a mix of healthy traffic, unhealthy traffic at the transport layer, and unhealthy traffic at the application layer can be observed in the network.
4. Application-specific knowledge has to be taken into account as well, if the precise identification of all availability problems is necessary. Such advanced monitoring requires flow exporters that can measure and export customized information per application.
5. A new idea can be taken into account regarding how to monitor problems in Cloud services. Most traditional methods rely on active measurements to analyze the performance of services or to benchmark Cloud providers. Other methods focus on measuring the performance of the infrastructure providing the services. The new approach differs

from those in several aspects, because it do not focus on a specific service, but instead propose a generic method applicable to a variety of Cloud services. It is a passive approach to Cloud monitoring because the active alternatives lack the ability to capture the impact of problems on actual end users. On the other hand, in contrast to active methods, a passive method can only identify problems in a Cloud service if end users are interacting with it. Consequently the use of NetFlow for measuring the availability of a remote service is central to this method.

6. Usually incoming/outgoing flow records are matched and alerts are triggered when the number of records without a match increases. We consider that is better to apply the differentiation when dealing with nonsampled data or with sampled data. When dealing with nonsampled data, we rely on a technique for aggregating flow records, using the normalization of the flow data, thus removing the effects of timeout parameters. For postprocessing packet-sampled flows, the recommendation is to estimate several properties of the original data stream, such as the flow length distribution, using packet-sampled flow records only.
7. There are limiting factors for the use of NetFlow to monitor Cloud services. These factors include the difficulties for filtering traffic and the need for more information in flow records.

As final conclusion, let note that the method discussed in this chapter for determining the health of a remote service solely from flow measurements is by no means a comprehensive solution to all monitoring needs of Cloud customers. However, by focusing on serious performance problems and by relying on flows exported by a widely deployed technology, it delivers an essential and immediate first layer of monitoring to Cloud customers, while being applicable to a wide range of services and scalable to high-speed networks.

6.4.5 Implementing Cloud services in the automation domain

The latest challenge in the industrial sector is oriented to developing Cloud architectures and services that will provide not only increased storage space and increased reliability but also high computing power and accessibility to advanced process control applications that can be used as services for modeling and optimization of industrial plants. The main challenges identified in this domain come from the variable access times and possible accessibility problems that arise from the virtualization of the Cloud components.

When integrating a process control application in a Cloud environment, we need to identify a generic, flexible solution adaptable to different application domains, programming environments, and hardware equipment. This section presents as original solution a model for integrating an advanced process control library in a Cloud-based environment.

Cloud computing is a technology that provides computational resources as a commodity. It enables seamless provisioning of both the computing infrastructure and software platforms. It is also an efficient way to manage complex deployments, as most of the installation and configuration task are automatically handled by the Cloud provider.

There are currently three main models for providing Cloud services: Infrastructure-as-a-Service (IaaS), Platform-as-a-Service (PaaS), and Software-as-a-Service (SaaS). IaaS Cloud solutions provide basic building blocks of the infrastructure—computing, storage, and network as a service. The PaaS Cloud solutions provide a high-level stack of services that enables application-specific environments. SaaS applications are hosted on the physical or virtual infrastructure and the computing platform is managed by the Cloud provider. They can run on the PaaS and IaaS layers as this is transparent to the customer who is only able to access and use the SaaS application.

The proposed system represents a Cloud-based support platform for process control engineers in the use and integration of advanced algorithms and control strategies. The architecture of the proposed system is illustrated in Figure 6.20. The system functionality is organized in relation to three levels: the user level, the Cloud level, and the process level.

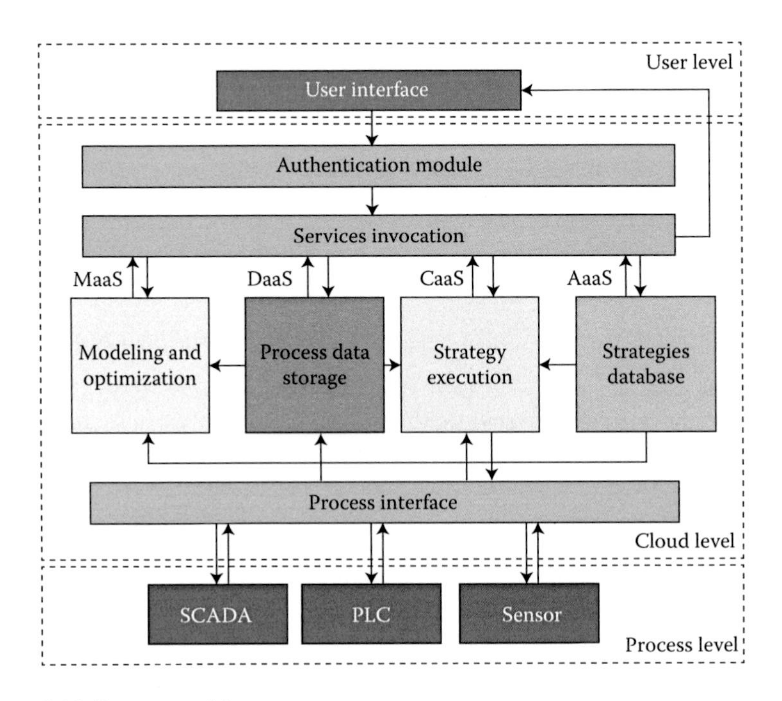

Figure 6.20 System architecture.

The library is organized as a modular structure, one for each service provided, with a clear decomposition of the functionality encapsulated at each component. The intermodule communication uses SOA services, while the communication with the process level uses OPC-UA as middleware for an easier integration with existing hardware equipment and SCADA applications.

Based on the authentication credentials, user's rights vary from requesting the download or execution of available strategies, to access to Cloud historian, to use the Cloud collaborative platform to upload new strategies or update the existing ones.

Four types of services are defined in this structure: (1) Model as a Service (MaaS): the user can request a model of the process based on specific data retrieved from the plant; (2) Data as a Service (DaaS): the platform can be used as a process data historian, providing increased reliability, security, and a large storage space; (3) Control as a Service (CaaS): the platform enables the remote execution of any algorithm or strategy available in the library, providing benefits in implementing advanced control algorithms; and (4) Algorithm as a Service (AaaS): the algorithms and strategies available in the library can be downloaded as standardized function blocks.

The strategies database stores all the available functions, algorithms, and strategies as XML files.

Their representation is based on the IEC 61499 standard. The execution module provides the functionality of a superior control level. For each CaaS service requested for a specific algorithm the execution module implements a virtual controller responsible with the execution that sends the results to a device in the controlled plant. A strategy can call the execution of another strategy.

Also, because of the use of the IEC 61499 standard, the execution can be distributed between several nodes of the Cloud architecture. A strategy execution takes place in real time but the virtualization of the Cloud environment can influence the execution time. For that reason, this functionality can be used only in noncritical control applications. The process data storage module can be accessed as a DaaS service by a library user and is also accessed by adjacent modules in case of modeling algorithms execution or process model optimization.

The modeling and optimization module runs complex computing applications for offline process model identification and performance optimization.

The process level implements standard automation control systems: sensor and actuator elements, PLC, DCS, or RTU controllers and SCADA applications. The interaction with the process is done using secured channels. This level is responsible for local control of an industrial plant and for implementing intelligent communication interfaces for the integration with the Cloud level (Ionescu 2014).

In order to take advantage of the main benefits that are provided by the Cloud infrastructures, such as the dynamic provisioning of computing resources, scalability, on-demand access, reliability, and elasticity, it is proposed to perform the deployment of the virtual controllers in a PaaS infrastructure.

The remote execution module can be used for real-time model identification or for offline model improvement. Considering the process parameters as input variables in the model execution, we compute the model response and compare it to the real process information to see if the accuracy indicators are met. Figure 6.21 illustrates the implementation details of the remote execution module.

When a user accesses the Web library web interface for executing an algorithm, the web server receives his request, parses the information regarding the execution details (sequence ID, remote device communication parameters, algorithm execution), and creates a new instance of the execution container.

Each user is assigned a different container instance. A container contains one application manager module and a number of algorithm execution modules associated to all the execution started by a specific user. If the processor effort on a container is overloaded, new instances can be further deployed. A container is terminated when all sequences finished their execution. A service orchestration module is responsible for container creation, termination, and management based on a user-IP table that keeps track of running containers and their associated users.

The communication with the process level is done using OPC-UA for easier integration with existing SCADA applications. The architecture for a library of process control strategies managed in a Cloud environment offer an extension of the current research in the automation domain by moving advanced control applications to a Cloud environment for increased processing power and storage space.

6.5 *Developing and validating strategies for traffic monitoring on RLS models*

In last part of this work, the behavior of the most used planning algorithms is analyzed in terms of guaranteeing the performance and the fairness between the traffic flows. As it was already shown in previous chapters, resource planning is an important element in QoS assurance. Both differentiated (DS) and integrated (IS) services use to some extent various planning algorithms. Features particular to each type of service impose using different algorithms. For example, in the case of IS, fair planning algorithms (FQ, DRR) are preferred and in the case of SD, while in the case of DS, active queue algorithms are chosen.

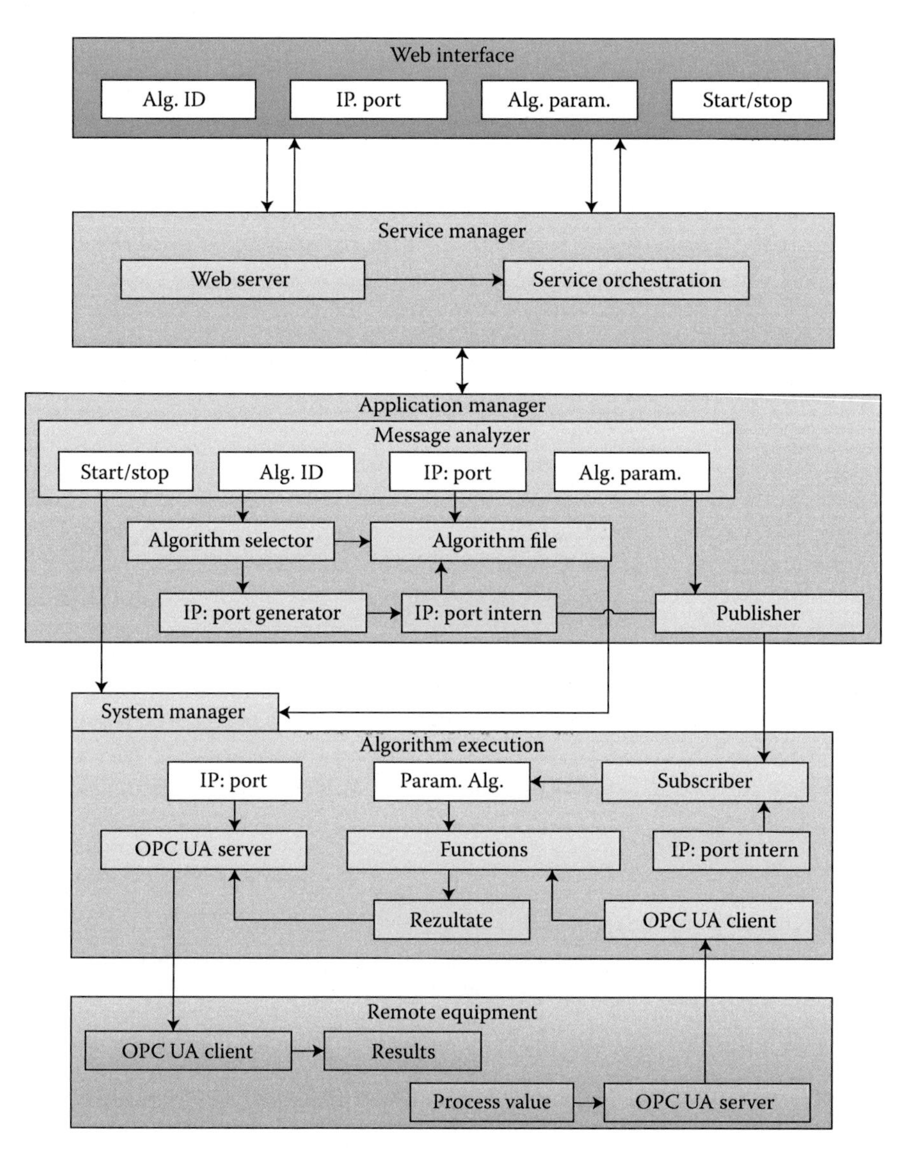

Figure 6.21 Architecture of the strategy execution module.

6.5.1 *Simulation framework*

For simulating, the open source network simulator ***ns*** was used, running on a Linux distribution (openSuse). ***ns*** is one of the most popular network simulators through the academic environment because of its extendibility (open source model) and its vast available documentation. It can be used for simulating routing and multicast protocols, both in fixed and in mobile networks. The development of ***ns*** started in 1989 as a variant to the network simulator REAL. Until 1995, ***ns*** received the support of DARPA, VINT project of LBL, Xerox PARC, UCB, and USC/ISI.

6.5.1.1 *Network topology*

The chosen topology is that of a star (Figure 6.22), in which the nodes 0 through 4 are directly connected to the output node 5 that provides a link with the outside through node 6. This topology is typical for a small size network that shares one Internet gateway. The problem of resource planning manifests at node 5 that must receive all the network-generated traffic and send it to node 6. As it can be seen, the speed of the internal links is much higher than for the outer link, resulting in it being easily saturated.

6.5.1.2 *Traffic sources*

In order to comparatively test the algorithm performance, four traffic sources have been used that should accomplish a high enough load. Two of them generated constant traffic (CBR) over UDP and the other two acted in bursts over TCP. The UDP traffic simulates a streaming application (voice, video) that is sensitive to delays and delay variations. Every one of the UDP traffic sources generate fixed size packets at a rate of 200 kbit/s. Considering the external link to be of 1 Mbit/s and the presence of four traffic sources, the fair share of each flux is of 250 kbit/s. So, the traffic generated by the two sources is lower than their share and, in the case of a correct bandwidth planning, should not experience delays or transfer rate variations.

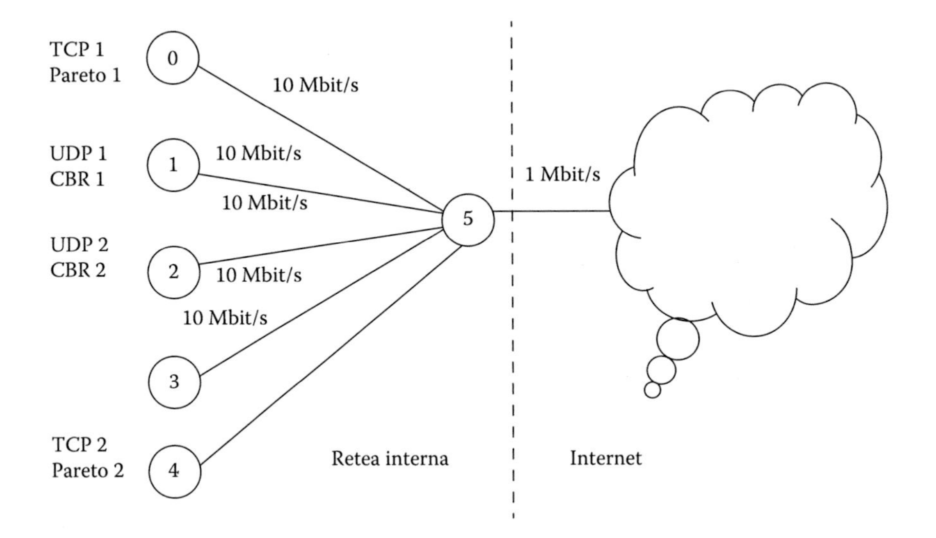

Figure 6.22 Network topology.

The TCP traffic sources simulate Web-based traffic, which is defined by bursts of data and a rather constant transfer rate. For generating this traffic, a Pareto distribution source was used. Each of the TCP sources is configured to send packets on the output link at a mean transfer rate of 200 kb/s. Because of the traffic characteristics, sometimes the transfer rate can reach higher values, especially when sending burst packets.

6.5.1.3 Simulation code

For simulation, the following script was used to create the network and the traffic sources, plan the network traffic and monitor the queue of node 5 using various algorithms. For each algorithm, the changes were minor, replacing only the queue type from the initial declaration between nodes 5 and 6 and the size of the queue so that it is the same for each algorithm.

These scripts were run using ***ns***, and the results were processed to obtain information related to delay, average transfer rate and delay variation for each data flow.

```
#Create a new simulator object
set ns [new Simulator]

#Define a different colour for each flow (for NAM)
$ns color 1 Blue
$ns color 2 Red
$ns color 3 Green

#Open the log file for NAM
set nf [open out.nam w]
$ns namtrace-all $nf

#Open log file
set tf [open out.tr w]
$ns trace-all $tf
set f0 [open cbr1.tr w]
set f1 [open tcp1.tr w]
set f2 [open tcp2.tr w]
set f3 [open cbr2.tr w]

#Define exit procedure
proc finish {} {
     global ns nf tf
     $ns flush-trace
     close $nf
     close $tf
     exec nam out.nam &
             exec xgraph cbr1.tr tcp1.tr tcp2.tr cbr2.tr &
     exit 0
}
```

```
#Define the recording procedure for the average data rates
proc record {} {
     global null2 sink1 sink0 null1 f0 f1 f2 f3
          set ns [Simulator instance]
     set time 0.5
     set bw0 [$null2 set bytes_]
     set bw1 [$sink1 set bytes_]
     set bw2 [$sink0 set bytes_]
          set bw3 [$null1 set bytes_]
     set now [$ns now]
     puts $f0 "$now [expr $bw0/$time*8/1000000]"
     puts $f1 "$now [expr $bw1/$time*8/1000000]"
     puts $f2 "$now [expr $bw2/$time*8/1000000]"
     puts $f3 "$now [expr $bw3/$time*8/1000000]"
     $null2 set bytes_0
     $sink1 set bytes_0
     $sink0 set bytes_0
     $null1 set bytes_0
     $ns at [expr $now+$time] "record"
}

#Create nodes
set n0 [$ns node]
set n1 [$ns node]
set n2 [$ns node]
set n3 [$ns node]
set n4 [$ns node]
set n5 [$ns node]
set n6 [$ns node]

#Create links between nodes
$ns duplex-link $n0 $n5 10Mb 10ms DropTail
$ns duplex-link $n1 $n5 10Mb 10ms DropTail
$ns duplex-link $n2 $n5 10Mb 10ms DropTail
$ns duplex-link $n3 $n5 10Mb 10ms DropTail
$ns duplex-link $n4 $n5 10Mb 10ms DropTail
$ns duplex-link $n5 $n6 1Mb 20ms DropTail

$ns queue-limit $n5 $n6 30

$ns duplex-link-op $n0 $n5 orient down
$ns duplex-link-op $n1 $n5 orient right-down
$ns duplex-link-op $n2 $n5 orient right
$ns duplex-link-op $n3 $n5 orient right-up
$ns duplex-link-op $n4 $n5 orient up
$ns duplex-link-op $n5 $n6 orient right

$ns duplex-link-op $n5 $n6 queuePos 0.5
```

```
#Create traffic sources

#The first TCP traffic source
set tcp0 [new Agent/TCP]
$tcp0 set class_2
$ns attach-agent $n0 $tcp0
set sink0 [new Agent/TCPSink]
$ns attach-agent $n6 $sink0
$ns connect $tcp0 $sink0
$tcp0 set fid_1

set par [new Application/Trafic/Pareto]
$par attach-agent $tcp0
$par set packetSize_210
$par set burst_time_50ms
$par set idle_time_50ms
$par set rate_200kb
$par set shape_1.5

#The second TCP traffic source
set tcp1 [new Agent/TCP]
$tcp1 set class_3
$ns attach-agent $n4 $tcp1
set sink1 [new Agent/TCPSink]
$ns attach-agent $n6 $sink1
$ns connect $tcp1 $sink1
$tcp1 set fid_1

set par1 [new Application/Trafic/Pareto]
$par1 attach-agent $tcp1
$par1 set packetSize_210
$par1 set burst_time_50ms
$par1 set idle_time_50ms
$par1 set rate_200kb
$par1 set shape_1.5

#The first UDP traffic source
set udp1 [new Agent/UDP]
$ns attach-agent $n1 $udp1
set null1 [new Agent/LossMonitor]
$ns attach-agent $n6 $null1
$ns connect $udp1 $null1
$udp1 set fid_2

set cbr1 [new Application/Trafic/CBR]
$cbr1 attach-agent $udp1
$cbr1 set type_CBR
$cbr1 set packet_size_100
```

```
$cbr1 set rate_200kb
$cbr1 set random_false

#The second UDP traffic source
set udp2 [new Agent/UDP]
$ns attach-agent $n2 $udp2
set null2 [new Agent/LossMonitor]
$ns attach-agent $n6 $null2
$ns connect $udp2 $null2
$udp2 set fid_2

set cbr2 [new Application/Trafic/CBR]
$cbr2 attach-agent $udp2
$cbr2 set type_CBR
$cbr2 set packet_size_100
$cbr2 set rate_200kb
$cbr2 set random_false

#Planning events
$ns at 0.0 "record"
$ns at 0.1 "$cbr1 start"
$ns at 0.3 "$cbr2 start"
$ns at 1.0 "$par start"
$ns at 1.1 "$par1 start"
$ns at 4.0 "$par1 stop"
$ns at 4.1 "$par stop"
$ns at 4.2 "$cbr2 stop"
$ns at 4.5 "$cbr1 stop"

$ns at 4.5 "$ns detach-agent $n0 $tcp0; $ns detach-agent $n6
    $sink0"
$ns at 4.5 "$ns detach-agent $n4 $tcp1; $ns detach-agent $n6
    $sink1"

$ns at 5.5 "finish"

#Run the simulation
$ns run
```

6.5.2 *Algorithms ran in simulation*

6.5.2.1 *DropTail*

DropTail is one of the simplest algorithms, also named FIFO (first in first out). It is also one of the most implemented and used algorithms. Additionally, it is the de facto standard routing method in the Linux kernel and in the majority of all the routers in the world. It employs a single tail to serve packets. The processing applied to flows is equal. Packets are extracted from the tail in the order of arrival (Figure 6.23).

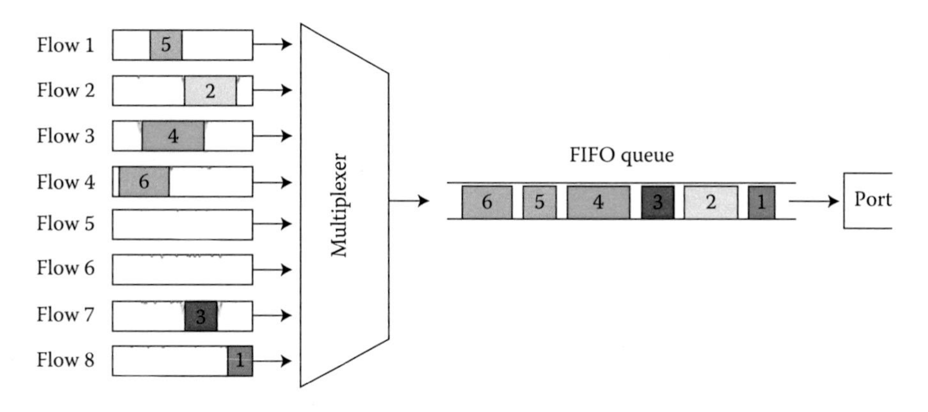

Figure 6.23 DropTail algorithm.

6.5.2.1.1 Advantages

For software routing, DropTail needs very few resources compared to more elaborate algorithm.

The behavior is very predictable, the packets are not reordered and the maximum delay depends on the depth of the tail.

While the size of the tail is small, DropTail provides a simple management of resources without a significant delay at each node.

6.5.2.1.2 Disadvantages

Having a single tail for all incoming packets, it does not allow for organizing stored packets and applying different processing to traffic flows.

The single tail affects all the flows because the medium delay of each one increases when congestion appears.

DropTail planning increases the delay, the delay variation, and the rate of lost packets for real-time applications that cross the tail.

During congestion periods, DropTail tends to favor UDP flows to the disadvantage of TCP flows. When a TCP application detects massive packet losses due to congestion, it will try to reduce the data rate, while UDP transmissions are impassable to such events, thus keeping their nominal transmission rate.

Because the reduction of transfer rate for TCP applications in order to cope with the network conditions, the congestion avoidance mechanism can produce increasing delays, delay variation, and decreasing bandwidth for TCP applications that pass through the tail.

A burst data flow can easily fill the tail, and from that moment on the node rejects packets from other flows while the burst is processed. This leads to an increasing delay and delay variation for all other TCP and UDP applications that behave correctly.

6.5.2.2 DRR

Deficit round robin, known also as deficit-weighted round robin (DWRR), is a modified form of the weighted round robin (WRR) planning algorithm. It was proposed by Shreedhar and Varghese in 1995.

WRR is a variant of the generalized processor sharing (GPS) algorithm. While GPS serves a small quantity of each tail that is not empty, WRR serves a number of packets from each tail that is not empty.

The number of packets = norming (weight/average packet size)

In order to obtain a normalized set of weights, the average packet weight must be known. Only then can WRR correctly estimate GPS. It is best for this parameter to be known before hand. This is not easy to do in real networks, so it must be estimated, which is quite tough to do in the GPS context.

An improved variant of WRR is DRR that can serve variable size packets without needing to know an average size. The maximum dimension of the packets is extracted from the packets in the tail. Packets that are even larger and arrive after determining this value must wait for a new pass of the planner.

While WRR serves every tail that is not empty, DRR serves each tail that is not empty and has a deficit higher than the packet size. If the packet is greater than the deficit, then it is incremented with a value named quantum and the packet is kept waiting. For each flow, the algorithm sends so many packets as the value of the quantum allows. The quantum that remains unused is added to the deficit of the specific tail. The quanta can be allocated based on a tail weight in order to emulate WFQ, and when a flow is waiting and has no packets to send, the quantum is canceled because it had the chance to send packets and it has lost it.

A simple variation of DRR is credit round robin (CRR). In the case of CRR, when a flow has a packet in the tail, but does not have enough quantum to send the packet, the planner sends but it will record the extra quantum in the credit of the flow.

6.5.2.3 FQ

Fair queuing (FQ) was proposed by John Nagle in 1987. FQ is the foundation for an entire class of planning disciplines oriented toward assuring that each flow has a fair access to the network resources and preventing the burst traffic from consuming more than its fair share from the output bandwidth. In FQ, the packets are firstly classified in flows and assigned to a tail dedicated to a certain flow. The tails are then served one packet at a time. Empty tails are not served. FQ is also referred to as flow-based planning.

FQ is sometimes compared to WFQ (weighted FQ). The WFQ algorithms are explained by the fluid model. In this model, it is assumed that

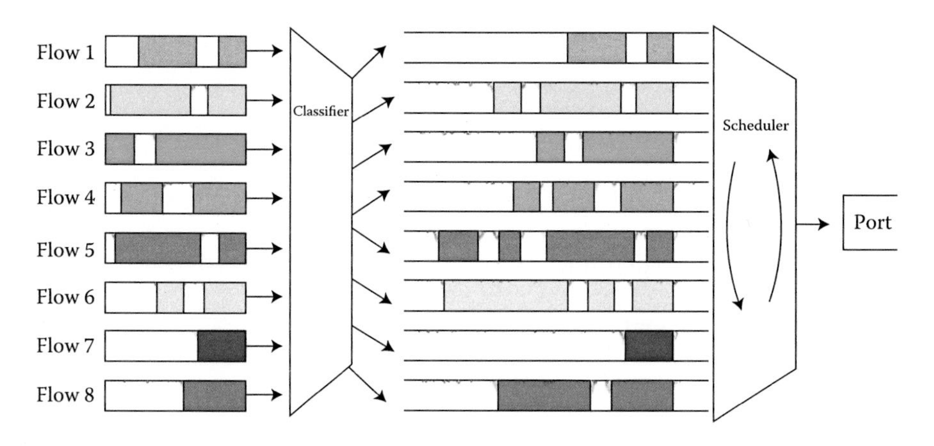

Figure 6.24 FQ algorithm.

the traffic is infinitely dividable and a node can serve simultaneously several flows. In real networks, packets are processed one at a time. So the size of the packets will affect the planning system. Nevertheless, the simplification from the fluid model allows for viewing more easily the planning disciplines operations and, in many cases, is accurate enough to extend the results to a packet system from a fluid model (Figure 6.24).

6.5.2.4 SFQ

SFQ is part of the class of algorithm based on the FQ algorithm. This algorithm was proposed by John Nagle in 1987. SFQ is a simple implementation of fair queuing algorithm. It is more accurate than other but it needs fewer computations, while still being almost perfectly fair.

The key point in SFQ is the conversation or the data flow, that corresponds to a TCP session or a UDP flow. The traffic is divided into a large enough FIFO tails, one for each conversation. It is then sent in a round robin manner, giving each session the chance of sending data one at a time.

This leads to a very fair behavior and avoids situations when a conversation suffocates the rest of the traffic. SFQ is deemed stochastic because it really does not allocate a tail for each session, rather it employs an algorithm that splits the traffic into a finite number of tails using a hashing method.

Because of this hash, several sessions can be allocated to the same tail, which also cuts in half the change of each session to send packets, thus reducing available speed. In order to prevent this to become noticeable, SFQ changes the hashing function frequent enough, so that two colliding sessions will only do it for a limited time.

It is important to note that SFQ is useful only if the output interface is under a heavy load. Otherwise, no tail will exist and its effects will not be visible.

6.5.2.5 RED

RED is an active tail management algorithm. It is also a congestion avoidance method (Figure 6.25).

RED was developed in conjunction with TCP to provide early warning on a potential congestion. RED computes the length of a tail Q_{mean} using an average weight, as follows:

$$Q_{\text{mean}} = (1 - \text{weight}) \times Q_{\text{mean}} + \text{weight} \times Q_{\text{current}}$$

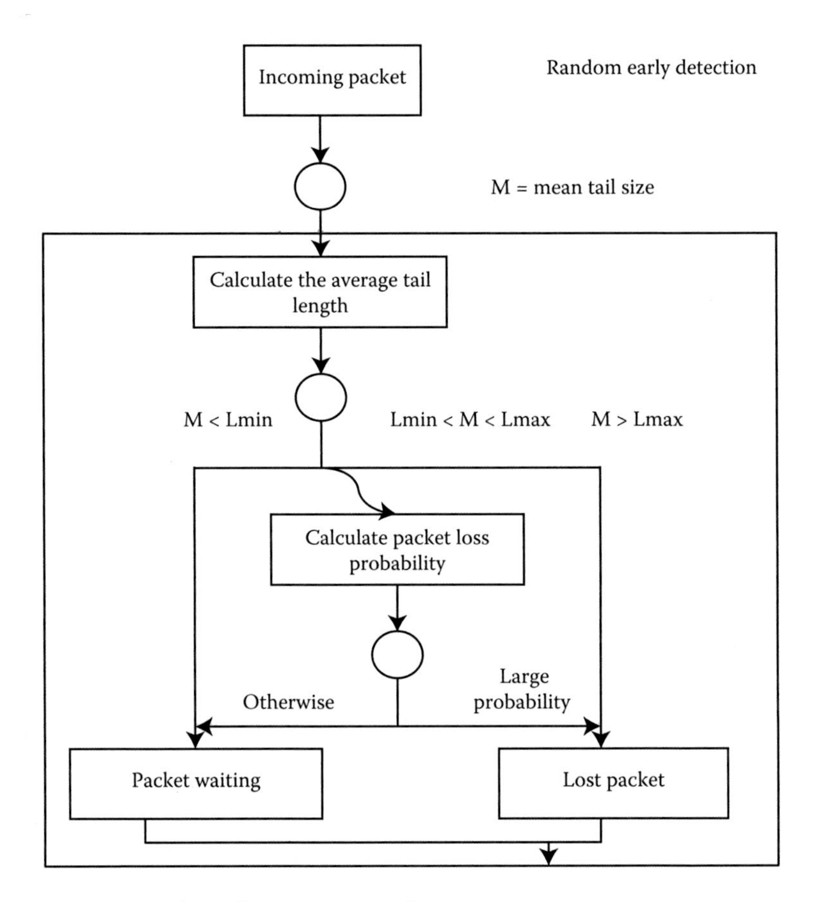

Figure 6.25 RED algorithm processing flow.

The averaging process smoothens the temporal traffic fluctuations and offers a stable indicator for traffic trends on the long term. RED keeps two limit values per tail: MinThreshold and MaxThreshold. Once the tail length drops under MinThreshold, incoming packets are rejected with a probability of *p*. The rejection probability increases linearly towards 1 as the average size of the queue over MaxThreshold. These parameters provide the service providers with fine tune mechanisms for acting against congestions.

When consecutive packets are rejected due to the overloading of the buffer memory, one can observe the synchronizing of the transfer rates for the systems involved in the congestion, as all detect the congestion quasi-simultaneous. Such a synchronization may lead to fluctuations of the queue size. The random rejection in RED helps reduce the synchronization phenomenon. Also, it increases the fairness in rejecting the packets; since the rejection is random, the emitter that sends more packets will be refused more packets in case of a congestion.

6.5.3 *Performance analysis*

Transfer rate, delay, and delay variation were measured in order to analyze the performance of each algorithm. The same simulation conditions are used for all the five algorithms. The results of these measurements are commented in the following pages.

6.5.3.1 *Transfer rate analysis*

The transfer rate is an important parameter to be monitored in the field of QoS. It is important for the majority of applications that use the Internet. Thus, a proper usage of available bandwidth is essential.

In general practice, the max–min planning policy is respected, meaning the maximizing of the minimum rate that a flow receives when participating in a connection. This implies that each flow obtains at least its fair part of the physical bandwidth, and when the load is small and there is some amount of unused bandwidth, the flow is allowed to occupy these resources.

In the case of this analysis, the average transfer rate on the five to six links is monitored for every flow. The graphics are drawn using the Xgraph utility. On the *OX* is shown the simulation time (0–5 s), and on the *OY* is shown the scaled transfer rate value.

The curves on the graph represent:

- Red—traffic to be generated by node 1 of type CBR1
- Green—traffic to be generated by node 2 of type CBR2
- Yellow—traffic to be generated by node 0 of type web TCP1
- Blue—traffic to be generated by node 4 of type web TCP2

6.5.3.1.1 DropTail case As already mentioned in the description of the algorithm, DropTail is not fair and does not offer equal chances to traffic flows that compete for the bandwidth. This can be seen from the significant variation registered for the UDP traffic. The UDP traffic is generated at a constant rate of 200 kB/s, adding to 400 kB/s on the outgoing link. This traffic can be compared to a video/audio conference with a constant rate. Note that when the TCP sources start to emit in bursts and the tail is full, the UDP flows are affected. Their transfer rate begins to oscillate and drops under 200 kB/s. In the moments of "peace" for the network, the transfer rate manages to return to its nominal value (Figure 6.26).

Such a behavior is inacceptable for a loss-sensitive transfer. Each time the data rate drops, the difference up to the nominal value is made of lost packets. In conclusion, DropTail does not manage to provide minimum guarantees for QoS in a network, and thus it is inappropriate for modern applications. Nevertheless, it constitutes a reference point for studying more complicated algorithms due to its simplicity.

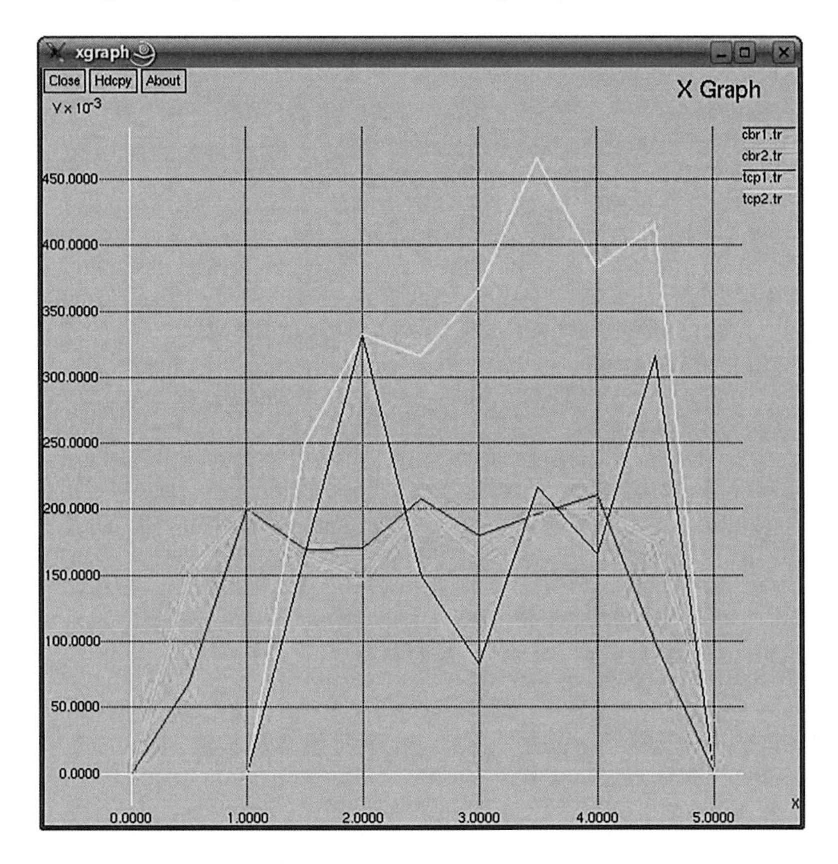

Figure 6.26 DropTail transfer rate.

6.5.3.1.2 DRR case Compared to DropTail, DRR offers fairness in traffic planning. Note that the UDP-generated traffic reaches and remains at the nominal transfer rate (Figure 6.27). Also, for an equal sharing of bandwidth among the four flows, each should get around 250 kB/s, so the 200 kB/s generated by each of the UDP sources are under this threshold and should not be affected by the traffic variation of the other sources. This happens in the case of DDR, which has a good fairness stability.

Another notable thing is that the TCP flows are equal. Each flow manages to occupy its own space without affecting the behavior of the other. The occasional bandwidth spikes are followed by periods when each flow occupies only its share of bandwidth, 300 kB/s.

6.5.3.1.3 FQ case FQ is one of the algorithms that offer and guarantee a fair planning. As DRR, FQ manages to fairly share the bandwidth between all four sources. The UDP traffic remains at a value of 200 kB/s for each source. This time, the variations are smaller than in the case of

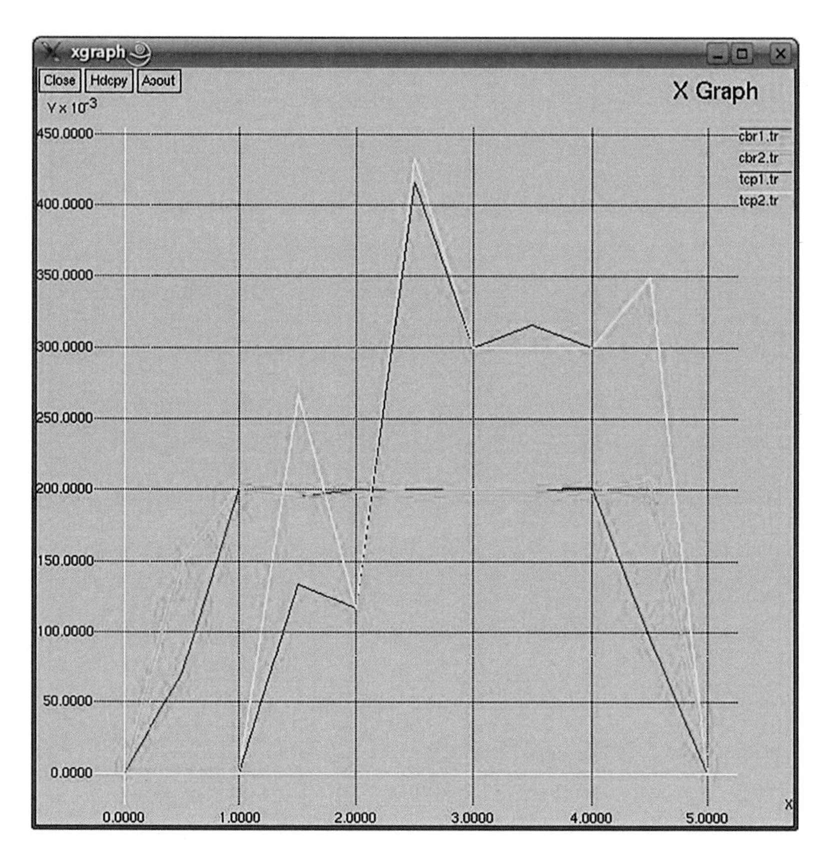

Figure 6.27 DDR transfer rate.

DRR. Note that for the TCP traffic the flows are equal, each one receiving 300 kB/s. An interesting detail is the symmetry between the TCP flows.

In general, the behavior of FQ is better than all other algorithms, managing to reach the best resource sharing in the fairest way. Nevertheless, FQ is quite a complex algorithm (Figure 6.28).

6.5.3.1.4 RED case RED is an active tail management algorithm that rejects packages with a specified probability. In the test, RED managed to shut down completely one of the TCP traffic sources, in the favor of the other one. This is due to the early trying of the source to increase significantly the transfer rate and the traffic burst sent at that moment. The congestion reduction algorithm gradually lowered the transferred rate, allowing the second TCP source to seize all the bandwidth without causing a significant tail increase that would have been penalized by RED. Note that the UDP flows are affected because of the rejection of

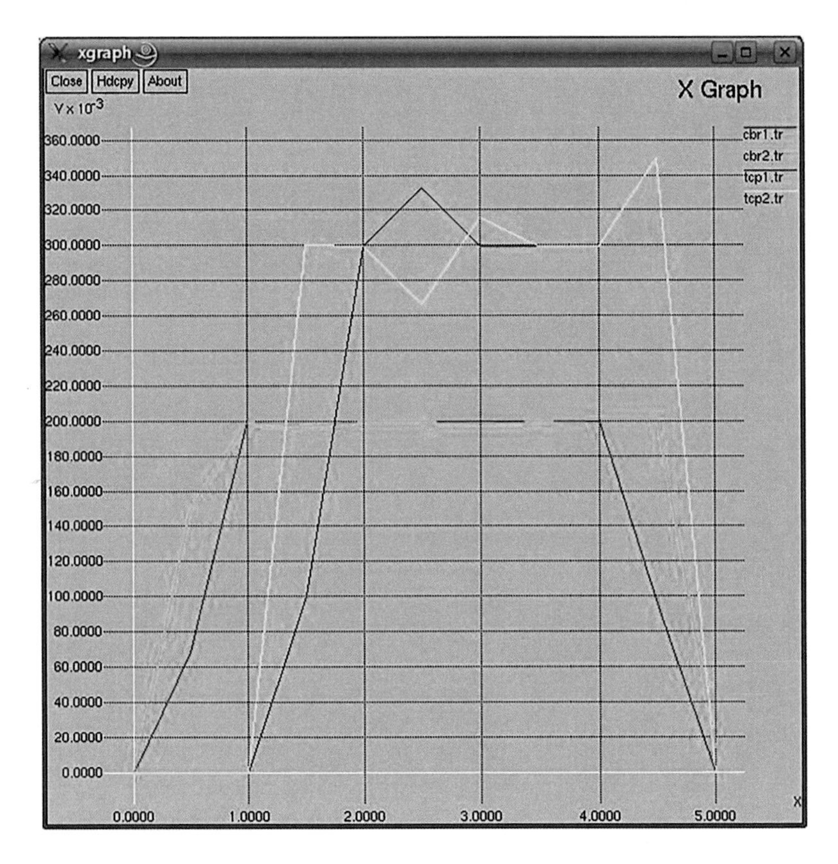

Figure 6.28 FQ transfer rate.

packages with a higher probability as the tail size increases. The moments when the UDP rate drops coincide with the moments when the TCP traffic increases, when the tail is full and packets begin to be rejected, including UDP packets (Figure 6.29).

Compared to FQ and DDR, RED does not accomplish to eliminate the interdependency between flows. The suppression of the TCP2 traffic is unacceptable.

6.5.3.1.5 SFQ case Although part of the FQ algorithms, SFQ displays the worst behavior regarding the UDP bandwidth. This can be due to a collision: the two UDP traffic sources end up in the same SFQ tail and the bandwidth for each one is halved. Although the SFQ algorithm has protection mechanism to prevent such events by changing the hashing method once at a couple of seconds, in our example, the time interval was too short to allow for a reconfiguration (Figure 6.30).

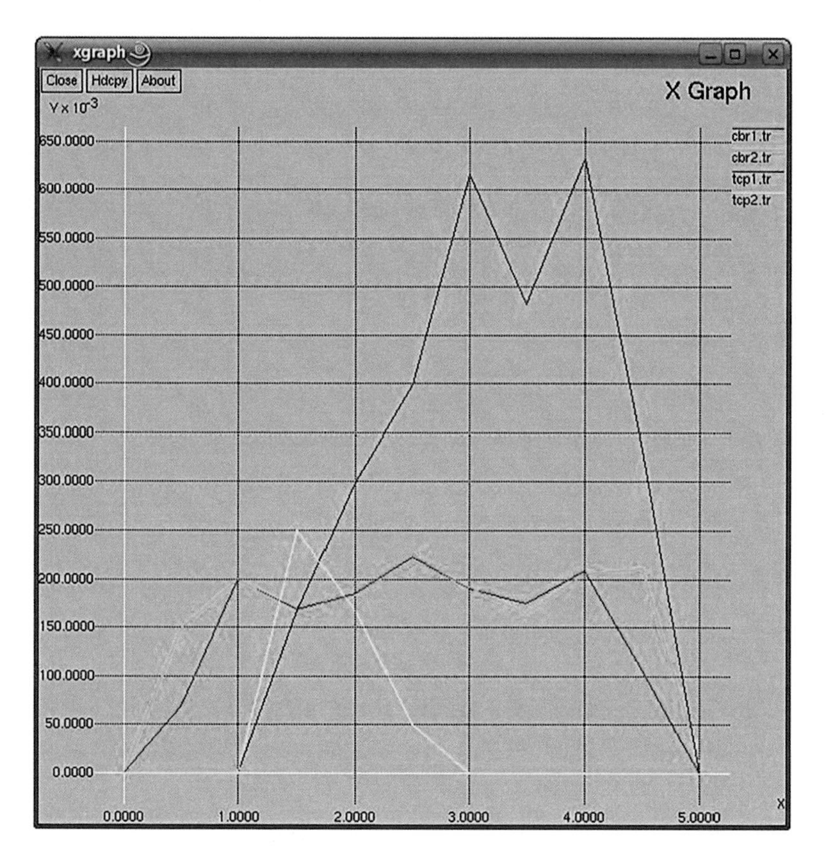

Figure 6.29 RED transfer rate.

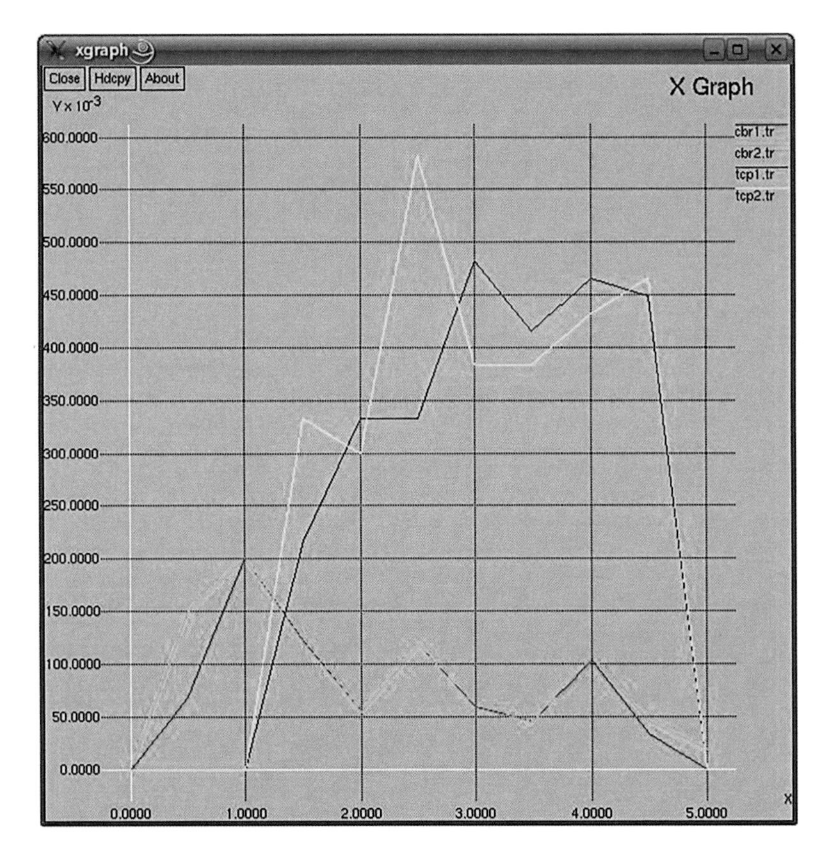

Figure 6.30 SFQ transfer rate.

Regarding the TCP traffic, there is some fairness among the sources. They share the available bandwidth equally. In this case of traffic, the hashing method works.

6.5.3.2 *UDP delay analysis*

UDP flow delays is important for applications that require a QoS. They need, beside a minimum bandwidth, a guaranteed delay. The delay of packets is due to a fixed component (processing, propagation time) and to time spent waiting in the tail. For providing quality services, it is good for this delay to be a priori known in terms of the maximum and average values (Figures 6.31 and 6.32).

For the analysis of the UDP delay, Figures 6.31 and 6.32 present side by side the delay of the two UDP flows in the five cases. On the *OX* axis is the packet number and on the *OY* axis the delay of the respective packet. Because of the difference between the total numbers of packets transferred in each case, the length of the graphs is not equal.

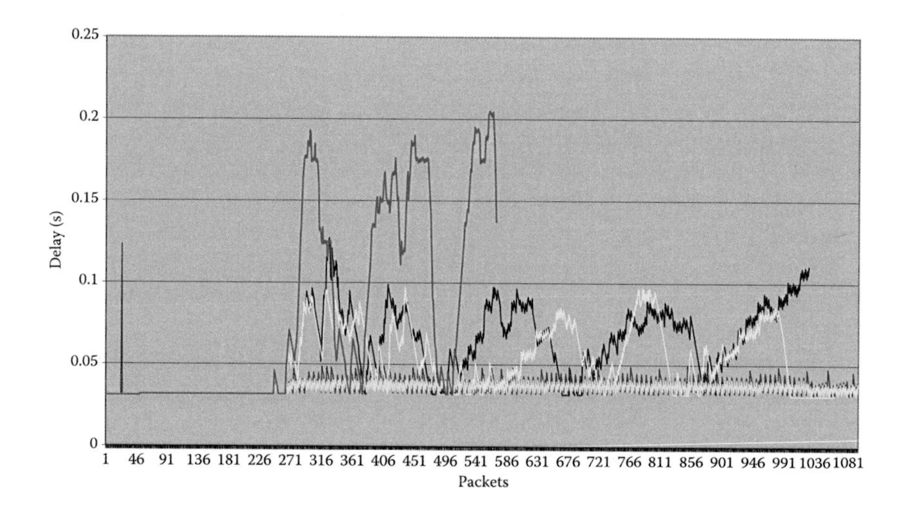

Figure 6.31 UDP 1 source delay.

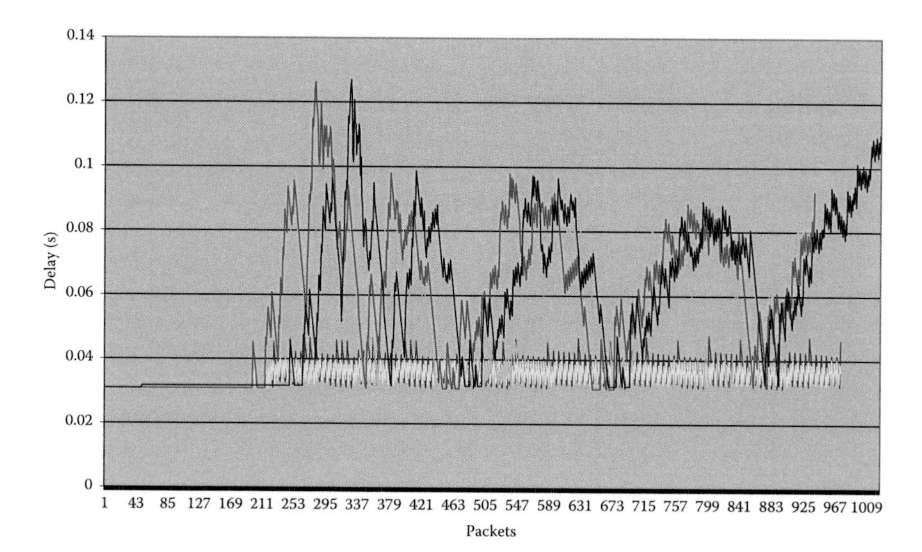

Figure 6.32 UDP2 source delay.

Consider the following legend:

DropTail—dark blue
DRR—red
FQ—yellow
RED—light blue
SFQ—green

As expected, in the analysis made on the transfer rate the algorithms that behaved best were the FQ and the DRR, with an advantage for FQ. The average delay for these algorithms was approximately 0.03 s. The small oscillations that appear in the graph are due to the synchronization of the UDP sources. Both of them transmit a packet at the same time, but the packets cannot be sent simultaneous, so one of them is delayed with a time necessary for sending one packet. This phenomenon alternates between flows. Note that the TCP flows, traffic bursts, do not affect the delay of the UDP packets. This is confirmed by the simulation when in the tail of node 5 there were mostly TCP packets.

In the case of the other algorithms, the delay becomes quite large at some points and it varies strongly. The worst behavior is the one of SFQ. Because of the phenomenon that also affects the transfer rate the UDP packets are delayed for a long time. Note that for these algorithms the TCP bursts also affect the delay of UDP. If for RED and DropTail the delay has a maximum of 0.1 s, for SFQ it can reach even 0.2 for the UDP1 source. A strange thing is the different treatment applied by SFQ to the UDP sources, in favor of the second source. In fact, the behavior of SFQ for source 2 is similar to some extent to RED and DropTail.

6.5.3.3 *TCP delay analysis*

For the TCP traffic analysis, the same way of presenting results will be used, by figuring in a structured manner the delay on both TCP flows in all the five cases (Figures 6.33 and 6.34).

Consider the same legend

DropTail—dark blue
DRR—red
FQ—yellow
RED—light blue
SFQ—green

The first noticeable thing for both traffic sources is that the algorithms that offer the best fairness and are the most correct when allocating bandwidth and packet priority are also the same that introduce the greatest delays in the TCP traffic. This is explainable because in their attempt to occupy as much bandwidth as possible, these flows penalize themselves by large delays and packet losses. Because of the nature of the applications that generate such a traffic, Web browsing, ftp, and so on, the large delay is not a major disadvantage. Note that DRR accomplishes a better distribution of the delay even if it reaches a higher fairness level than FQ.

Regarding the other planning algorithms, DropTail, RED, and SFQ their delay is quite small for the TCP traffic because they do not differentiate so strongly as FQ and DRR do. The lack of packets for the TCP1 flow

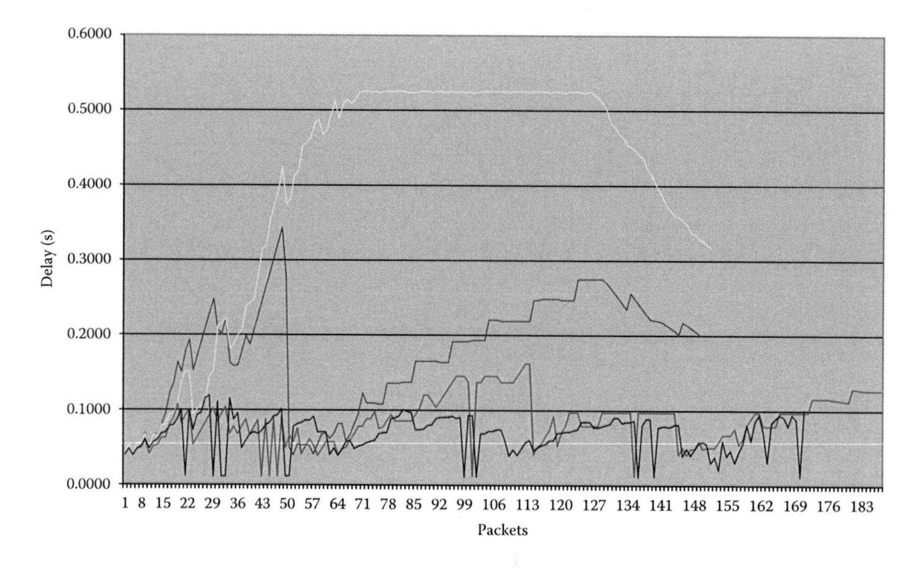

Figure 6.33 TCP1 source delay.

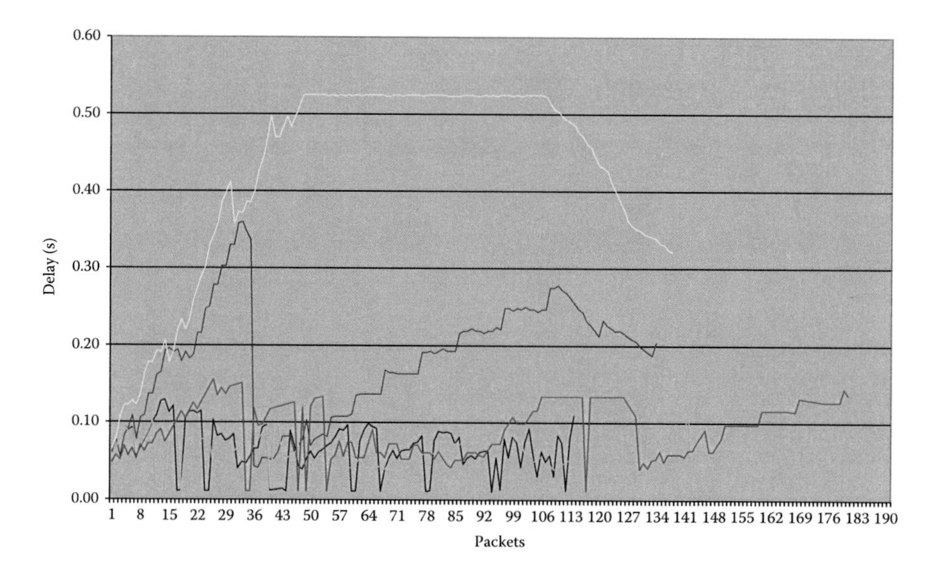

Figure 6.34 TCP2 source delay.

on the RED algorithm is caused by the traffic choking phenomenon, presented in the transfer rate analysis. Another noteworthy detail is that the delay introduced by these algorithms for the TCP traffic is approximately just as large as the one for UDP, thus highlighting the lack of a preferential treatment of flows.

6.6 Conclusions

Considering the way the Internet is growing now, service providers are barely managing to cope with the high demand. The proliferation of broadband connections like DSL, cable, or fiber optics increased the bandwidth for accessing the Internet tenfold. Service providers must cope with this increase by adding more bandwidth to their own networks. The short life of network equipment and the continuous upgrade of network infrastructure help implement QoS on the long term. This will facilitate next implementations and the service providers will experiment and develop new services using the capabilities of the new equipment.

Bibliography

Abry, P. and D. Veitch. Wavelet analysis of long range dependent traffic, *IEEE Transactions on Information Theory*, 44(1), pp. 2–15, 1998.

Adamova, K., D. Schatzmann, B. Plattner, and P. Smith. Network anomaly detection in the cloud: The challenges of virtual service migration, in: *IEEE International Conference on Communications (ICC)*, Sydney, Australia, pp. 3770–3775, 2014.

Adas, A. Traffic models in broadband networks, *IEEE Communications Magazine*, pp. 82–89, 1997.

Adelman, I. Long cycles–fact or artifact, *American Economics Review*, 55–3, pp. 444–463, 1965.

Ahuja, N. and B.J. Schachter. *Pattern Models*, John Wiley and Sons, New York, 1983.

Akimaru, H. and K. Kawashima. *Teletraffic, Theory and Applications*, Springer-Verlag, Berlin, 1999.

Albert, R. and A.L. Barabási. Topology of evolving networks: Local events and universality, *Physical Review Letters*, 85, pp. 5234–5237, 2000.

Albert, R. and A.L. Barabási. Statistical mechanics of complex networks, *Reviews of Modern Physics*, 74(1), pp. 47–97, 2002.

Albert R., H. Jeong, and A.L. Barabási Error and attack tolerance of complex networks, *Nature*, 406, pp. 378–382, 2000.

Altman, E., K. Avrachenkov, C. Barakat, A. Kherani, and B. Prabhu. Analysis of MIMD congestion control algorithm for high speed networks, *Computer Networks*, 48, pp. 972–989, 2005.

Amaral, L.A.N., A. Scala, M. Barthelemy, and H.E. Stanley. Classes of small-world networks, *Proceedings of the National Academy of Sciences of the United States of America*, 97, p. 11149, 2000.

Arbaugh, W.A., N. Shankar, and Y.C.J. Wan. *Your 802.11 Wireless Network has No Clothes*, Department of Computer Science University of Maryland, 2001.

Arrowsmith, D., M. di Bernardo, and F. Sorrentino. Effects of variations of load distribution on network performance, in: *IEEE International Symposium on Circuits and Systems, ISCAS*, pp. 4616–4621, 2005.

Assenza, S., J. Gomez-Gardenes, and V. Latora. Enhancement of cooperation in highly clustered scale-free networks, *Physical Review E*, 78, p. 017101, 2008.

Aukia, P., M. Kodialam, P. Koppol, T. Lakshman, H. Sarin, and B. Suter. Rates: A server for Mpls traffic engineering, *IEEE Network Magazine*, March–April, pp. 34–41, 2000.

Baccelli, F. and S. Zuyev. Poisson-Voronoi spanning trees with applications to the optimization of communication networks, *Operations Research*, 47(4), pp. 619–631, 1999.

Barabási, A., Z. Dezso, E. Ravasz, S.H. Yook, and Z. Oltvai. Scale-Free and hierarchical structures in complex networks, in *Modeling of Complex Systems*, Seventh Granada Lectures 661(1), pp. 1–16, 2003.

Barabási, A.L. and E. Bonabeau. Scale-free networks, *Scientific American*, 2003.

Barabási, A.L. and R. Albert. Emergence of scaling in random networks, *Science*, 286, pp. 509–512, 1999.

Barabási, A.L., R. Albert, and H. Jeong. Mean-field theory for scale-free random networks, http://arxiv.org/pdf/cond-mat/9907068v1.pdf, 1999.

Barakat, C., P. Thiran, G. Iannaccone, C. Diot, and P. Owezarski. Modelling Internet backbone traffic at the flow level, *IEEE Transactions on Signal Processing*, 51, pp. 2111–2124, 2003.

Barford, P., J. Kline, D. Plonka, and A.A. Ron. Signal analysis of network traffic anomalies, in: *Proceedings of the 2nd ACM SIGCOMM Workshop on Internet Measurement*, pp. 71–82, 2002.

Beran, J. *Statistics for Long-Memory Processes*, Chapman & Hall, London, 1994.

Bestavros, A., K, Harfoush, and J. Byers. Robust identification of shared losses using end-to-end unicast probes, in: *Proceedings of IEEE International Conference on Network Protocols*, Osaka, Japan, pp. 22–33, November 2000.

Bianconi, G. and A.L. Barabási. Competition and multiscaling in evolving networks, *Europhysics Letters*, 54, pp. 436–442, 2001.

Bollobas, B. *Random Graphs*, London Math. Society Monographs, Academic Press, London, 1985.

Box, G.E. and G.M. Jenkins. *Time Series Analysis: Forecasting and Control*, 2nd ed., Holden-Day, New York, 1976.

Brenner, P.A. Technical Tutorial on the IEEE 802.11 Protocol, BreezeCom Wireless Communications, 1997.

Breunig, M., H.P. Kriegel, and R.N.J. Sander. LOF: Identifying density-based local outliers, in: *Proceedings of ACM SIG-MOD International Conference on Management of Data*, pp. 93–104, 2000.

Broder, A.R., F. Kumar, P. Maghoul, S. Raghavan, R. Rajagopalan, A. Stata, A. Tomkins, and J. Wiener. Graph structure in the web, *Computer Networks*, 33, pp. 309–320, 2000.

Brutlag, J.D. Aberrant behavior detection in time series for network service monitoring, in: *Proceedings of 14th Systems Administration Conference*, pp. 139–146, 2000.

Cao, J., D. Davis, S. Wiel, and B. Yu. Time-varying network tomography: Router link data, *The Journal of American Statistics Association*, 95(452), pp. 1063–1075, 2000.

Casal, C.R., F. Schoute, and R. Prasad. *Evolution towards Fourth Generation Mobile Multimedia Communication*, Delft University of Technology, The Netherlands, 1999.

Casas, P., S. Vaton, L. Fillatre, and I. Nikiforov. Optimal volume anomaly detection and isolation in large-scale IP networks using coarse-grained measurements, *Computer Networks*, 54(11), pp. 1750–1766, 2010.

Chao H.J. and X. Guo. *Quality of Service Control in High-Speed Networks*, Wiley-Interscience, New York, 2001.

Chen, T. Network traffic modelling, in: *The Handbook of Computer Networks*, Hossein Bidgoli (Ed.), Wiley, New York, 2007.

Chung, J. NS by Example. Available online: http://www.isi.edu/nsnam/ns/, 2011.

Chung, F. and L. Lu. Connected components in random graphs with given degree sequences, *Annals of Combinatorics*, 6(2), pp. 125–145, 2002.

Claise, B. *Cisco Systems NetFlow Services Export Version 9*, RFC 3954 (Informational), October 2004.

Clark, C.R. and D.E. Schimmel. Scalable multi-pattern matching on high-speed networks, in: *IEEE Symposium on Field-Programmable Custom Computing Machines*, Napa, California, pp. 249–257, 2004.

Clegg, R.G. *The Statistics of Dynamic Networks*, PhD thesis, University of York, 2004.

Cleveland, W.S., D. Lin, and D.X. Sun. Ip packet generation: Statistical models for TCP start times based on connection rate super position, in: *Proceedings ACM SIGMETRICS*, pp. 166–177, 2000.

Coates, M. and R. Nowak. Network delay distribution inference from end-to-end unicast measurement, in: *Proceedings of IEEE International Conference on Acoustics, Speech and Signal Proceedings*, May 2001.

Coates, M., A. Hero, R. Nowak, and B. Yu. Internet tomography, *IEEE Signal Processing Magazine*, 19, pp. 47–65, 2002.

Cohen, R., K. Erez, D. Ben-Avraham, and S. Havlin. Resilience of the internet to random breakdown, *Physical Review Letters*, 85, pp. 4626–4628, 2000.

Colesniuc, D. *Contribuţii privind Implementarea Tehnologiilor de Bandă Largă în Reţeaua Militară Naţională de Comunicaţii*, Teză de doctorat, 2004. (In Romanian.)

Costa, L. da F., M.S. Barbosa, V. Coupez, and D. Stauffer. *Brain and Mind*, 4, p. 1, 2003.

Costa, L. da F. On the separability of attractors in grandmother dynamic systems with structured connectivity, *Physics*, arxiv-physics0701089, 2007.

Crovella, M.E. and A. Bestavros. *Self-Similarity in World Wide Web Traffic: Evidence and Possible Causes*, Computer Science Department, Boston University, Boston, 1996.

Crovella, M.E. and A. Bestavros. Self-similarity in world wide web traffic: Evidence and possible causes, *IEEE/ACM Transactions Networking*, 5(6), pp. 835–846, 1997.

Crovella, M. and M. Taqqu. Estimating the heavy tail index from scaling properties, *Methodology and Computing in Applied Probability*, 1, pp. 55–79, 1999.

Deane, J.H.B., C. Smythe, and D.J. Jefferies. Self-similarity in a deterministic model of data transfer, *Journal of Electronics*, 80(5), pp. 677–691, 1996.

Deri, L., M. Martinelli, and A. Cardigliano. Realtime high-speed network traffic monitoring using ntopng, in: *Proceedings of the 28th Conference on Large Installation System Administration*, pp. 69–79, 2014.

Dimitropoulos, D.K. and G. Riley. Graph annotations in modeling complex network topologies, *ACM Transactions on Modeling and Computer Simulation*, 19(4), pp. 17-1–17-29, 2009.

Dobrescu, M., R. Dobrescu, and St. Mocanu. A video traffic feedback control mechanism for ATM networks, *WSEAS Transactions on Computers*, 3(6), pp. 1746–1751, 2004a.

Dobrescu, M., S. Mocanu, R. Ulrich, and M. Rothenberg. An analysis of traffic characteristics in a WWAN network, *Scientific Bulletin of UPB, Bucharest*, 67(C, 2), pp. 9–20, 2005b.

Dobrescu, R. and M. Dobrescu. Reactive congestion control for multipoint video services, in: *8th International Workshop on Systems, Signals and Image Processing*, Bucharest, RO, pp. 95–99, 2001a.

Dobrescu, R. and M. Dobrescu. Dedicated hardware and software architecture for multipoint services over ATM Networks, in: *5th International Conference on Knowledge-Based Intelligent Information Systems*, Osaka, Japan, pp. 288–292, September 2001b.

Dobrescu, R., M. Dobrescu, and St. Mocanu. Using self-similarity to model network traffic, *WSEAS Transactions on Computers*, 3(6), pp. 1752–1757, 2004b.

Dobrescu, R., D. Hossu, and R. Ulrich. Procedure of Pointing out Self-Similarity of Internet Traffic, Romanian Patent RO125569-A0, 2013 (approved by OSIM decision 6/156/29.11.2013).

Dobrescu, R. and M. Rothenberg. A method for inferencing of network characteristics based on traffic similarities, *Scientific Bulletin of UPB, Bucharest*, 67(C, 1), pp. 23–32, 2005a.

Dorogovtsev, S.N., J.F.F. Mendes, and A.N. Samukhin. Structure of growing networks with preferential linking, *Physics Review Letters*, 85(21), pp. 4633–4636, 2000a.

Dorogovtsev, S.N., J.F.F. Mendes, and A.N. Samukhin. Structure of growing networks: Exact solution of the Barabási-Albert's model, *Physics Review Letters*, 85, p. 4633, 2000b; Cond-mat/0004434.

Dorogovtsev, S.N. and J.F.F. Mendes. Natural scale of scale-free networks, *Physical Review E.*, 63, p. 62101, 2001.

Dorogovtsev, S.N. and J.F.F. Mendes. Evolution of networks, *Advances in Physics*, 51(4), pp. 1079–1187, 2002.

Drago, I. Understanding and monitoring cloud services, PhD thesis, Centre for Telematics and Information Technology, 2013.

Duffield, N.G. Sampling for passive internet measurement: A review, *Statistical Science*, 19(3), pp. 472–498, 2004.

Embrechts, P. and M. Maejima. *Self-Similar Processes*, Princeton University Press, Princeton, 2002.

Erramilli, A., M. Roughan, D. Veitch, and W. Willinger. Self-similar traffic and network dynamics, *Proceedings of IEEE*, 90(5), pp. 810–819, 2002.

Erdös, P. and A. Rényi. On the evolution of random graphs, *Publications of the Mathematical Institute of the Hungarian Academy of Sciences*, 5, pp. 17–61, 1960.

Esker, H. van den. A geometric preferential attachment model with fitness, *Mathematics—Combinatorics*, 2008, arXiv:0801.1612v1.

Estan, C., S. Savage, and G. Varghese. Automatically inferring patterns of resource consumption in network traffic, in: *Proceedings of ACM SIGCOMM*, pp. 30–33, August 2003.

Etoh, M. *Next Generation of Mobile Systems: 3G & Beyond*, John Wiley & Sons, New York, 2005.

Fall, K. and K. Varadhan. *The NSs Manual*, The VINT Project: A Collaboration between Researchers at UC Berkeley, LBL, USC/ISI, and Xerox PARC, 2008.

Falconer, K. *Techniques in Fractal Geometry*, John Wiley & Sons, Chichester, 1997.

Faloutsos, M., P. Faloutsos, and C. Faloutsos. On power-law relationships of the internet topology, *Computer Communications Review*, 29, pp. 251–262, 1999.

Fayoumi, A. Performance evaluation of a cloud based load balancer severing Pareto traffic, *Journal of Theoretical and Applied Information Technology*, 32(1), pp. 28–34, 2011.

Field, T., U. Harder, and P. Harrison. Measurement and modelling of self-similar traffic in computer networks, *IEEE Proceedings—Communications*, 151(4), pp. 355–363, 2004.

Floyd, S. and V. Paxson. Difficulties in simulating the internet, *ACM/IEEE Transactions on Networking*, 9(4), pp. 392–403, 2001.

Fraleigh, C., S. Moon, B. Lyles, C. Cotton, M. Khan, D. Moll, R., Rockell, T. Seely, and C. Diot. Packet-level traffic measurements from the sprint IP backbone, *IEEE Network*, 17, pp. 6–16, 2003.

Franceschi, A.S., L.F. Kormann, and C.B. Westphall. Performance evaluation for proactive network management, in: *Proceedings of IEEE ICC*, pp. 22–26, 1996.

Frost, V.S. and B. Melamed. Traffic models for telecommunications networks, *IEEE Communications Magazine*, pp. 70–81, 1994.

Garcia-Teodoro, P., J. Dıaz-Verdejo, G. Macia-Fernandez, and E. Vazquez. Anomaly-based network intrusion detection: Techniques, systems and challenges, *Computers and Security*, 28(1–2), pp. 18–28, 2009.

Gast, M. 802.11 *Wireless Networks: The Definitive Guide—Creating and Administering Wireless Networks*, O'Reilly, Sebastopol, CA, 2002.

Grama, A., A. Gupta, G. Karpys, and V. Kumar. *Introduction to Parallel Computing*, 2nd Ed, Pearson Prentice Hall, Englewood Cliffs, NJ.

Granger, C.W.J. Long memory relationships and the aggregation of dynamic models, *Journal of Econometrics*, 14(2), pp. 227–238, 1980.

Grossglauser, M. and J.C. Bolot. On the relevance of long-range dependence in network traffic, *IEEE/ACM Transactions on Networking*, 7(5), pp. 629–640, 1999.

Hernandez, P.C. *Statistical Analysis of Network Traffic for Anomaly Detection and Quality of Service Provisioning*, PhD thesis, ENSTB, 2010.

Heyman, D., A. Tabatabai, and T.V. Lakshman. Statistical analysis and simulation study of video teletraffic in ATM networks, *IEEE Transactions on Circuits and Systems for Video Technology*, 2(1), pp. 49–59, 1992.

Hurst, H.E. Long-term storage capacity of reservoirs, *Transactions of the American Society of Civil Engineers*, 116, pp. 770–799, 1951.

Ihler, A., J. Hutchins, and P. Smyth. Adaptive event detection with time-varying Poisson processes, in: *Proceedings of ACM SIGKDD International Conference on Knowledge Discovery and Data Mining (KDD)*, Philadelphia, PA, August 2006.

Ionescu, F., C. Constantin, M. Nicolae, and V. Kotev. Object oriented virtual manufacturing lines, in: *Proceedings of 5th International Conference on Manufacturing Engineering*, pp. 119–132, 2014.

Jagerman, D.L., B. Melamed, and W. Willinger. *Stochastic Modelling of Traffic Processes, Frontiers in Queueing: Models, Methods and Problems*, CRC Press, Boca Raton, FL, 1997.

Jeong, H., B. Tombor, R. Albert, Z.N. Oltvai, and A.L. Barabási. The large-scale organization of metabolic networks, *Nature*, 407, pp. 651–654, 2000.

Jeong, H.D.J., D. McNickle, and K. Pawlikowski. *A Comparative Study of Generators of Synthetic Self-Similar Teletraffic*, Technical Report TR-COSC 10/98, University of Canterbury, New Zealand, 1998.

Jiang, M., M. Nikolic, S. Hardy, and L. Trajkovic. *Impact of Self-Similarity on Wireless Data Network Performance*, Simon Frasier University Vancouver, Canada, 2001.

Karagiannis, T. and M. Faloutsos. *SELFIS: A Tool for Self-Similarity and Long-Range Dependence Analysis*, Edmonton, Canada, 2002.

Karagiannis, T., M. Faloutsos, and M. Molle. A user-friendly self-similarity analysis tool, *ACM SIGCOMM Computer Communication Review*, 33(3), pp. 81–93, 2003.

Kalden, R. and S. Ibrahim. *Searching for Self-Similarity in GPRS*, Ericsson Research Aachen, Germany, 2004.

Karonski, M. and A. Rucinski. The origins of the theory of random graphs, in: *Mathematics of Paul Erdos*, Springer, Berlin, pp. 311–336, 1997.

Katiyar, S. and J. Grover. Agent based dynamic load balancing in cloud computing, in: *Proceedings of the International Conference on Human Computer Interactions*, pp. 1–6, 2013.

Kato, M., K. Cho, M. Honda, and H. Tokuda. Monitoring the dynamics of network traffic by recursive multi-dimensional aggregation, in: *Workshop on Managing Systems Automatically and Dynamically*, pp. 1–7, 2012.

Kelly, F. *The Mathematics of Traffic in Networks*, Cambridge, UK, 2005.

Kitsak, M., S. Havlin, G. Paul, M. Riccaboni, F. Pammolli, and H.E. Stanley. Betweenness centrality of fractal and non-fractal scale-free model networks and tests on real networks, *Physical Review E*, 75(5), p. 056115, 2007.

Kleinberg, J.M., R. Kumar, P. Raghavan, S. Rajagopalan, and A. Tomkins. The web as a graph: Measurements, models and methods, in: *Proceedings of the 5th Annual International Conference, COCOON'99*, Springer-Verlag, Berlin, 1999.

Kleinberg, J. The small-world phenomenon: An algorithmic perspective, in: *32nd ACM Symposium on Theory of Computing*, 2000.

Krapivsky, P.L. and S. Redner. Organization of growing random networks, *Physical Review E*, 63, p. 066123, 2001.

Krapivsky, P.L., S. Redner, and F. Leyvraz. Connectivity of growing random networks, *Physical Review Letters*, 85(21), pp. 4629–4632, 2000.

Kumar, R., P. Raghavan, S. Rajalopagan, D. Sivakumar, A.S. Tomkins, and E. Upfal. Stochastic models for the web graph, in: *Proceedings of 41st IEEE Symposium on Foundations of Computer Science*, pp. 57–65, 2000.

Lakhina, A., M. Crovella, and C. Diot. Diagnosing network-wide traffic anomalies, in: *Proceedings of SIGCOMM'04*, 2004.

Lamperti, J. Semi-stable stochastic processes, *AMS Transactions*, 104, pp. 62–78, 1962.

Lan, K. and J. Heidemann. *On Utilizing the Correlations between User Populations for Traffic Inference*, Technical Report ISI-TR-544, USC/Information Sciences Institute, September 2001.

Lan, K. and J. Heidemann. Rapid model parameterization from traffic measurement, *ACM, Transactions on Modelling and Computer Simulation*, 12(3), pp. 201–229, 2002.

Lan, K. and J. Heidemann. *On the Correlation of Internet Flow Characteristics*, Technical Report ISI-TR-574, USC/Information Sciences Institute, July 2003.

Lan, K. and J. Heidemann. On the feasibility of utilizing correlations between user populations for traffic inference, in: *Proceedings of IEEE International Conference on Local Computer Networks*, pp. 132–139, 2005.

Lazar, A., W. Wang, and R. Deng. Models and algorithms for network fault detection and identification: A review, in: *Proceedings of IEEE International Contribution Conference*, Singapore, pp. 999–1003, 1992.

Leinen, S. *Evaluation of Candidate Protocols for IP Flow Information Export (IPFIX)*, RFC 3955, October 2004.

Leland, W., M. Taqqu, W. Willinger, and D. Wilson. On the self-similar nature of Ethernet traffic, in: *Proceedings of ACM SIGCOMM '93*, pp. 183–193, 1993.

Leland, W.E., W. Willinger, M.S. Taqqu, and D.V. Wilson. On the self-similar nature of ethernet traffic (extended version), *IEEE/ACM Transactions on Networking*, 2(1), pp. 1–15, 1994.

Li, M., S. Yu, and L. He. Detecting network-wide traffic anomalies based on spatial HMM, in: *IFIP International Conference on Network and Parallel Computing*, pp.198–203, 2008.

Li, L., D. Alderson, R. Tanaka, J.C. Doyle, and W. Willinger. Towards a theory of scale-free graphs: Definition, properties, and implications (Extended Version), *Internet Mathematics*, 2005.

Li, X., F. Bian, M. Crovella et al. Detection and identification of network anomalies using sketch subspaces, in: *Proceedings of 6th ACM SIGCOMM Conference on Internet Measurement*, Rio de Janeriro, Brazil, pp. 147–152, 2006.

Liu, J., Y. Shu, L. Zhang, F. Xue, and O. Yang. Traffic modelling based on FARIMA models, in: *Proceedings of IEEE Canadian Conference on Electrical and Computer Engineering*, 1, pp. 162–167, 1999.

Lopez-Ardao, J.C., C. Lopez-Garcia, A. Suarez-Gonzalez, M. Fernandez-Veiga, and R. Rodriguez-Rubio. On the use of self-similar processes in network simulation, *ACM Transactions on Modeling and Computer Simulation*, 10(2), 2000.

Ma, J. and S. Perkins. On-line novelty detection on temporal sequences, in: *Proceedings of ACM SIGKDD International Conference on Knowledge Discovery and Data Mining (KDD)*, New York, NY, pp. 613–618, 2003.

Mandelbrot, B.B. and J.W. Van Ness. Fractional Brownian motions, fractional noises and application, *SIAM Review*, 10(4), pp. 422–437, 1968.

Mardani, M. and G. Giannakis. Robust network traffic estimation via sparsity and low rank, in: *Proceedings of IEEE International Conference on Acoustics, Speech and Signal Processing*, pp. 4529–4533, 2013.

Maslov, S., K. Sneppen, and A. Zaliznyak. Detection of topological patterns in complex networks: Correlation profile of the Internet, *Physica A, 333*, pp. 529–540, 2004.

Melamed, B. The empirical TES methodology: Modelling empirical time series, *Journal of Applied Mathematics and Stochastic Analysis*, 10(4), pp. 333–353, 1997.

Mell, P. and T. Grance. *The NIST Definition of Cloud Computing, National Institute of Standards and Technology (NIST)*, Technical Report Special Publication, 800-145, Sept 2011.

Meng, S. *Monitoring as a Service in the Cloud*, PhD thesis, Georgia Institute of Technology, 2012.

METIS—*Serial Graph Partitioning and Fill-Reducing Matrix, METIS stable version: 5.1.0*, http://glaros.dtc.umn.edu/gkhome/metis/metis/overview, 2013.

Mocanu, S. and S. Ţarălungă. Cluster based simulations of scale-free networks immunization strategies, *WSEAS Transactions on Computers*, 6(2), p. 268, 2007.

Molloy, M. and B. Reed. The size of the giant component of a random graph with a given degree sequence, *Combinatory Probability Computation*, 7(3), pp. 295–305, 1998.

Molnar, S. and G. Miklos. On burst and correlation structure of teletraffic models, in: *5th IFIP Workshop on Performance Modelling and Evaluation of ATM Networks*, West Yorkshire, UK, 1997.

Nahrstedt, K. and R. Steinmetz. *Resource Management in Multimedia Networked Systems*, Technical Reports (CIS), Paper 331, University of Pennsylvania, 1994.

Newman, M.E.J., S.H. Strogatz, and D.J. Watts. Random graphs with arbitrary degree distributions and their applications, *Physical Review E*, 64, p. 026118, 2001.

Nicolaescu, Ş.V. *Comunicaţii Mobile, Generaţiile 3G şi 4G*, Institutul Naţional de Studii şi Cercetări pentru Comunicaţii, 2003. (In Romanian.)

Norros. I. On the use of fractional Brownian motion in the theory of connectionless networks, *IEEE Journal on Selected Areas in Communications*, 13(6), pp. 953–962, 1995.

O'Mahony, M.J., D. Simeonidou, D. Hunter, and A. Tzanakaki. The application of optical packet switching in future communication networks. *IEEE Communications Magazine*, 39, pp. 128–135, 2001.

Ostring, S.A.M. and H. Sirisena. The influence of long-range dependence on traffic prediction, in: *Proceedings of IEEE ICC'01*, 4, Helsinki, Finland, pp. 1000–1005, 2001.

Park, K. and W. Willinger (Eds.). *Self-Similar Network Traffic and Performance Evaluation*, Wiley, New York, 2000.

Park, K., G. Kim, and M. Crovella. On the effect of traffic self-similarity on network performance, in: *Proceedings of 1997 SPIE International Conference on Performance and Control of Network Systems*, 1997.

Paxson, V. and S. Floyd. Wide-area traffic: The failure of Poisson modelling, in: *Proceedings of SIGCOMM '94*, September 1994.

Pfeiffenberger, Th., U. Hofmann, I. Miloucheva, and A. Nassri. Inferencing of interdomain path characteristics based on active end-to-end QoS monitoring—Scenario studies in emulation environment, in: *First International Workshop IPS*, Salzburg, 2003.

Phaal, P.S. Panchen, S., and S. McKee. *InMon Corporation's sFlow: A Method for Monitoring Traffic in Switched and Routed Networks*, RFC 3176, September 2001.

Popescu, A. Traffic self-similarity, in: *Proceedings of IEEE International Conference on Telecommunications, ICT2001*, June 2001.

Porekar, J. *Random Networks, Course: Faculty of Mathematics and Physics*, Ljubljana University, Ljubljana, 2002.

Pruthi, P. *An Application of Chaotic Maps to Packet Traffic Modelling*, PhD dissertation, Royal Institute of Technology, Stockholm, Sweden, 1995.

Ramamurthy, G. and B. Sengupta. Modelling and analysis of a variable bit rate video multiplexor, in: *Proceedings of INFOCOMM '92*, Florence, Italy, pp. 817–827, 1992.

Rabkin, A., M. Arye, S., Sen, V. Pai, and M. Freedman. Aggregation and degradation in jet stream: Streaming analytics in the wide area, in: *Proceedings of 11th USENIX Symposium on Networked Systems Design and Implementation*, pp. 275–288, 2014.

Riley, G., M. Ammar, R. Fujimoto, A. Park, K. Perumalla, and D. Xu. A federated approach to distributed network simulation, *ACM Transactions on Modelling and Computer Simulation (TOMACS)*, 14(2), pp. 116–148, 2004.

Rothenberg, M. *Access Net Traffic Analysis*, Technical Report, http://www.access-net.ro, 2006.

Roberts, J. Traffic theory and the Internet, *IEEE Communication Magazine*, pp. 94–99, 2001.

Roberts, J. Internet traffic, QoS and pricing, *Proceedings of the IEEE*, 92(9), pp. 1389–1399, 2004.

Sadasivan, G., N. Brownlee, B. Claise, and J. Quittek. *Architecture for IP Flow Information Export*, RFC 5470, 2009.

Sahinoglu, Z. and S. Tekinay. *On Multimedia Networks: Self-Similar Traffic and Network Performance*, New Jersey Institute of Technology, New Jersey, 1999.

Saroiu, S., K. Gummadi, R. Dunn, S. Gribble, and H. Levy. An analysis of Internet content delivery systems, *ACM SIGOPS Operating Systems Review*, 36, pp. 315–327, 2002.

Saru, D. and Ş. Mocanu. *Arhitecturi Hardware/Software pentru Sisteme de Comunicaţie Wireless*, Universitatea "Politehnica" Bucureşti, 2003. (In Romanian.)

Scherrer, A., N. Larrieu, P. Borgnat, P. Owezarski, and P. Abry, Non Gaussian and long memory statistical modelling of internet traffic, *Transactions of IEEE on Dependable and Secure Computing*, 4(1), pp. 56–70, 2007.

Sen, P., B. Maglaris, N.E. Rikli, and D. Anastassiou. Models for packet switching of variable-bit-rate video sources, *IEEE Journal on Selected Areas in Communications*, 7(5), pp. 865–869, 1989.

Serfozo, R. *Introduction to Stochastic Networks*, Springer-Verlag, New York, 1999.

Sheridan, P., Y. Yagahara, and H. Shimodaira. A preferential attachment model with Poisson growth for scale-free networks, *Annals of the Institute of Statistical Mathematics*, 60(4), pp. 747–761, *arXiv: 0801.2800v2*, Tokyo, 2008.

Sikdar, B. and K.S. Vastola. On the contribution of TCP to the self-similarity of network traffic, in: *Evolutionary Trends of the Internet*, Lecture Notes in Computer Science, pp. 596–613, September 2001, Springer-Verlag, Berlin.

Soule, A., K. Salamatian, and N. Taft. Combining filtering and statistical methods for anomaly detection, in: *Proceedings of 5th ACM SIGCOMM Conference on Internet Measurement*, pp. 31–45, 2005.

Sriram, I. and D. Cliff. Hybrid complex network topologies are preferred for component-subscription in large-scale data-centres, in: *Complex Networks*, Costa L. da F., A. Evsukoff, G. Mangioni, and R. Menezes, (Eds.), Vol. 116, Communications in Computer and Information Science, pp. 130–137, Springer, Berlin, Heidelberg, 2011.

Stallings, W. *SNMP, SNMPv2, and CMIP: The Practical Guide to Network Management Standards*, Addison-Wesley, New York, 1994.

Takine, T., B. Sengupta, and T. Hasegawa. An analysis of a discrete-time queue for broadband ISDN with priorities among traffic classes, *IEEE Transactions on Communications*, 42(234), pp. 1837–1845, 1994.

Tanaka, T., M. Inayoshi, and N. Mizuhara. Overview of communication network evolution, *Hitachi Review*, 47, 1998.

Ţarălungă, S., I. Hangiu, M. Hangiu, and R. Ulrich. Distributed simulation for traffic analysis of large scale-free networks, in: *Proceedings of the 3rd IAFA Symposium*, 2007.

Terdik, G. and T. Gyires. Does the internet still demonstrate fractal nature?, in: *Proceedings of the 8th International Conference on Networks*, pp. 30–34, 2009.

Thottan, M. and C. Ji. Using network fault predictions to enable ip traffic management, *Journal of Network and Systems Management*, 9(3), 327–346, 2001.

Thottan, M. and C. Ji. Anomaly detection in IP networks, *IEEE Transactions on Signal Processing*, 51(8), 2191–2204, 2003.

Toroczkai, Z., K. Gyorgy, M. Novotny, and H. Guclu. Virtual time horizon control via communication network design, in: *Computational Complexity and Statistical Physics*, A. Percus, G. Istrate, C. Moore (Eds.), Oxford University Press, Oxford, 2006.

Ulrich, R. and M. Rothenberg. Self-similarity in wireless networks, in: *Proceedings of the 2nd IAFA Symposium*, 2005.

Vardi, Y. Network tomography: Estimating source-destination traffic intensities from link data, *Journal of American Statistics Association*, 91(433), pp. 365–377, 1996.

Veres, A., Zs. Kenesi, S. Molnar, and G. Vattay. On the propagation of long-range dependence in internet, in: *Proceedings of SIGCOMM*, Stockholm, Sweden, 2000.

Veres, A. *Modelling TCP Dynamics and Engineering Service Differentiation in TCP/IP Network*, PhD dissertation, Department of Telecommunications and Media Informatics, Budapest University of Technology and Economics, Budapest, Hungary, 2004.

Vetsigian, K. *Models of Complex Networks, Course: Statistical Physics*, Biological Information and Complexity, University of Illinois, 2001.

Wang, H., D. Zhang, and K.G. Shin. Detecting SYN flooding attacks, in: *Proceedings of IEEE INFOCOM*, 2002.

Watts, D.J. and S.H. Strogatz. Collective dynamics of "small-world" networks, *Nature*, 393, pp. 440–442, 1998.

Waxman, B. Routing of multipoint connections, *IEEE Journal on Selected Areas in Communications*, SAC-6, pp. 1617–1622, 1988.

Wilkinson, B. and M.A. Pearson. *Parallel Programming*, 2nd ed, Prentice-Hall, Englewood Cliffs, NJ, 2004.

Willinger, W., V. Paxon, R.H. Riedi, and M.S. Taqqu. Long-range dependence and data network traffic, in: *Long-Range Dependence: Theory and Applications*, P. Doukhan, G. Oppenheim, and M.S. Taqqu (Eds.), Boston, Brikhauser, pp. 373–406, 2001.

Willinger, W., M.S. Taqqu, and A. Erramilli. *A Bibliographical Guide to Self-Similar Traffic and Performance Modelling for Modern High-Speed Networks*, Bellcore, Oxford University Press, Oxford, 1996.

Willinger, W., M.S. Taqqu, R. Sherman, and D.V. Wilson. Self-similarity through high-variability: Statistical analysis of Ethernet LAN traffic at the source level, *IEEE/ACM Transactions on Networking*, 5(1), pp. 71–86, 1997.

Wolman, A., G.M. Voelker, N. Sharma, N. Cardwell, M. Brown, T. Landray, D. Pinnel, A.R. Karlin, and H.M. Levy. Organization-based analysis of web-object sharing and caching, in: *USENIX Symposium on Internet Technologies and Systems*, USENIX, 1999.

Xiong, Y.J., M. Vandenhoute, and H.C. Cankaya. Control architecture in optical burst-switched WDM networks, *IEEE Journal on Selected Areas in Communications*, 18(10), pp. 1838–1851, 2000.

Yao, L., M. Agapie, J. Ganbar, and M. Doroslovacki. Long-range dependence in Internet backbone traffic, in: *IEEE ICC*, 3, pp. 1611–1615, 2003.

Yu, X., Y. Chen, and C. Qiao. A Study of traffic statistics of assembled burst traffic in optical burst switched networks, in: *Proceedings of SPIE Optical Networking and Communication Conference*, pp. 149–159, 2002.

Zang, H., J.P. Jue, and B. Mukherjee. A review of routing and wavelength assignment approaches for wavelength-routed optical WDM networks, *SPIE Optical Networks Magazine*, 1(1), pp. 47–60, 2000.

Zhang, Y., M. Roughan, C. Lund, and D. Donoho. An information-theoretic approach to traffic matrix estimation, in: *ACM SIGCOMM*, pp. 301–312, 2003.

Zseby, T., E. Boschi, N. Brownlee, and B. Claise. *IP Flow Information Export (IPFIX) Applicability*, RFC 5472, 2009.

Index

D

O

P

Z